T0398945

VOLUME SEVENTY

SOLID STATE PHYSICS

Recent Advances in Topological Ferroics and their Dynamics

VOLUME SEVENTY

SOLID STATE PHYSICS

Recent Advances in Topological Ferroics and their Dynamics

Edited by

ROBERT L. STAMPS
Department of Physics and Astronomy,
University of Manitoba, Canada

HELMUT SCHULTHEIß
Institute for Ion Beam Physics and Materials Research,
Helmholtz-Center Dresden-Rossendorf,
Dresden, Saxony, Germany

Academic Press is an imprint of Elsevier
50 Hampshire Street, 5th Floor, Cambridge, MA 02139, United States
525 B Street, Suite 1650, San Diego, CA 92101, United States
The Boulevard, Langford Lane, Kidlington, Oxford OX5 1GB, United Kingdom
125 London Wall, London, EC2Y 5AS, United Kingdom

First edition 2019

ISBN: 978-0-08-102920-6
ISSN: 0081-1947

For information on all Academic Press publications
visit our website at https://www.elsevier.com/books-and-journals

Publisher: Zoe Kruze
Acquisition Editor: Jason Mitchell
Editorial Project Manager: Joanna Collett
Production Project Manager: Abdulla Sait
Cover Designer: Miles Hitchen

Typeset by SPi Global, India

Contents

Contributors

Chandrima Banerjee
School of Physics, Trinity College Dublin, College Green, Dublin, Ireland

Anjan Barman
Department of Condensed Matter Physics and Material Sciences, S. N. Bose National Centre for Basic Sciences, Kolkata, India

Biswanath Bhoi
National Creative Research Initiative Center for Spin Dynamics and Spin-Wave Devices, Nanospinics Laboratory, Research Institute of Advanced Materials, Department of Materials Science and Engineering, Seoul National University, Seoul, Republic of Korea

Eugene A. Eliseev
Institute for Problems of Materials Science, National Academy of Sciences of Ukraine, Kyiv, Ukraine

Pawel Gruszecki
Faculty of Physics, Adam Mickiewicz University in Poznan, ul. Uniwersytetu Poznańskiego, Poznań, Poland

Olav Hellwig
Institute of Physics, Chemnitz University of Technology, Chemnitz; Institute of Ion Beam Physics and Materials Research, Helmholtz-Zentrum Dresden-Rossendorf, Dresden, Germany

Jiri Hlinka
Institute of Physics of the Czech Academy of Sciences, Prague, Czech Republic

M. Benjamin Jungfleisch
Department of Physics and Astronomy, University of Delaware, Newark, DE, United States

Victoria V. Khist
National Technical University of Ukraine "Igor Sikorsky Kyiv Polytechnic Institute"; Institute of Magnetism, National Academy of Sciences of Ukraine and Ministry of Education and Science of Ukraine, Kyiv, Ukraine

Sang-Koog Kim
National Creative Research Initiative Center for Spin Dynamics and Spin-Wave Devices, Nanospinics Laboratory, Research Institute of Advanced Materials, Department of Materials Science and Engineering, Seoul National University, Seoul, Republic of Korea

Maciej Krawczyk
Faculty of Physics, Adam Mickiewicz University in Poznan, ul. Uniwersytetu Poznańskiego, Poznań, Poland

Sergi Lendinez
Department of Physics and Astronomy, University of Delaware, Newark, DE, United States

Anna N. Morozovska
Institute of Physics; Bogolyubov Institute for Theoretical Physics, National Academy of Sciences of Ukraine, Kyiv, Ukraine

Michal Mruczkiewicz
Institute of Electrical Engineering, Slovak Academy of Sciences, Bratislava, Slovakia

Petr Ondrejkovic
Institute of Physics of the Czech Academy of Sciences, Prague, Czech Republic

Victor Polinger
Department of Chemistry, University of Washington, Seattle, WA, United States

Jan Seidel
School of Materials Science and Engineering; ARC Centre of Excellence in Future Low-Energy Electronics Technologies, UNSW Sydney, Sydney, NSW, Australia

Joseph Sklenar
Department of Physics and Astronomy, Wayne State University, Detroit, MI, United States

Preface

It is our great pleasure to present the 70th edition of *Solid State Physics*. The vision statement for this series has not changed since its inception in 1955, and *Solid State Physics* continues to provide a "mechanism ... whereby investigators and students can readily obtain a balanced view of the whole field." What has changed, is the field and its extent. As noted in 1955, the knowledge in areas associated with solid state physics has grown enormously, and it is clear that boundaries have gone well beyond what was once, traditionally, understood as solid state. Indeed, research on topics in materials physics, applied and basic, now requires expertise across a remarkably wide range of subjects and specialties. It is for this reason that there exists an important need for up-to-date, compact reviews of topical areas. The intention of these reviews is to provide a history and context for a topic that has matured sufficiently to warrant a guiding overview.

The topics reviewed in this special, thematic volume highlight recent advances in ferroic materials along lines of unusual and interesting topological excitations. These include domain walls and skyrmions, as well as other exotic examples of noncolinear ferroic orderings. What is perhaps most interesting is that such structures and their analogues can be found in a variety of systems and in this volume we review in particular examples found in magnetic, ferroelectric, multiferroic, and lithographically patterned arrays. The many chapters in this volume highlight features that are of particular importance and interest for applications, with a primary focus on the various dynamic properties that have been predicted and observed in these material systems. The editors and publishers hope that readers will find the introductions and overviews useful and of benefit both as summaries for workers in these fields, and as tutorials and explanations for those just entering.

Robert L. Stamps and Helmut Schultheiß

Photon-magnon coupling: Historical perspective, status, and future directions

Biswanath Bhoi, Sang-Koog Kim*
National Creative Research Initiative Center for Spin Dynamics and Spin-Wave Devices, Nanospinics Laboratory, Research Institute of Advanced Materials, Department of Materials Science and Engineering, Seoul National University, Seoul, Republic of Korea
*Corresponding author: e-mail address: sangkoog@snu.ac.kr

Contents

Solid State Physics, Volume 70
ISSN 0081-1947
https://doi.org/10.1016/bs.ssp.2019.09.001

1. Introduction

In the last few decades, technologies engineered from quantum systems have flourished and found actual and potential applications in the fields of quantum information and communication science and technology. An exciting strategy has emerged in the last few years that enables the combining of the individual advantages of distinct physical systems as well as the development of new innovations by integration into hybrid quantum systems [1–3]. Accordingly, light-matter interactions are an interesting and important subject in condensed matter physics, providing physical insight into coupled physical properties while enabling the design of new-concept devices. In this context, photon-magnon (P-M) coupling is a newly developing, interdisciplinary field that brings microwave and optical communities together with spintronics and magnetism researchers [4–6] to develop the integration of quantum information and spintronic technologies. Controllable P-M interactions offer a powerful means of accessing and controlling quantum states as well as new insights into mutual manipulation of magnon and photon modes that are applicable for quantum information communication technologies [4–6].

The field of P-M coupling started around 2014, when it was found that the quanta of collective spin excitations in ferromagnetic crystals (i.e., magnons) can be coupled with both microwaves and optical photons through light-matter interaction, as illustrated in Fig. 1A [7–9]. Since then, the field has attracted broad interest from different groups working in the respective research fields including spintronics, electrodynamics, photonics, and

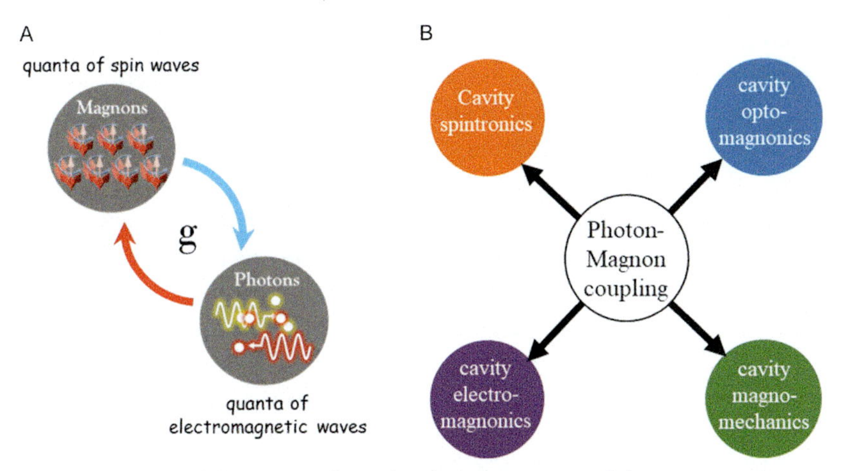

Fig. 1 (A) Conceptual illustration of coupling between quanta of electromagnetic waves (photons) and quanta of spin waves (magnons), where g is the coupling strength. (B) Schematic illustration of different emerging hybrid quantum systems incorporating photon-magnon (P-M) coupling.

electro-mechanical systems. Subsequently, new research areas based on the underlying physics of P-M coupling have emerged, including quantum electrodynamics, cavity spintronics, as well as optomagnonic and cavity magnomechanics (Fig. 1B) [6,10–15]. Magnon-photon coupling links some of the most exciting concepts in modern physics, such as quantum information and quantum optics, along with one of the oldest subjects, which is to say magnetism, via the quantum physics of spin-photon entanglement on the one hand and, on the other, classical electrodynamic coupling. This review aims to provide an introduction to this new frontier of condensed matter physics for researchers working in the diverse research fields of magnetism, spintronics, quantum information, and microwave technologies.

From a theoretical viewpoint, understanding P-M coupling requires simultaneous solving of Maxwell's equations and the Landau-Lifshitz-Gilbert (LLG) equation [16]. On the other hand, in the experimental aspect, realizing strong P-M coupling requires a well-defined, high-quality microwave cavity to confine photons for a sufficiently long time period, and a large sample-to-cavity filling factor to increase the number of spins in the microwave cavity. The latter can be realized using high-spin-density, low-loss ferrimagnetic materials, such as yttrium-iron-garnet (YIG), commonly used in magnonic devices [17]. The former has been realized for several decades in the field of cavity quantum electrodynamics, where the quantum nature of light is brought to the forefront [18].

To this end, much work has recently been devoted to the observation of strong magnon-photon interactions between low-loss magnetic materials and high-quality microwave cavities (or planar resonators) [19–30] in order to achieve large coherent coupling in magnetically ordered systems. At the beginning of the research, initial experiments demonstrated such phenomena only at low temperatures [7–9,19], whereas strong magnon-photon coupling at room temperature has only very recently been realized [19–30]. Such work has provided a foundation for many exciting new possibilities, such as cavity-mediated coupling of spatially separated magnetic moments [31], cavity-mediated qubit-magnon coupling [32], the merging of quantum optics and spintronics through the use of whispering-gallery modes [13,33], possible integration of microwave, mechanical, optical and magnonic systems [6], and the development of magnon dark-mode memory architectures [34].

This book chapter starts with an historical review that traces this new field back to some of the most innovative work in the field of magnon-photon coupling. Then, recent experiments that focus on the development of new cavity-mediated techniques, such as coupling of magnetic moments, distant manipulation of spin current, and conversion between optical and microwave photons, will be highlighted along with non-linear dynamic phenomena in P-M coupling. This paper is organized as follows. Sections 2 and 3 address the different basic elements required to build a P-M coupled system. Section 4 explains the concepts, origin and fundamental characteristics of P-M coupling. An overview of the theoretical progress, from a classical oscillator model to a semi-classical electromagnetic model, is provided. Following this succinct discussion of the fundamentals and theoretical framework of P-M coupling, Sections 5 and 6 highlight the progress achieved so far in integrating the basic elements into hybrid systems. The experimental design for observation of P-M coupling in different hybrid systems and the different parameters for control of P-M coupling are discussed. Section 7 describes the nonlinear effects observed in P-M coupling, while Section 8 addresses the applicability of P-M coupling to technical devices. Finally, Section 9 presents several potential systems for P-M coupling that extend the capability of existing systems to entirely new functionalities.

2. Photon modes

Photons are defined as quantized electromagnetic waves manifesting elementary excitations [35–37]. A concept of the quantization of electromagnetic waves was first introduced by Planck in 1900 in order to explain the spectra of black-body radiations [35–37]. Einstein also adopted the concept, specifically to explain the photoelectric effect in 1905 [37], as did Compton in 1923, to explain the shift of wavelengths of incident X-rays scattered by massive particles [38]. The term "photon" was first introduced by Lewis in 1926 to represent quanta of energy bundles of electromagnetic waves [39]. The formal quantization of electromagnetic waves was first performed by Dirac in 1927 [40]. Electromagnetic waves for all possible radiation types are categorized according to their wavelength and frequency in Fig. 2.

Meanwhile, current research into photon excitation sources and their coupling with other quantized particles has given rise to the explosive growth of the field of quantum information science and technologies over the last few decades. Photonic qubits, whereby information can be encoded in the quantum state of a photon using degrees of freedom such as polarization, momentum, energy, etc., are an ideal choice for many of these applications, since (a) photons travel at the speed of light and interact weakly with their environment over long distances, which results in lower noise and loss, and (b) photons can be controllable in the linear optics regimes.

Since photon coupling with magnons has been experimentally studied only in the microwave (GHz) frequency regime (due to the lack of efficient THz radiation sources and their detection electronics), this review paper restricts the discussion of photon-mode excitation to the microwave frequency range. In the following section, many different geometries of microwave photon resonators will be summarized.

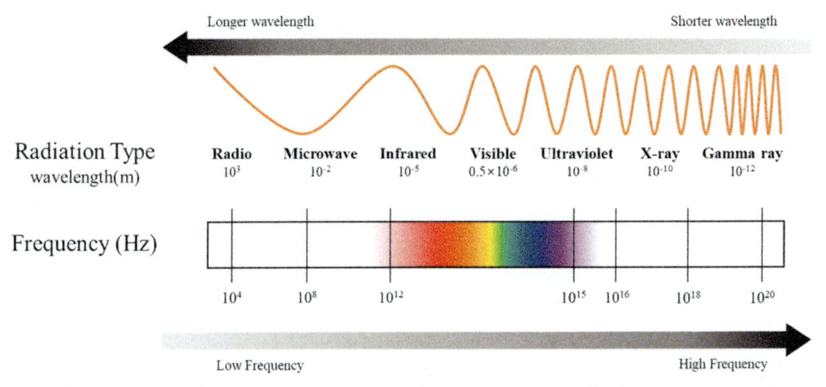

Fig. 2 Illustration of electromagnetic waves for all possible radiation types according to wavelength and frequency.

2.1 Three-dimensional (3-D) cavity resonator

A cavity resonator is a space surrounded by a rectangular- or cylindrical-shape metal conductor that can confine electromagnetic waves (RF waves) by reflecting them back and forth between the cavity's boundaries. The cavity is excited by a coaxial cable carrying signals from a network analyzer. Currents in the walls build up standing electromagnetic waves that form specific resonant modes of certain frequencies, and thereby, the cavity stores electromagnetic energy. For example, for a cylindrical cavity of radius r and height d (see Fig. 3A), the frequency of resonance modes is given as

$$f_{mnl} = \frac{c}{2\pi\sqrt{\mu_r \varepsilon_r}} \sqrt{\left(\frac{p'_{mn}}{r}\right)^2 + \left(\frac{l\pi}{d}\right)^2} \tag{1}$$

where ε_r and μ_r are the relative permittivity and permeability, respectively, p'_{mn} is the m-th root of the Bessel function J'_n of the first kink, and n, m, and l refer to the number of half-wavelength variations in the standing-wave patterns in the radial, axial, and longitudinal directions, respectively [41]. Since the resonant frequency depends only on the geometric parameters of the cavity, it can be tuned by moving one of the cavity's walls inward or outward to change its size.

Standing electromagnetic waves' electric and magnetic field components are exactly out of phase, under which condition, the magnetic field amplitude is the maximum while the electric field amplitude is the minimum,

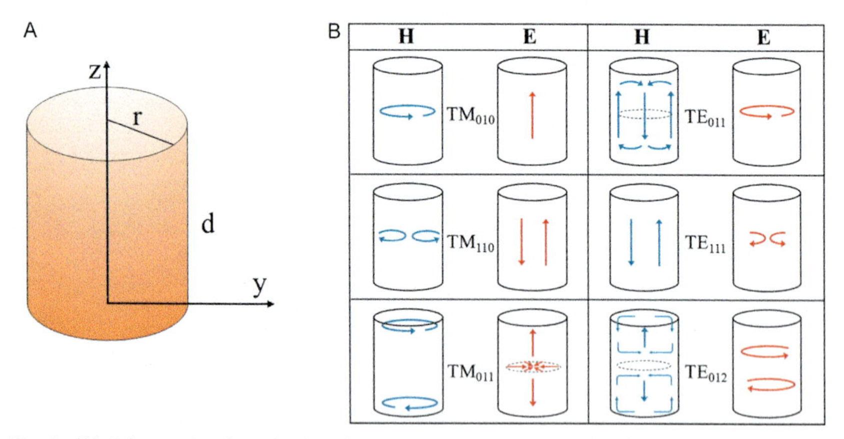

Fig. 3 (A) Schematic of a cylindrical cavity resonator. (B) Mode chart table showing the lowest-order TE and TM modes. For each mode, the magnetic field (left) is shown in blue and the electric field (right) in red.

or vice versa. Fig. 3B shows several examples of the most common transverse magnetic (TM_{nml}) and transverse electric (TE_{nml}) modes. In the *TM* modes, the magnetic field encloses an electric field that "travels" back and forth through the cavity. In the *TE* modes, the electric field either encloses the magnetic field or terminates on the induced surface charge. Since the microwave magnetic field can be absorbed in a magnetic material, the magnetic material is usually placed at the position where the microwave electric field and magnetic field are at the minimum and the maximum, respectively. The spatial distributions of the amplitudes of the electric and magnetic fields for the most commonly used FMR cavity are those for the TE_{012} cavity, as shown in Fig. 3B (right side: bottom), which correspond to a high spatial uniformity of microwave magnetic field right at the middle of the cavity [42]. Such microwave field uniformity is very crucial for accurate interpretation of experimental results. Also, the quality factor of a cavity is an important parameter that represents $Q=f_{res}/\Delta f$, where f_{res} is the resonant-mode frequency of the cavity and Δf is the full width at half maximum of the resonance peak [42]. The quality factor generally is a measure of the ability of a resonator to store energy in relation to time-averaged power dissipation. The higher the Q factor, the lower the rate of energy loss in the cavity; hence the very high sensitivity of the cavity resonator.

2.2 Planar resonator

Besides the above mentioned 3-D microwave cavities, some other planar resonators loaded with microstrip feeding lines have been explored to realize strong P-M coupling. These planar resonators are more suitable for potential practical applications devised using artificially engineered structures such as a -split-ring resonator (SRR), as shown in Fig. 4A. The SRR design first proposed by Pendry et al. [43] is a small metal loop with a splitting gap in it. When electronic currents circulate in the metallic ring, a magnetic field perpendicular to the plane of the ring is generated. This is a kind of LC resonator in which the loop acts as inductance and the splitting gap corresponds to capacitance. Due to the lumped-element approach, an SRR may have resonance frequencies noticeably smaller than those emitted by standing electromagnetic waves in a given photon resonator of the same size as that of the SRR. In order to excite the SRR in a planar device, a microstrip feeding line is necessary so that microwave currents flowing through the line can induce a current flow in the metallic SRR. The directions of both the electric and magnetic fields generated from the microstrip line without an SRR are illustrated in Fig. 4B.

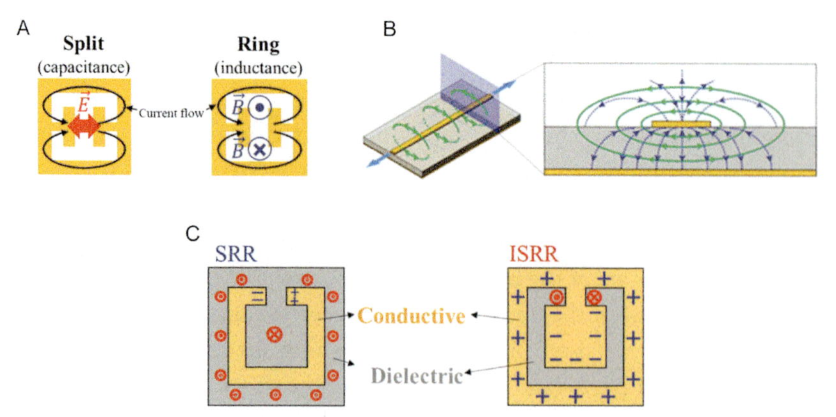

Fig. 4 Schematic of (A) electric current flows (black arrows) and possible generations of electric (wide red arrow) field in the plane and magnetic (blue) field along the out-of-plane axis in a certain-type split-ring resonator (SRR). (B) Electric (blue) and magnetic field (green) distributions generated by electric current flow along a microstrip line (dark yellow color strip). (C) Comparison of SRR and its inverted pattern (ISRR) structures along with schematic drawing of distributions of electric charges (left) and magnetic charges (right) at split gap. The SRR structure has a strong magnetic dipole perpendicular to the metallic plane, while the ISRR structure has a strong magnetic dipole near the split gap.

Similarly, to the SRR structure, an inverted pattern of SRR (i.e., ISRR) has been realized by replacing the conductive (rings) and dielectric parts with dielectric and conductive materials, respectively, which results in the ISRR shown at the right of Fig. 4C. In the ISRR, the generation of the electric and magnetic fields are exchanged with those in the SRR. The ISRR structure can be fabricated by etching some part of the metallic surface. Electric fields originate from the central microstrip line and terminate perpendicularly on the ground plane when a dielectric material is inserted between the microstrip line and the ISRR (or SRR) (Fig. 4B). Owing to the presence of the dielectric substrate, the electric fields are tightly concentrated just below the central conductor, and the electric flux density reaches its maximum just below the microstrip line. To achieve strong electrodynamic coupling, the ISRR accordingly can be designed usually on the ground plane just below the microstrip line. When microwave currents are flowing through the microstrip feeding line, the ISRR can be excited by the axial electric field that is generated from the microstrip line. In this regard, it behaves like a parallel LC resonant circuit, thereby leading to a quasi-static resonant effect. This planar structure of the microstrip line as combined with the LC resonator, along with its capability of localizing microwave electric and magnetic fields, is very convenient for excitation of higher-order spin-wave modes in magnetic films especially when the LC resonator is coupled with a magnetic thin film. Therefore, nowadays in P-M coupling, a microstrip-line-loaded planar resonator is used to replace 3-D microwave cavities in P-M coupling measurements. Since the resonance frequency of such structures depends on the inductance and capacitance of the circuit, it can be readily controllable (tuned) by changing the dimensions of the SRR (or ISRR) and its split gap.

3. Magnon (spin-wave) modes

Felix Bloch first introduced the concept of a magnon in 1930 [44] to explain the temperature dependence of magnetizations in ferromagnetic materials. In spin lattices where spins are exchange- and dipolar-coupled, if certain external perturbations are locally applied, the disturbance of spin dynamics can propagate in waveforms such as collective excitations of the precessions of individual spins in a ferromagnetic medium. The quanta of spin waves are called magnons, which are quasi-particles of definite amounts of energy obeying boson characteristics. The classical description of dynamic precession motions of individual spins is governed by the Landau-Lifshitz-Gilbert (LLG) equation [16]

$$\frac{d\mathbf{m}}{dt} = -\gamma\mathbf{m}\times\mathbf{H}_{\text{eff}} + \alpha\mathbf{m}\times\frac{d\mathbf{m}}{dt} \tag{2}$$

where $\mathbf{m} = \mathbf{M}/M_s$, γ is the gyromagnetic ratio, M_s is the saturation magnetization, α is the Gilbert damping, and $\mathbf{H}_{\text{eff}}$ is the effective magnetic field. The first term on the right of Eq. (2) corresponds to the precessional motion of the magnetization vector $\mathbf{m}$ about the effective magnetic field direction. The second term is the damping term that is responsible for the alignment of the magnetization vector in the direction of $\mathbf{H}_{\text{eff}}$. Schematic representations of magnetization precession with and without damping are shown in Fig. 5A and B, respectively. Since individual spins are magnetically coupled, collective excitations of magnetizations form certain spin-wave modes of a given frequency and wavelength, as shown in a one-dimensional (1D) array of magnetic moments (see Fig. 6).

Generally, spin-wave modes are of two different types in terms of interaction energy. For a short-wavelength regime, the angle variation between the neighboring spins is large enough, and thus, the exchange interaction is dominant. Such spin-wave modes are called exchange-dominated modes. On the other hand, for a long-wavelength regime, the tilting angle between

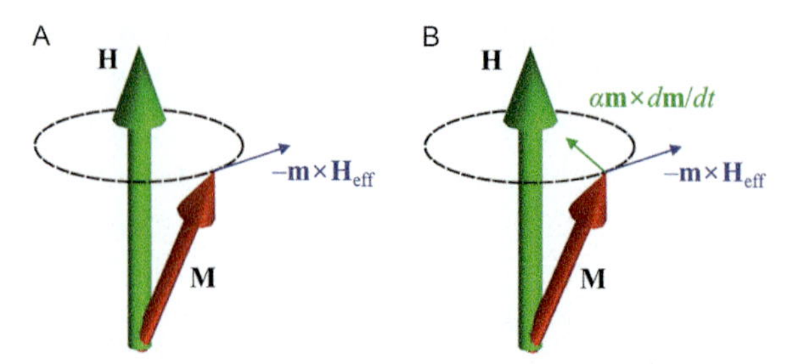

Fig. 5 Sketch of magnetization precession around an applied magnetic field (H) (A) without damping and (B) with Gilbert damping.

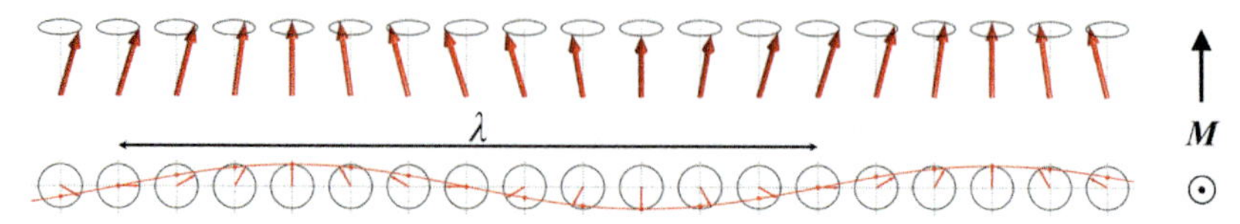

Fig. 6 Schematic representation of collective magnetization excitations (a certain mode of spin wave) in a one-dimensional (1D) array of magnetic moments (M): side view (top) and top view (bottom).

the neighboring spins is so small that long-range dipolar interaction can play a dominant role in the spin-wave characteristics. Such a spin-wave type is called the dipolar mode. In an intermediate-wavelength regime, both dipolar and exchange interactions are dominant in spin-wave excitations, which is the so-called dipolar-exchange spin-wave mode. The frequency of a certain specific spin-wave mode thus depends on the geometry and dimensions of the sample, externally applied magnetic field, the given material parameters as well as the microwave frequency range.

3.1 Ferromagnetic resonance (FMR)

In cases of uniformly oriented magnetizations under a magnetic field, individual magnetic moments process coherently about the field axis at the same frequency and in the same phase, and thus they can be considered as a single macro-spin. In this case, the exchange energy of this system does not change with the orientation of informally aligned magnetizations. This behavior is referred to as ferromagnetic resonance (FMR), the mode of which is a spin wave of infinite wavelength. The frequency of this FMR mode for an ellipsoid magnetized along the z-axis is expressed as the Kittel equation [45,46]

$$\omega_r = \gamma\sqrt{\left[H + \mu_0 M_S\left(N_{xx} - N_{zz}\right)\right]\left[H + \mu_0 M_S\left(N_{yy} - N_{zz}\right)\right]} \tag{3}$$

where N_{xx}, N_{yy}, and N_{zz} are the demagnetization factors along the x-, y-, and z-axes for any shape of ellipsoid. For a thin film magnetized in the film plane ($N_{xx} = 1$, $N_{yy} = N_{zz} = 0$ in SI unit), this formula reduces to [46,47]

$$\omega_r = \gamma\sqrt{H(H + \mu_0 M_s)} \tag{4}$$

while for magnetizations oriented in the axis normal to the film plane ($N_{xx} = N_{yy} = 0$, $N_{zz} = 1$), it turns out to be

$$\omega_r = \gamma(H - \mu_0 M_s) \tag{5}$$

In Eqs. (3)–(5), $\omega_r/2\pi$ is the resonant frequency and H is the applied dc magnetic field. This mode can be resonantly excited if the frequency of microwave uniform magnetic fields equals the mode resonance frequency $\omega_r/2\pi$. A detailed discussion of the rigorous quantum mechanical derivation of the Kittel formula is found in Ref. [47].

3.2 Spin waves in isotropic and infinite-size ferromagnetic media

The property of the propagating spin waves can be characterized by the dispersion, which is to say, by the relationship between the frequency and the wave vector $\vec{k}_{sw}$. For the cases of $\vec{k}_{sw} \neq 0$, the phase of the precessions of the neighboring spins differs, and thus the spins are no longer parallel with each other, and thus too, the exchange interaction cannot be neglected. Furthermore, the dynamic part of the magnetization is a function of the position. For small-angle precession, $\mathbf{m}\left(\vec{r}, t\right)$ can be written as the sum of a series of plane waves $\mathbf{m}\left(\vec{r}, t\right) = \sum_{\vec{k}_{sw}} \mathbf{m}_{\vec{k}_{sw}}(t)e^{i\vec{k}_{sw}\cdot\vec{r}}$, where the summation is done over all of the wavenumbers of the reciprocal space. Under this condition, the nonlinear LLG equation can be linearized and solved. Herrings and Kittel derived the dispersion relation of dipole-exchange spin waves in a ferromagnetic material of infinite size as given by the formula [48]:

$$\omega(k_{sw}) = \gamma\sqrt{\left(H_{\text{eff}} + Dk_{\text{sw}}^2\right)\left(H_{\text{eff}} + Dk_{\text{sw}}^2 + \mu_0 M_s \sin^2\varphi\right)} \tag{6}$$

where φ is the angle between the direction of the spin-wave vector and the static magnetization; H_{eff} is the effective magnetic field as defined by the vector sum of an externally applied magnetic field and the demagnetization field in the magnetic film, and the parameter D is the exchange stiffness constant that is given in terms of the exchange constant A as $D=2g_L\mu_B A/M_s$, where g_L is the Landé factor and μ_B is the Bohr magneton.

3.3 Dispersion characteristics of spin waves in thin films

The dispersion relations of spin-waves in isotropic and infinite ferromagnetic media can be determined by the Herring-Kittel equation, Eq. (6). However, for very thin films of finite size, the dispersion relation cannot be explained with Eq. (6), but was solved by Damon and Eshbach [49,50]. In their work, they solved the LLG equation by considering Maxwell's equations in the magnetostatic limit for an in-plane magnetized thin film. Since then, many examples have been reported [16,51–53].

In thin films, three different types of spin wave, namely, magnetostatic surface-spin waves (MSSWs), forward-volume magnetostatic spin waves (FVMSWs), and backward-volume magnetostatic spin waves (BVMSWs), can be excited according to the relative orientation between the wave vector and the magnetization orientation, as shown in Fig. 7. In in-plane magnetized films, BVMSWs are excited in the case of the wave vector being parallel to the applied magnetic bias field (magnetization orientation), and MSSWs are excited in the case of the wave vector being perpendicular to the field direction. On the other hand, FVMSWs are excited in the case where the magnetization is oriented to the field direction perpendicular to the film plane [17]. These three different types of spin wave show distinct characteristic dispersion curves analytically derived by Kalinikos and Slavin as shown below [54],

$$\omega(k_{sw}) = \gamma\sqrt{\left(H_{\text{eff}} + Dk_{\text{sw}}^2\right)\left(H_{\text{eff}} + Dk_{\text{sw}}^2 + \mu_0 M_s F\right)} \tag{7}$$

where $\omega(k_{\text{sw}})$ is the spin-wave frequency, and factor F is the field angle (θ, φ) dependence, as given by

$$F = P + \sin^2\theta_{\text{eff}}\left\{1 - P\left(1 + \cos^2\varphi\right) + P(1-P)\frac{\mu_0 M_s}{H_{\text{eff}} + Dk_{\text{sw}}^2}\sin^2\varphi\right\} \tag{8}$$

with $P = 1 - (1 - e^{-k_{sw}d})/k_{sw}d$ for the case of unpinned surface spins and d being the thickness of the film. The static magnetic fields applied are defined as $\mathbf{H} = H_x\hat{\mathbf{x}} + H_y\hat{\mathbf{y}} + H_z\hat{\mathbf{z}}$ with arbitrary orientation denoted as (θ, φ) where θ and φ are the polar and azimuthal angles from the $+z$ and $+x$ directions, respectively (Fig. 7A), and H_{eff} is the effective magnetic field. For cases of uniform magnetizations in a given static field orientation, $H_{\text{eff}}\cos\theta_{\text{eff}} = H\cos\theta - \mu_0 M_s \cos\theta_{\text{eff}}$ and $\mu_0 M_s \sin 2\theta_{\text{eff}} = 2H\sin(\theta_{\text{eff}} - \theta)$, where H equals

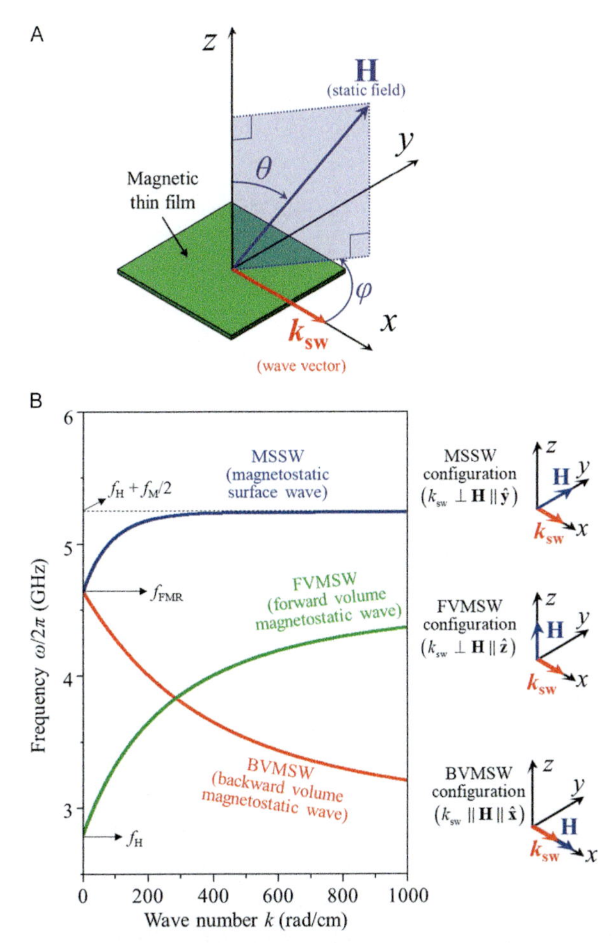

Fig. 7 (A) The wave-vector k_{sw} of spin waves and the direction of static magnetic fields H are indicated with respect to the x-, y-, and z-axes. The polar angle θ and the azimuthal angle φ are defined from the $+z$ and $+x$ directions, respectively. (B) Calculated dispersion characteristics using formulae (7–8) for forward-volume magnetostatic waves (FVMSW), backward-volume magnetostatic waves (BVMSW), and magnetostatic surface-spin waves (MSSW). The values were chosen for calculation as: the applied magnetic field $H=0.1$ T, saturation magnetization $\mu_0 M_s=0.174$ T, film thickness $=25\,\mu$m and exchange stiffness constant $D=5.33\times10^{-17}$ T/m^2. The excitation geometries of the three different types of spin waves also are shown (right side).

to $|\mathbf{H}|$ and θ_{eff} is the angle of the effective field from the $+z$-direction [54]. According to the relative orientation of the saturation magnetization with respect to k_{sw}, the BVMSW, MSSW, and FVMSW modes are excited at $(\theta, \varphi)=(90°, 0°)$, $(90°, 90°)$, and $(0°, 0°)$, respectively. The analytical forms of the dispersion curves for the BVMSW, MSSW, and FVMSW modes can be derived by inserting their corresponding values of (θ, φ), into Eqs. (7) and (8). The resultant explicit forms are given as [17,54]:

$$f_{\mathrm{BVMSW}} = \frac{\gamma}{2\pi}\sqrt{H_{\mathrm{eff}}\left(H_{\mathrm{eff}} + \mu_0 M_s \frac{1 - e^{-k_{sw}d}}{k_{sw}d}\right)}, \tag{9}$$

$$f_{\mathrm{MSSW}} = \frac{\gamma}{2\pi}\sqrt{\left(H_{\mathrm{eff}} + \frac{\mu_0 M_s}{2}\right)^2 - \left(\frac{\mu_0 M_s}{2}\right)^2 e^{-2k_{sw}d}}, \tag{10}$$

$$f_{\mathrm{FVMSW}} = \frac{\gamma}{2\pi}\sqrt{H_{\mathrm{eff}}\left[H_{\mathrm{eff}} + \mu_0 M_s\left(1 - \frac{1 - e^{-k_{sw}d}}{k_{sw}d}\right)\right]}, \tag{11}$$

As shown in Fig. 7B, the MSSW (blue line) and FVMSW (green line) modes have positive slopes in the dispersion curves, while the BVMSW (red line) mode has negative slopes. The frequency of FVMSWs approaches $f_H = (\gamma/2\pi)H_{\mathrm{eff}}$ as the wavenumber k_{sw} comes to zero. With increasing k_{sw}, the frequency goes to $f_{FMR} = (\gamma/2\pi)\sqrt{H_{\mathrm{eff}}(H_{\mathrm{eff}} + \mu_0 M_s)}$. However, the BVMSW's dispersion curve is mirrored with respect to the FVMSW's: it spans the region between f_{FMR} and f_H. For both volume modes (FVMSWs and BVMSWs), the amplitude of the magnetization precession has a cosinusoidal distribution across the film thickness. The negative slope of the BVMSWs (i.e., negative group velocity) in their dispersion curve implies that their phase and group velocities are counter-propagating and that an increase in k_{sw} is associated with a decrease in frequency. On the other hand, unlike FVMSWs and BVMSWs, MSSWs are localized to one surface of the film on which they propagate. The distribution of precessional amplitude across the film thickness is exponential, with a maximum at one surface of the film, and can be switched to the other surface by reversal (i.e., 180° rotation) of either the field or the propagation direction.

3.4 Yttrium iron garnet (YIG)

One of the most significant challenges in the technological application of spin waves (magnons) to real devices is how to increase the magnons' lifetime (order of nano-sec) in pure iron and in commonly used polycrystalline permalloy ($Ni_{81}Fe_{19}$) [55,56]. Such short lifetimes along with the low speed of magnons' propagation (four orders of magnitude less than the speed of light) result in mean free paths of spin waves typically $<10\,\mu m$ [57,58].

Thus, searching for low-damping magnetic materials is crucial to the development of magnonic circuits.

Monocrystalline yttrium iron garnet ($Y_3Fe_5O_{12}$: YIG) [59,60] was discovered by Bertaut and Forrat in 1956 [59]. The first study of the crystallographic and magnetic properties of YIG material was done by Geller and Gilleo in 1957 using a single-crystal YIG sphere of 0.23 mm diameter [60]. The unit cell of the garnet structure has cubic symmetry with a lattice constant of 12.376 Å and an X-ray density of 5.17 g/cm^3 at room temperature. Each unit cell contains 160 atoms (8 formula units of $Y_3Fe_5O_{12}$) and, thus, 24 Y^{3+} ions, 40 Fe^{3+} ions, and 96 O^{2-} ions. The coordination of the cations in YIG is complex, and therein, there are three types of crystallographic sites that are occupied by these metal ions, as schematically shown in Fig. 8A. Tetrahedral (*d*) sites are surrounded by four O^{2-} ions located at the corners of a tetrahedron, and there are 24 of those sites occupied by Fe^{3+} ions. Octahedral (*a*) sites are surrounded by six O^{2-} ions located at the corners of an octahedron, and there are 16 of them that are occupied by Fe^{3+} ions. Dodecahedral (*c*) sites have 8 oxygen ions located at the corners of a polyhedron with 12 faces and 18 edges, and Y^{3+} ions occupy the 24 sites of this type. The basic arrangement of the ions in the unit cell of YIG is shown in Fig. 8B. Each O^{2-} ion is surrounded by one *a*-site Fe^{3+} ion, one *d*-site Fe^{3+} ion, and two *c*-site Y^{3+} ions [61,62].

In such a garnet, antiferromagnetic coupling occurs through super-exchange interaction between the cations mediated by oxygen ions. Since Y^{3+} has no 4f electrons, it does not possess any magnetic moment. The magnetization in YIG originates from the super-exchange interactions between the *a*-site Fe^{3+} ions and the *d*-site Fe^{3+} ions. According to the Neel's theory, the magnitude of the super-exchange interactions between the two magnetic ions depends strongly on the bonding angle of the magnetic ion-oxygen ion-magnetic ions. The strongest interaction occurs at the bond angle of ~180°, while the weakest occurs at ~90°. In the YIG material, the Fe^{3+} (*a*-site)–O^{2-}–Fe^{3+} (*d*-site) bonding angle is about 126.6° [61]. Thus, the strongest super-exchange interaction in the material occurs between the *a*-site Fe^{3+} ion and the *d*-site Fe^{3+} ion. The net magnetic moment of YIG thus comes from one ferric ion ($5\mu_B$) per each formula unit. Since each unit cell contains eight formula units of $Y_3Fe_5O_{12}$, each unit cell has a net magnetic moment of $40\,\mu_B$ ($8 \times 5\,\mu_B$). This net magnetic moment corresponds to a theoretical saturation value of $4\pi M_s \sim 2470$ G, which is very close to YIG films' measured value of 2463 G at 4.2 K [63]. The room-temperature value of $4\pi M_s$ reportedly is in the range of 1730–1780 G [64].

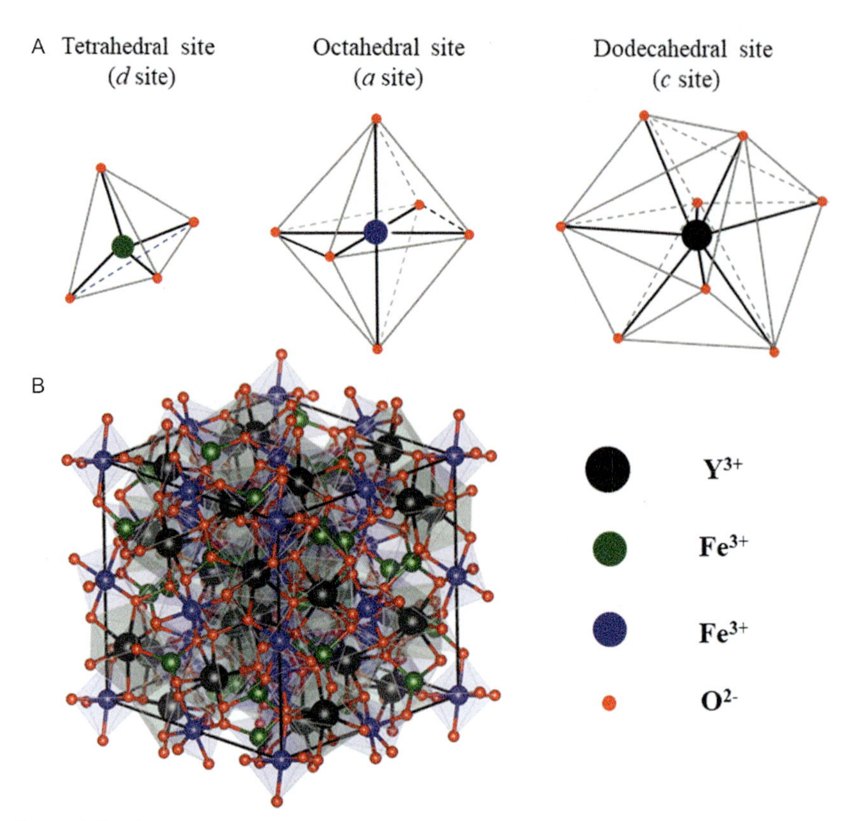

Fig. 8 (A) Schematic diagrams of 3 different sites occupied by cations in yttrium iron garnet (YIG), a trivalent ion of iron at a site surrounded by 6 oxygen ions in octahedral symmetry, a divalent ion of iron at a site surrounded by 4 oxygen ions in tetrahedral symmetry, and a yttrium ion surrounded by 8 oxygen ions, forming an 8-cornered 12-sided polyhedron. (B) Schematic diagrams representing one YIG unit cell with cation arrangement.

YIG also has a cubic magnetocrystalline anisotropy of (111) easy axis. The first- and second-order cubic anisotropy constants at room temperature are $K_1 = -6100$ erg/cm^3 and $K_2 = -260$ erg/cm^3, respectively [16]. YIG thin films are usually grown on (111)-oriented gadolinium gallium garnet ($Gd_3Ga_5O_{12}$, GGG) substrates, due to their extremely small lattice mismatch on the order of 0.001 Å. (111)-oriented YIG films have an out-of-plane effective anisotropy field ($2K_1/M_s$) of ~85 Oe, which is much smaller than external magnetic fields used in typical experiments and, therefore, is usually ignored in data analysis.

In YIG crystals, the FMR linewidth is about 0.2 Oe at 10 GHz (Fig. 9) [65]. This linewidth corresponds to an intrinsic Gilbert damping constant of $\sim 3 \times 10^{-5}$, which is two orders of magnitude smaller than typical values in

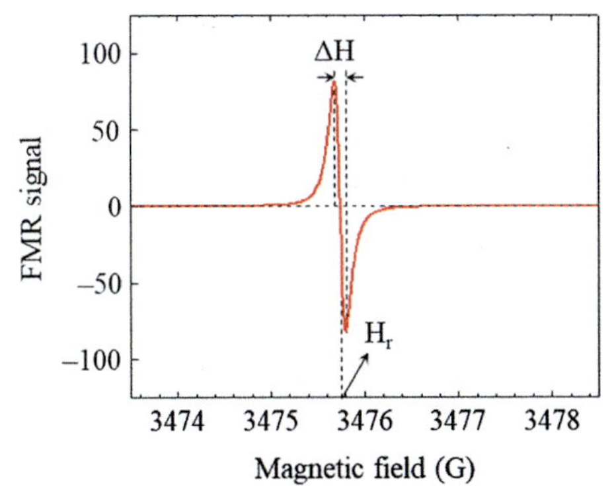

Fig. 9 FMR absorption curve for a YIG sphere. H_r, resonance field; ΔH, linewidth at 9.8 GHz.

ferromagnetic metals (FMs) [66,67]. Such an extremely small damping constant and very high spin density ($\sim 10^{27}\,m^{-3}$) [65,68] make YIG material well suited for studies of both spin waves themselves and magnons coupled with photon modes, as will be explained later.

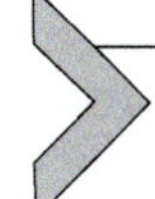

4. Photon-magnon (P-M) coupling

4.1 Research trends

The field of quantum information has achieved significant progress in recent decades, and its continued success is driven by a strong focus on quantum technology. Ongoing development of the devices and experimental techniques that utilize the quantum nature of the system to perform storage, transfer, and processing of quantum information is critical to achieving the ambitious goal of workable quantum computation. In this framework, a high-fidelity technology must be able to exchange information with preserved coherence [19,69,70]. To this end, a system that consists of two sub-systems has to operate in a regime called "strong coupling." The strong-coupling regime is characterized by the strength of coupling between the subsystems, which is larger than the mean energy loss in either of them. A straightforward way to decrease energy loss is to make use of resonant systems [62,71–73].

The first step toward the physical realization of a scheme for manipulating quantum information requires a proper choice of subsystems to be fabricated. While one of the subsystems is typically an electromagnetic cavity, the other should demonstrate quantum behaviors, preferably with long coherence times. Many promising solid-state implementations of the latter subsystem have been demonstrated, including trapped ions [74], superconducting qubits [75], quantum dots [76], photonic nanostructures [77], ultracold atoms and ensembles [78], and electronic and nuclear spins in solids [79]. In particular, systems that exploit plasmonic resonances make use of strong coupling between electric dipoles and optical fields localized on the sub-wavelength scale. This makes possible the creation of solid-state sources of quantum states of light (single photons, indistinguishable or entangled photons) such as those based on semiconductor quantum dots or in nitrogen-vacancy centers in diamonds, as embedded in optical microcavities or precisely placed in close proximity to optical nanoantenna [1,3,80–82].

On the other hand, the direction in which to achieve strong coupling using large spin ensembles was pointed out by Imamoğlu [83], who claimed that the collective excitations of spin ensembles can be incorporated into hybrid systems with a nonlinear element such as a Josephson junction. In general, the interaction of a single spin with the electromagnetic field mode is very weak; however, collective enhancement, which scales as the square root of the total number of spins, makes the effective coupling strong enough. Therefore, large-spin ensembles are considered to be of great potential for application to hybrid quantum systems and quantum information manipulation [84–86].

The utilization of spin-doped 3-D cavities makes operation in the strong-coupling regime possible, though the coupling strength is always limited due to the fact that an increase in the number of spins usually leads to a broadening of both the spin and cavity resonances, thus resulting in a trade-off with electromagnetic coupling. This problem may be solved by using dielectrics that exhibit ferromagnetism instead of dilute paramagnetic impurities. Because ferro- or ferrimagnetic materials are perfectly ordered systems, they do not suffer from excessive loss due to spin-spin interactions when the spin density is increased. At the same time, ferrite materials have much larger magnetic susceptibilities than paramagnetic systems, due to the largest volumetric density of spins; thus, the offer the potential for much stronger P-M coupling per unit volume.

A natural choice for a ferromagnetic system is yttrium iron garnet [YIG: $Y_3Fe_5O_{12}$], which is a ferrite material of unique microwave properties. YIG exhibits a very low microwave magnetic-loss parameter and excellent dielectric properties at microwave frequencies. The collective character of the precession, the large density of spins, and the extremely small magnetic loss result in the very strong response of YIG to an external microwave stimulus.

As a result, only very small volumes of a bulk YIG material may be needed in order to achieve strong quantum electrodynamics in coupled-resonator systems based on ferro- and ferrimagnetic materials. Thus, the combination of a specially designed microwave cavity and YIG represents a promising path toward the realization of an ultra-strong-coupling regime, and has been extensively studied for the purposes of P-M coupling.

4.2 Sample design of hybrid P-M systems

FMR behavior is excited by placing a sample in a uniform (or quasi-uniform) microwave magnetic field. The onset of FMR is easily seen as applied-field-dependent absorption of the microwave power by the magnetic material. Microwave cavities are more suitable for driving FMR in bulk magnetic materials. Another more convenient way to excite FMR in ferro- and ferrimagnetic films is by placing them on top of a microstrip or coplanar microwave transmission line [46] or forming a microstrip line directly on top of the film [87,88]. The idea of using collective spin excitations in a ferro/ferrimagnet for P-M coupling is to allow for coherent interaction between the microwave optical photons and spin systems, thereby efficiently enhancing the light-matter interaction. In principle, a spatially uniform magnetostatic mode, called the Kittel mode, in YIG, manifests itself as a macro magnetic dipole in precession. The largeness of the magnetic dipole moment and the long-time coherence of the magnons in the Kittel mode make it possible to couple them strongly to the microwave photons in a microwave cavity mode and hybridize them. The possibility of realizing hybridized P-M excitations in ferromagnetic systems was first discussed theoretically by Soykal and Flatté in 2010 [89,90], but was realized experimentally only in 2013. In this first experiment, Huebl et al. [7] employed a YIG sphere coupled to a planar superconducting resonator at ultra-low temperatures, using microwave transmission to measure the key coupling signature of anti-crossing in the eigenspectrum.

In recent years, P-M coupling has often been experimentally studied in the microwave (GHz) regime, because efficient electronics for generation and detection of terahertz (THz) radiation are still under development. Different hybrid systems have been proposed to demonstrate the coupling between photon and magnon modes where a microwave resonator (3-D cavity) [8–10,14,19–21,25] and planar configuration [11,24,26–30] was loaded with a ferrimagnetic insulator such as YIG (film or bulk). Broadly, these hybrid systems can be categorized into three types: YIG sphere/3-D cavity resonator, YIG sphere/planar resonator, and YIG film/planar resonator.

The first experimental demonstration of the coupling of a magnetostatic mode (the Kittel mode) in a sphere of YIG and a 3-D microwave cavity mode at room temperature was reported by Tabuchi et al. in 2014 [8]. This was achieved by inserting a YIG sphere into a high-quality 3-D microwave cavity wherein the microwave magnetic field amplitude was the maximum. Thereafter, different characteristic features of P-M coupling have been realized by designing high-quality 3-D cavities [Fig. 10B] with confined microwave fields. Since a YIG sphere or cavity is essentially 3-D, it may be incompatible with the current Complementary metal-oxide-semiconductor (CMOS) platforms incorporated in planar geometry. Therefore, a planar microwave resonator such as the split-ring resonator (SRR) or T-type resonator [Fig. 10C and D] loaded with a YIG sphere or film has been proposed [11,26,27,29]. Similarly, a newly found planar hybrid system that consists of a YIG film positioned on top of a microstrip line, being capacitive-coupled to an inverted pattern of SRR (i.e., ISRR), demonstrate strong P-M coupling along with a nearly five-times-higher gain and a wider frequency band relative to those of the SRR of the same dimensions as the ISRR [26]. In this case, to achieve strong electrodynamic coupling, the ISRR was designed on the ground plane just below the microstrip line, as shown in Fig. 10E. Also, the planar structure and the easy localization of microwave fields both allow for much excitation of the higher-order spin-wave modes, which also can contribute to P-M coupling.

Apart from that, studies using transition-metal thin films such as FeCo [91] or Py [92] coupled with planar resonators also have been done in order to demonstrate the possibility of achieving YIG-type functionalities and to overtake the working frequency limitation of YIG. In those cases, the sample consisted of patterned FeCo films and copper SRRs on a MgO(001) substrate. The 30 nm-thick FeCo film was subsequently patterned into 100 μm disks using standard lithographic techniques. An 8 μm-thick insulating SU-8 layer was subsequently spun and cured on top of the FeCo disks. Next, a 300-nm-thick Cu film was deposited on top of the SU-8 layer and patterned into an SRR. Their proximity ensured excellent coupling between the FeCo disks and the SRR (see Fig. 10F).

4.3 Theory of P-M coupling

Condensed matter physics is a science geared to technological development, and as such, it requires active interplay between theory and experimentation. In order to understand the complex and nuanced behavior of P-M coupling,

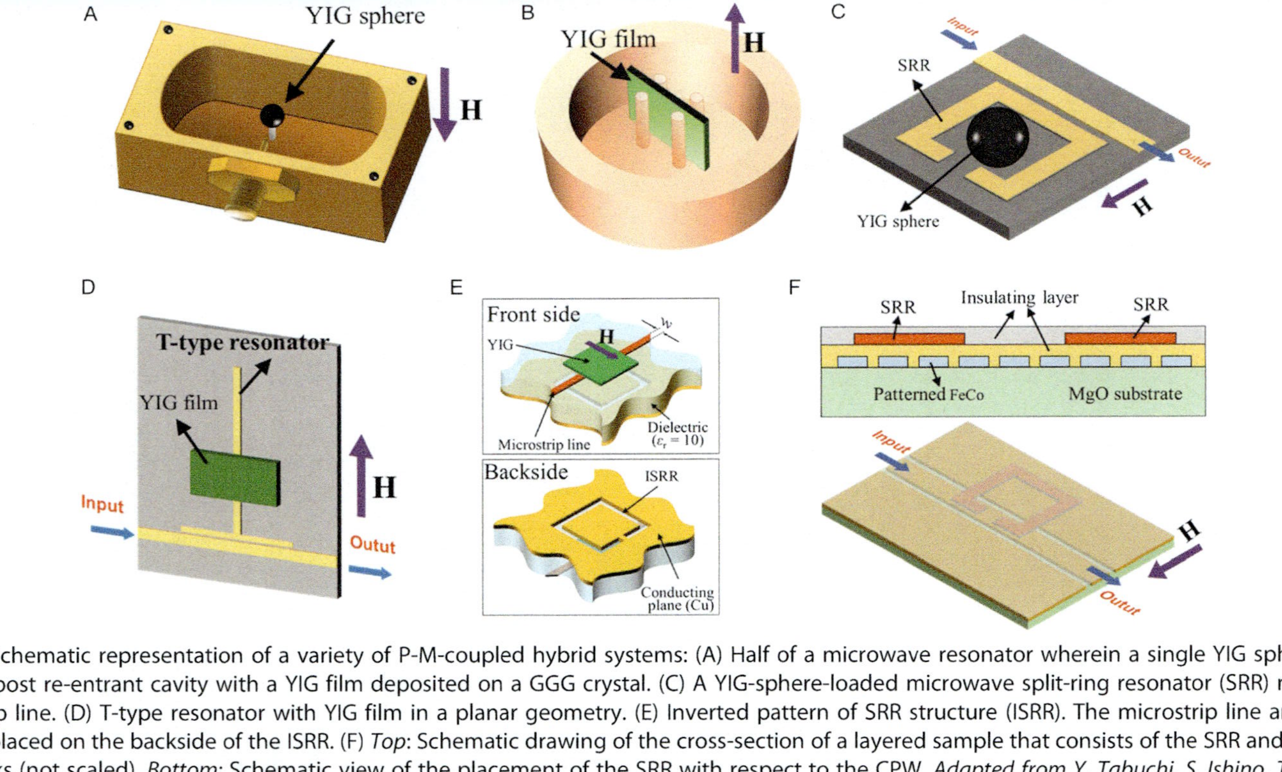

Fig. 10 Schematic representation of a variety of P-M-coupled hybrid systems: (A) Half of a microwave resonator wherein a single YIG sphere exists. (B) Four-post re-entrant cavity with a YIG film deposited on a GGG crystal. (C) A YIG-sphere-loaded microwave split-ring resonator (SRR) next to the microstrip line. (D) T-type resonator with YIG film in a planar geometry. (E) Inverted pattern of SRR structure (ISRR). The microstrip line and the YIG film are placed on the backside of the ISRR. (F) *Top*: Schematic drawing of the cross-section of a layered sample that consists of the SRR and patterned FeCo disks (not scaled). *Bottom*: Schematic view of the placement of the SRR with respect to the CPW. *Adapted from Y. Tabuchi, S. Ishino, T. Ishikawa, R. Yamazaki, K. Usami, Y. Nakamura, Hybridizing ferromagnetic magnons and microwave photons in the quantum limit, Phys. Rev. Lett. 113 (2014) 083603; M. Goryachev, W.G. Farr, D.L. Creedon, Y. Fan, M. Kostylev, M.E. Tobar, High-Cooperativity Cavity QED with Magnons at Microwave Frequencies, Phys. Rev. Appl. 2 (2014) 054002; B. Bhoi, B. Kim, J. Kim, Y.-J. Cho, S.-K. Kim, Robust magnon-photon coupling in a planar-geometry hybrid of inverted split-ring resonator and YIG film, Sci. Rep. 7 (2017); V. Castel, R. Jeunehomme, J. Ben Youssef, N. Vukadinovic, A. Manchec, F. K. Dejene, et al., Thermal control of the magnon-photon coupling in a notch filter coupled to a yttrium iron garnet/platinum system, Phys. Rev. B 96 (2017) 064407; D. Zhang, W. Song, G. Chai, Spin-wave magnon-polaritons in a split-ring resonator/single-crystalline YIG system, J. Phys. D Appl. Phys. 50 (2107) 205003; S.A. Gregory, L.C. Maple, G.B.G. Stenning, T. Hesjedal, G. van der Laan, G.J. Bowden, Angular control of a hybrid magnetic metamolecule using anisotropic FeCo, Phys. Rev. Appl., 4 (2015) 054015.*

three different models have been proposed: coupled harmonic oscillators, microscopy theory (quantum model), and dynamic phase correlation [5–11,19–26,29]. The details of the derivations and significances of these models can be found in a recent review article by Harder and Hu [5] based on P-M coupling in 3-D hybrid structures. Herein we present, in particular, coupled harmonic oscillators and a dynamic phase correlation model for coupled systems in a planar geometry not yet discussed by Harder et al. [5,22]. Before proceeding to the modified models of P-M coupling for planar geometry, this section of this chapter first presents brief reviews and highlights each of those theoretical models.

In general, P-M coupling can be described using the coupled harmonic oscillator analogy, which assumes that the photon mode and the magnon mode can be modeled as two harmonic oscillators coupled to each other via a coupling constant. In this harmonic oscillator perspective, the dispersion and the linewidth evolution of coupled modes between the photon and the magnon can be calculated using either the quantum Hamiltonian or the classical equations of motion for the coupled harmonic oscillator system. A limitation of the harmonic oscillator model is that it cannot explain the physical origin of P-M coupling. On the other hand, Bai et al. [10] proposed a different model of dynamic phase correlation based on electromagnetic interactions between the magnon mode of magnetic materials and the photon mode of a 3-D cavity. This semiclassical model describes the origin of the P-M coupling from the phase correlation arising from both the Faraday's induction and Ampere's circuit laws. Although the above two models based on a classical point of view well describe the P-M coupling behaviors, further insights into the nature of the coupling have to be gained quantum-mechanically [1,3,10,22]. This approach, for example, reveals the microscopic origin of the coupling strength [1,3]. Quantum formalism also can be extended to explore quantum effects of hybridizations between two different systems.

All three of the models can accurately describe microwave transmission spectra, thereby providing an important tool for the analysis of strongly coupled P-M systems. However, the following basic question arises: Which model should be used by beginners studying in this new field of P-M coupling? To answer this question, it is useful to first consider that, in general, the models can be split into either of two categories, classical or quantum: the former entails solving the coupled LLG and Maxwell's equations, while the latter requires first defining a Hamiltonian and then determining the eigenfrequencies and transmission properties. In cases where an experimental system contains an obvious extension of the classical approach to multiple cavities or spin-wave modes [25,26], it is natural to use an analogous classical model. On the other hand, when an inherently quantum phenomenon is investigated, it is natural to extend the quantum mechanical approach by modifying the Hamiltonian through the inclusion of additional interaction terms [5,34,93].

The final equation for theoretical modeling of cavity/P-M coupling remains the same for all three of the models; near the anti-crossing center, all reduce to a set of coupled equations, leading to a 2×2 matrix:

$$\begin{pmatrix} \omega-\widetilde{\omega}_p & g \\ g & \omega-\widetilde{\omega}_r \end{pmatrix}\begin{pmatrix} h \\ m \end{pmatrix}=\begin{pmatrix} \omega_p h_0 \\ 0 \end{pmatrix}, \tag{12a}$$

where $\widetilde{\omega}_r=\omega_r-i\alpha\omega_r$ and $\widetilde{\omega}_p=\omega_p-i\beta\omega_p$ are the two complex resonance frequencies of the magnon and photon modes, respectively, h_0 is a driving field strength at the frequency ω, and g is the coupling rate. The top and bottom terms in Eq. (12a) describe the resonant behavior of the cavity (h) and magnetization (m), respectively. The dispersion of the hybridized modes can then be determined by solving the determinant of the matrix, as given by

$$\widetilde{\omega}_\pm=\frac{1}{2}\left[\left(\widetilde{\omega}_r+\widetilde{\omega}_p\right)\pm\sqrt{\left(\widetilde{\omega}_r-\widetilde{\omega}_p\right)^2+4g^2}\right]. \tag{12b}$$

The real and imaginary parts of the eigenvalues ($\widetilde{\omega}_\pm=\omega_\pm-i\Delta\omega_\pm$) are plotted in Fig. 11A and B, respectively. Fig. 11A shows the distinct anti-crossing of P-M-coupled modes typically observed. The horizontal and inclined dashed lines indicate the uncoupled cavity and uncoupled FMR modes, respectively. The strength of the P-M interaction determines the size of the mode splitting, and is given by the frequency gap ($\omega_{gap}=\omega_+-\omega_-$), as shown in Fig. 11A. For a strongly coupled P-M system with $\alpha,\beta\ll 1$, the effect of damping is negligible, and thus Eq. (12b) becomes

$$\omega_\pm=\frac{1}{2}\left[\left(\omega_r+\omega_p\right)\pm\sqrt{\left(\omega_r-\omega_p\right)^2+(2g)^2}\right] \tag{13}$$

This indicates that in the strong-coupling regime, the modes' frequency gap gives us the coupling strength as $\omega_{gap}=2g$, as shown in Fig. 12. This theory was extended later by Harder et al. [94], who took the energy difference between the two eigenmodes in order to obtain the mathematical expression of frequency splitting (gap) in terms of coupling strength as well the damping parameters β and α, as written by (for details see Refs. [94,95])

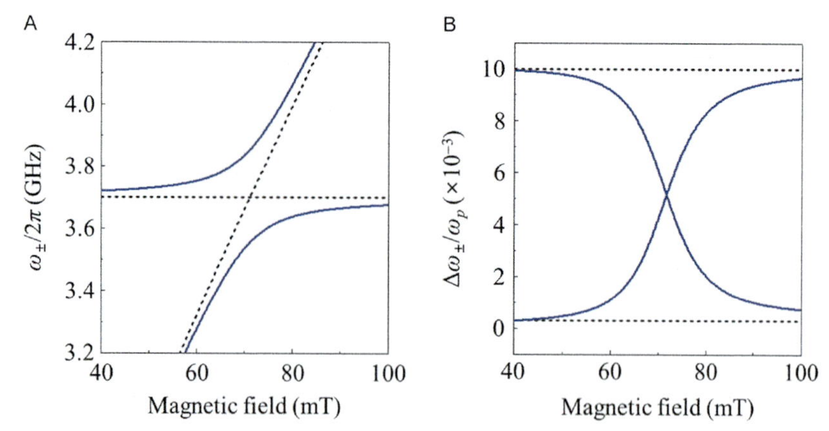

Fig. 11 Calculation of (A) the dispersion $\omega_{\pm}$ and (B) the line width $\Delta\omega_{\pm}$ using Eq. (12b). The horizontal and diagonal dashed lines in (A) show the uncoupled cavity and FMR dispersions, respectively, while the horizontal dashed lines in (B) indicate the damping limit owing to the intrinsic damping constants of FMR α and cavity β.

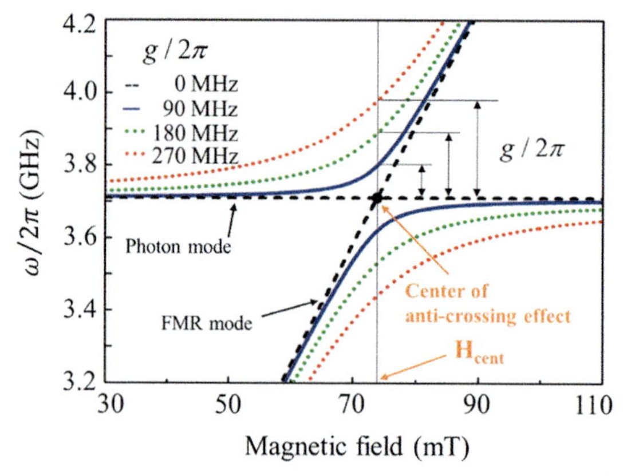

Fig. 12 Dispersion curves of coupled modes for different coupling strengths calculated according to a simple harmonic coupled oscillator model.

$$\omega_{gap} = (\omega_{+} - \omega_{-}) = \sqrt{4g^2 - \omega_p^2(\beta - \alpha)^2} \tag{14}$$

For the cases of $g > \omega_p|\beta - \alpha|/2$, then, the frequency gap is real, indicating that the dispersion spectra of the two coupled modes are anti-crossing with each other, and that their linewidths are crossing with each other, as found typically in strong P-M coupling. On the contrary, for cases of $g < \omega_p|\beta - \alpha|/2$, the frequency gap is imaginary, indicating that the dispersions of the two modes are crossing while their linewidths are anti-crossing, as demonstrated experimentally by Hardar et al. [96]. This behavior is known as level attraction, as will be discussed in Section 5.4. On the other hand, under the condition of $g = \omega_p|\beta - \alpha|/2$, the frequency gap in the coupled modes disappears and their linewidth merges. This special value of coupling strength defines exceptional points (EPs), where the eigenvectors of the system coalesce [96,97], as experimentally demonstrated by Zhang et al. [97]. Although the distinct phenomena originating from the EP singularity have been demonstrated in other electromagnetic, atomic and molecular physics systems [94,98–102], the related interesting properties in a magnon-photon coupled systems have yet to be explored.

4.4 Harmonic oscillator model

In the phenomenological coupled-oscillator model discussed in Refs. [5, 22], the external driving force acts only on one oscillator, under the assumption that there is no analogous driving force F acting on the other oscillator. This assumption is applicable to P-M coupling in a microwave 3-D cavity/YIG system where the feeding line excites only the cavity resonator and is located far from the YIG. However, the assumption no longer holds for the case of planar hybrid systems such as SRR-YIG (or ISRR-YIG), where the microstrip feeding line can excite both the resonator and the YIG. As schematically shown in Fig. 13, in the first pathway, the microwave magnetic field generated by ac currents flowing along the microstrip feeding line can induce an electromotive force, which is to say, currents in the SRR via Faraday's induction law, which force subsequently couples

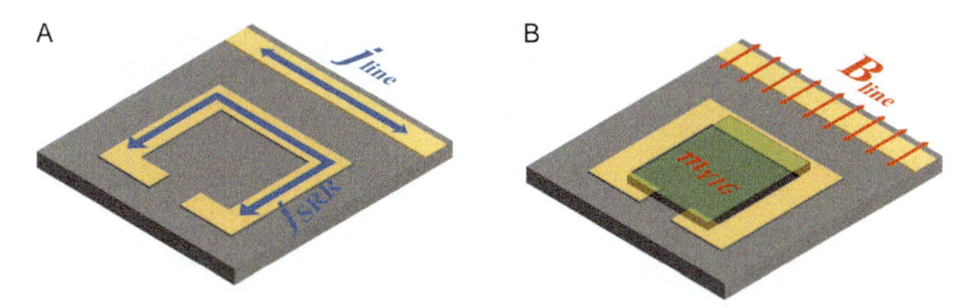

Fig. 13 Schematic representation of excitation pathways in a planar hybrid SRR-YIG system: Magnetic fields generated from the current flow in the microstripline can excite both magnetizations of YIG film and SRR via Faraday's induction law. (A) Coupling of currents between the microstrip feed line and the SRR. Blue lines indicate the direction of electric current. (B) Coupling of magnetic fields to the magnetization of the YIG. Red lines indicate the direction of magnetic fields. *Adapted from D. Zhang, W. Song, G. Chai, Spin-wave magnon-polaritons in a split-ring resonator/single-crystalline YIG system, J. Phys. D Appl. Phys. 50 (2107) 205003.*

with the YIG on the SRR. In the second pathway, the magnetic field produced by the microstrip feed line can excite magnetizations in the YIG, which subsequently couple with the SRR. Tay et al. [30] introduced a new parameter τ, which represents the relative strength of the two excitation pathways in the planar system described above. The coupled oscillator model results in

$$\begin{pmatrix} \omega^2 - \omega_p^2 + i\beta\omega\omega_p & -k\omega_r^2 \\ -k\omega_p^2 & \omega - \omega_r^2 + i\alpha\omega\omega_r \end{pmatrix} \begin{pmatrix} j_R \\ m_m \end{pmatrix} = \begin{pmatrix} 1 \\ \tau \end{pmatrix}, \tag{15}$$

$$\begin{pmatrix} 1 \\ \tau \end{pmatrix} \propto \begin{pmatrix} g_R & 0 \\ 0 & g_m \end{pmatrix} \begin{pmatrix} E_{line} \\ B_{line} \end{pmatrix}, \tag{16}$$

where j_{line} couples with j_R with a coupling constant g_R, while B_{line} couples with m_m with a coupling constant g_m; E_{line} and B_{line}, meanwhile, denote the electric and magnetic fields, respectively, produced by the microstrip line, and k denotes the general coupling constant between the magnon and photon modes. Since the strengths of j_{line} and B_{line} are given by the current in the microstrip line, τ depends on g_m/g_R. Tay et al. [30] calculated transmission spectra for different values of τ and k/α, as shown in Fig. 14. In the cases of the lower τ values ($g_R > g_m$), the microstrip primarily excites the SRR,

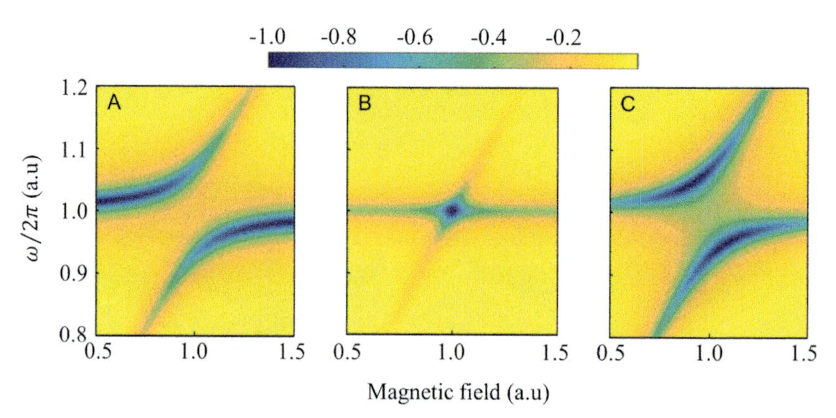

Fig. 14 Theoretical absorption spectra calculated for (A) $\tau = 2$, $k/\alpha = 4$, (B) $\tau = 10$, $k/\alpha = 0.5$, and (C) $\tau = 10$, $k/\alpha = 4$. The dark-blue and yellow colors indicate strong and weak absorptions of microwaves, respectively, as shown by the color-bar scale. *Reprinted with permission from Ref. [29] by Copyright Clearance Center Elsevier. A modified version of this figure was originally published in Z.J. Tay, W.T. Soh, C.K. Ong, Observation of electromagnetically induced transparency and absorption in Yttrium Iron Garnet loaded split ring resonator, J. Magn. Magn. Mater. 451 (2018) 235–242, https://doi.org/10.1016/j.jmmm.2017.11.029.*

which then is coupled to the YIG. The resulting dispersion spectra are consistent with the anti-crossing reported in P-M-coupled systems, where only one excitation pathway is considered. In this case, k/α does not affect the dispersion type. However, for the cases of higher τ values ($g_R < g_m$) and lower k/α values, the usual anti-crossing between the photon and magnon modes disappears, and the dispersion spectra show a strong absorption. In this case, the microstrip primarily excites the YIG, which is then coupled to the SRR. This additional excitation pathway, which thus far has not been considered in 3-D hybrid systems, gives rise to new behavior, as shown in Fig. 14B. On the other hand, the theoretical spectra for the high τ and high k/α parameter regime are given in Fig. 14C. In this regime, due to the high coupling constant, there is an anti-crossing region. However, the present study was experimentally limited to normal anti-crossing behavior. Further, this theory does not provide any information on the conditions under which the anti-crossing behavior would disappear. A more detailed comprehensive theory and experimental demonstration of P-M coupling were provided by Bhoi et al. [103] for a planar hybrid system.

4.5 Dynamic phase correlation model

The semi-classical picture describing P-M interaction, proposed by Bai et al. [10] for a 3-D cavity, is based on the combination of a microwave RLC and Landau-Lifshitz-Gilbert (LLG) equations [46]. The coupling of the magnetization dynamics with microwave photons is established via two electromagnetic interactions: one is known as Faraday induction, which induces a voltage in the RLC circuit due to the precessing magnetization; the other is governed by Ampere's circuit law, which provides magnetic fields to excite the magnetizations in magnetic materials. This model was later modified by Grigoryan et al. [104] for a hybrid 3-D system by considering an additional phase-shifted FMR driving force that acts on the magnetizations along with the magnetic component of the microwave electromagnetic fields in the cavity.

Bhoi et al. [103] recently provided a further extension of this model for a P-M-coupled hybrid system consisting of an ISRR and YIG film in a planar geometry, as schematically illustrated in Fig. 15. The hybrid system used in this study consists of three different physical systems: (1) the microstrip line to excite the magnon and photon modes as well as to probe those coupled modes; (2) the YIG, and (3) the ISRR, wherein the magnon and photon modes are to be excited, respectively. By considering all of the three interactions, namely, between ① and ②, ① and ③, and ② and ③, the analytical form of the coupling matrix can be written as

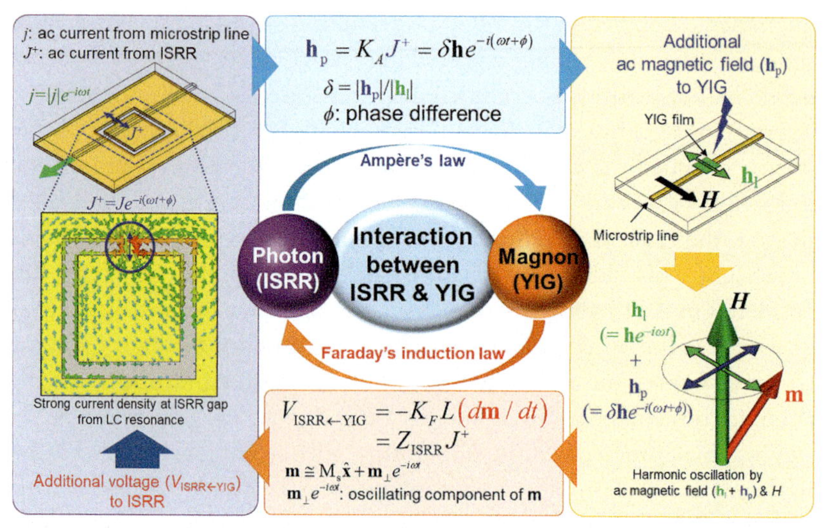

Fig. 15 Schematic illustration of involved mechanisms that cause P-M coupling in a planar hybrid system consisting of ISRR and YIG film. AC currents applied along the microstripline create two different microwave magnetic fields of $\mathbf{h}_l(=\mathbf{h}e^{-i\omega t})$ and $\mathbf{h}_p(=\delta\mathbf{h}e^{-i(\omega t+\phi)})$ around the feeding line and ISSR split gap, respectively. ϕ is the phase difference between these two microwave magnetic fields $\mathbf{h}_p$ and $\mathbf{h}_l$, and $\delta=|\mathbf{h}_p|/|\mathbf{h}_l|$. Both magnetic fields excite the magnetization of the YIG. Once the magnetizations are excited in the YIG, they can yield an additional voltage in the ISRR as $V_{ISRR\leftarrow YIG}=-K_F L(dm/dt)$, according to Faraday's induction law. The values of both ϕ and δ can be controlled by changing the ISRR's split-gap orientation/position with respect to the microstrip line axis.

$$\begin{pmatrix} \omega^2-\omega_p^2+2i\beta\omega\omega_p & ik_F\omega^2 \\ -i\omega_m\left(1+\delta e^{i\phi}\right)k_A & \omega-\omega_r+i\alpha\omega \end{pmatrix}\begin{pmatrix} J^+ \\ m^+ \end{pmatrix}=\begin{pmatrix} 0 \\ 0 \end{pmatrix}, \tag{17}$$

$$\Omega\begin{pmatrix} m^+ \\ J^+ \end{pmatrix}=\begin{pmatrix} 0 \\ 0 \end{pmatrix}, \tag{18}$$

where m^+ and J^+ are the magnetization in the YIG film and currents in the ISRR, respectively, $\omega_r=\gamma\sqrt{H(H+\mu_0 M_s)}$ is the FMR frequency of YIG film and $\omega_m=\gamma\mu_0 M_s$ with gyromagnetic ratio $\gamma/2\pi=28$ GHz/T. ω_p is the resonance frequency of the ISRR. Here, k is the coupling constant between the ISRR photon and the magnon modes, while β and α are the damping parameters of the ISRR and YIG film, respectively. The determinant Ω is expressed as $(\omega-\omega_r+i\alpha\omega_r)(\omega^2-\omega_p^2+2i\beta\omega\omega_p)-k_A k_F\omega_m\omega^2(1+\delta e^{i\phi})=0$ with $k^2\cong k_A k_F$; as such, it finally describes P-M coupling in the ISRR-YIG hybrid system. The coupling of the magnetizations in the YIG to currents in the ISRR is represented by the matrix's first line, which describes the RLC circuit of the ISRR as affected by the magnetization motions of the YIG film, while the effect of the net currents of the ISRR on the magnetization dynamics in the YIG film is described by the second line of the matrix. Here, the magnetizations in the YIG are influenced by the effective field, which is the sum of two time-dependent magnetic fields $\mathbf{h}_l(=\mathbf{h}e^{-i\omega t})$ from the feeding line and $\mathbf{h}_p(=\delta\mathbf{h}e^{-i(\omega t+\phi)})$ from the ISSR split gap, where ϕ is the phase difference between these two microwave magnetic fields $\mathbf{h}_p$ and $\mathbf{h}_l$, yielding $\mathbf{h}_p=\delta e^{-i\phi}\mathbf{h}_l$, with $\delta=|\mathbf{h}_p|/|\mathbf{h}_l|$. The two parameters of ϕ and δ can remarkably influence anti-crossing effects, including the dispersion type, the linewidth, and the net coupling strength of the two coupled modes. These effects can be manipulated readily by changing the position/orientation of the ISRR's split gap with respect to the microstrip line axis. Hereafter, the ϕ and δ parameters are referred to collectively as the geometry factor [103].

The complex eigenvalues $\widetilde{\omega}_\pm=\omega_\pm-i\Delta\omega_\pm$ of the coupled modes can be numerically calculated by solving the determinant of Eq. (18) using the experimentally observed numerical values of $\alpha=3.2\times10^{-4}$ and $\beta=0.02$, $k=0.03$, and $\omega_p/2\pi=3.7$ GHz. For two specific values of $(\delta, \phi)=(1/2, 0)$ and $(2, \pi)$, the resultant numerical calculations are shown in Fig. 16A and B, respectively. For $(\delta, \phi)=(1/2, 0)$, the ω_+ and ω_- branches (anti-crossing) repel each other at the crossing point (H_{cent}) between the isolated photon and magnon modes (top of Fig. 16A), while their linewidths $\Delta\omega_+$ and $\Delta\omega_-$ cross each other at H_{cent} (bottom of Fig. 16A). This anti-crossing behavior is observed typically in most of the P-M hybrid systems [7–31]. On the other hand, for the case of $(\delta, \phi)=(2, \pi)$, the ω_+ and ω_- branches attract each other, thereby resulting in opposite anti-crossing (top of Fig. 16B), while the linewidths $\Delta\omega_+$ and $\Delta\omega_-$ are repulsive without crossing each other (bottom of Fig. 16B).

For the ISRR/YIG hybrid system, the net coupling strength cannot be directly determined using the frequency gap in the anti-crossing center, but rather, it can be modified in terms of the geometry factor of ISRR, δ and ϕ, as well as materials parameters β and α, as shown by (for details see Ref. [103])

$$\Delta'=\frac{\sqrt{2}k}{4\pi}\sqrt{\omega_p\omega_m(1+\delta\cos\phi)-\omega_p^2(\beta-\alpha)^2/2k^2} \tag{19}$$

However, for cases of α, $\beta<<1$, Δ' becomes Δ, which is given as

$$\Delta=\frac{1}{4\pi}\sqrt{2k^2\omega_m\omega_p(1+\delta\cos\phi)}, \tag{20a}$$

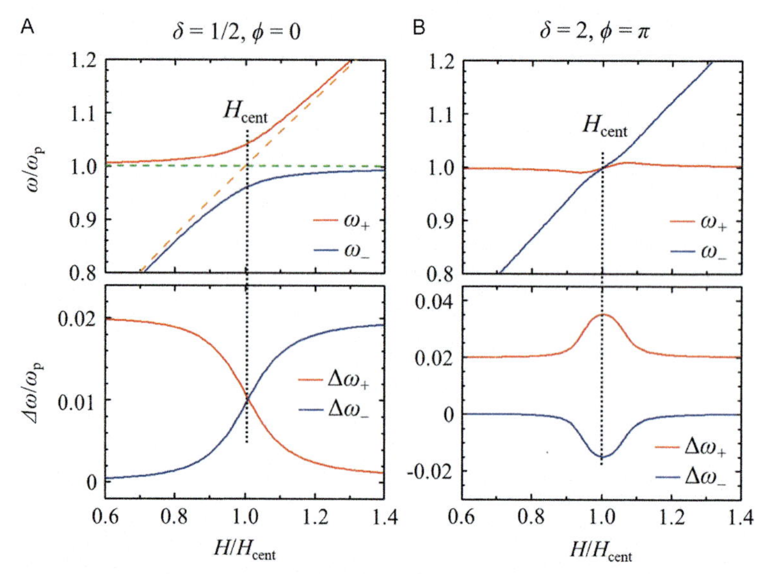

Fig. 16 Calculated frequencies (top) and linewidths (bottom) of P-M modes using Eq. (18) for (A) $\delta = 1/2$, $\phi = 0$ (normal anti-crossing) (B) $\delta = 2$, $\phi = \pi$ (opposite anti-crossing). The dashed lines in (A) show the pure photon (green line) and magnon (orange line) modes, while the vertical dotted line represents the center of anti-crossing. The vertical dotted line in (B) shows the center of the coupling region in opposite anti-crossing. *A modified version of this figure was originally published in B. Bhoi, B. Kim, S.-H. Jang, J. Kim, J. Yang, Y.-J. Cho, S.-K. Kim, Abnormal anticrossing effect in photon-magnon coupling, Phys. Rev. B 99 (2019) 134426, https://doi.org/10.1103/PhysRevB.99.134426.*

where Δ is the net coupling strength at $H = H_{cent}$ ($\omega_r = \omega_p$) in Hz units. Since ω_+ and ω_- can be expressed in terms of Δ, as

$$\omega_\pm \approx \frac{1}{2}\left[\left(\omega_r + \omega_p\right) \pm \sqrt{\left(\omega_r - \omega_p\right)^2 + (4\pi\Delta)^2}\right], \tag{20b}$$

the P-M net coupling strength can be directly estimated from the experimental data by fitting Eqs. (20a) and (20b) to experimentally observed coupled modes. The term $\sqrt{2k^2\omega_m\omega_p}/4\pi$ shown in Eq. (20a) exactly corresponds to the coupling strength $g/2\pi$, as previously discussed for generalized hybridization models of a 3-D cavity P-M-coupled system [5,22,26,27,29].

For general cases of β and α, the real and imaginary values of Δ' are determined by the sign of $\omega_m\omega_p(1 + \delta\cos\phi) - \omega_p^2(\beta - \alpha)^2/2k^2$, and thus approximately by the relative magnitude of the intrinsic parameter $\Delta_{mat} = (\beta - \alpha)^2/2k^2$ and the geometry parameter $\Delta_{geom} = (1 + \delta\cos\phi)$.

For the case of $\Delta_{mat} = (\beta - \alpha)^2/2k^2 = 0$, Eq. (19) becomes Eq. (20a). For the case of $\Delta_{geom} = (1 + \delta \cos\phi) \gg \Delta_{mat} = (\beta - \alpha)^2/2k^2$, the anti-crossing dispersion is not much varied with α and β, but rather is determined dominantly by ϕ and δ. Under this condition, Eq. (20b) represents that the net coupling strength can be obtained directly from the dispersion spectra. However, for the case of $\Delta_{geom} \sim \Delta_{mat}$, it leads to $\Delta' \approx 0$; thus, the frequency gap in the dispersion spectra disappears. On the other hand, for the case of $\Delta_{geom} < \Delta_{mat}$, Δ' should be imaginary, always leading to opposite anti-crossing dispersions. Therefore, for the last two cases, the net coupling strength can be determined by the full expression of Eqs. (20a) and (20b). Thus, both terms $\Delta_{mat} = (\beta - \alpha)^2/2k^2$ and $\Delta_{geom} = (1 + \delta \cos\phi)$ determine the net coupling strength, and consequently, the types of complex coupling dispersions as well.

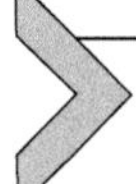

5. Experimental demonstration of P-M coupling

5.1 Measurement techniques

To characterize strong P-M coupling behaviors, it is necessary to measure both the field and frequency dependence of microwave responses from a hybrid system. The key requirements for such measurements are the following: (i) a low damping material with high spin density, (ii) a high-quality cavity or resonator, (iii) a static magnetic field to tune the magnon and photon resonances into coincidence where the influence of coupling is most notable, and (iv) a microwave source and detector. A variety of structures of cavity/resonator hybrids have been used, which include 3-D geometries [7–10,12–15,20,21,24], planar structures [11,23,24,26,27,29,103,105], waveguide resonators, and other special cavity configurations to enhance local field strengths [19].

Photon-magnon (P-M) coupling was observed for the first time by measuring the microwave transmission spectra of YIG samples in a superconducting resonator at ultra-low temperatures [7]. This technique has proven to be a useful probe, and has been extended for wide use both at cryogenic and room temperatures. Due to the potential of strong P-M coupling for spintronics applications, the interaction also has been remarkably detected electrically via spin pumping measurements [10,21]. In addition to microwave spectroscopy and electrical detection techniques, micro-focused Brillouin light scattering (BLS) has also been used to study P-M coupling in a system consisting of an SRR and a YIG thin film [106]. It is worth noting that the measurement techniques for P-M coupling can be categorized according to detection methods into (i) Microwave detection, (ii) DC electrical detection, and (iii) Optical detection, as described in the next section.

5.1.1 Microwave detection (vector network analyzer)

A typical experimental setup used to perform microwave transmission or reflection measurements of P-M coupling in planar-geometry hybrid systems is shown in Fig. 17. This setup is schematic, and many variations in the setup are possible according to the design of the resonator and YIG shape. This measurement is most easily performed using a vector network analyzer (VNA). However, other components, for example, a microwave generator and a spectrum analyzer, have also been used, according to the requirements of a given experiment.

Microwave propagation can be described by a transmission-line theory that can describe the propagation of waves in terms of voltage, current, and impedance. Therefore, this theory has been widely used in the radio and microwave frequency ranges to describe the propagations of waves through cavities, transmission lines, and resonators. Also, microwave propagation has been experimentally measured in terms of the scattering parameters [41].

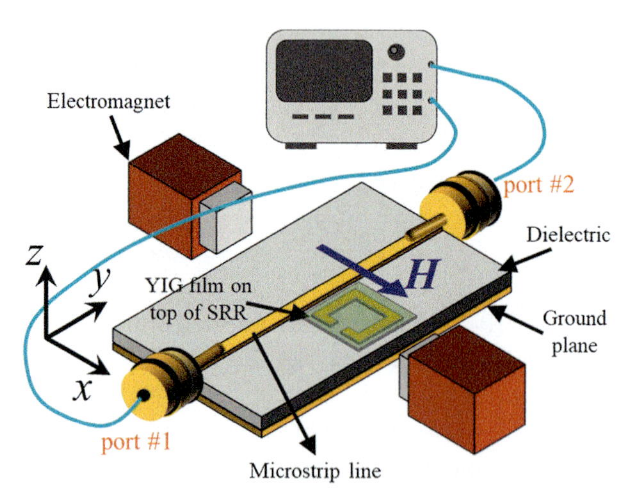

Fig. 17 Schematic drawing of an experimental setup for microwave transmission measurements of P-M coupling. The hybrid sample consists of a split ring resonator (SRR) and a YIG film in the given planar geometry. In the experiment, the input (port 1) and output (port 2) of the feeding line are connected to a vector network analyzer (VNA), and the split ring is inductively coupled to the microstrip feeding line. During the measurements, the YIG film is placed on the top of the SRR and static magnetic fields *H* are applied along the *x*-axis using an electromagnet.

In a microwave range, the magnitude and phase of the scattering parameters are usually measured using a VNA, which consists of two or four channels. For a two-port network, the scattering (or S-matrix) is given as

$$[S] = \begin{bmatrix} S_{11} & S_{12} \\ S_{21} & S_{22} \end{bmatrix} \tag{21}$$

with $S_{ij} = \frac{V_i^-}{V_j^+}\Big|_{V_i^+=0}$, where V_i^- is the voltage wave reflected from port i when voltage wave V_j^+ is incident toward port j. For a two-port network, S_{11} or S_{22} measures the signal reflected from the system, whereas S_{12} and S_{21} measure the transmitted signal through the system when microwaves are input at port 1 and port 2, and output at port 2 and port 1, respectively. For a reciprocal system, it turns out to be $S_{12}=S_{21}$, which case often occurs in a microwave cavity P-M-coupled system.

The advantage of using a VNA for S-parameter measurements is that it can make accurate measurements by taking into account a variety of possible errors via in-built error correction and calibration processes [41]. Also, in order to maximize the coupling strength, the magnetic materials are often placed at the location of the maximum rf magnetic field in the cavity resonator or, in the case of a planar resonator, on the microstrip line.

5.1.2 DC electrical detection

One of the important results gleaned from the study of P-M coupling is that electrical detection of strong coupling provides the foundation for cavity-based spintronics [10,21,31]. For example, from a YIG/Pt bilayer in a 3-D cavity, Bai et al. [10] demonstrated for the first time that the voltage generated due to spin pumping and the inverse spin Hall effect (ISHE) both could also monitor P-M coupling. Such hybridization influences the generation of spin current and can, therefore, be used as a new control mechanism for the development of spintronic devices. It also provides another avenue for probing of P-M coupling.

Most key requirements in electric measurements are the same as those in microwave transmission measurements. However, since hybridization will be detected via the spin current, a specific sample configuration, such as the combination of ferrimagnetic materials and normal metals (NM), is necessary, as is the case in typical spin pumping measurements [10,21]. A microwave generator provides magnetic fields to excite magnetizations' precession and measures the voltage generated through the combined effort of spin pumping and the ISHE, as shown schematically in Fig. 18A.

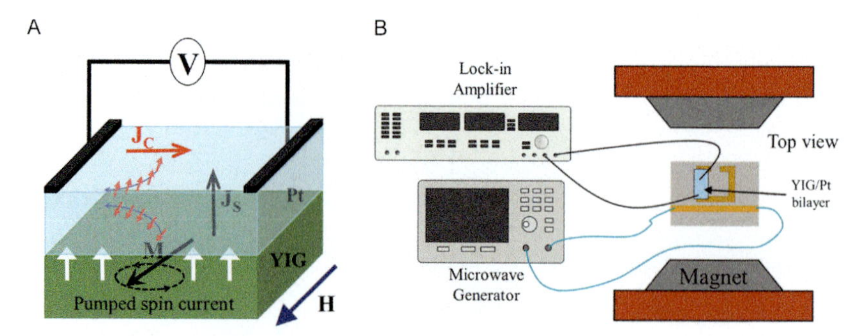

Fig. 18 Schematic drawing of an experimental setup for electrical detection of P-M coupling: (A) Scheme of spin-pumping process and inverse spin Hall effect (ISHE). During the spin pumping process, a ferrimagnetic YIG is excited by a microwave field, and the magnetization (M) precession of the YIG allows for the transfer of angular momentum in the YIG to conduction electrons in Pt, which results in the generation of pure spin currents without accompanying charge current. The pure spin current (J_S) is converted to a net charge current (J_C) in the Pt via the ISHE. (B) Schematic diagram of an electrical detection system using a lock-in technique. The microwave generator signal excites the SRR and consequently the magnetizations of YIG. The lock-in amplifier measures voltage generated via spin-pumping/ISHE in the YIG/Pt bilayer system.

An experimental setup of measurements for such planar hybrid systems is shown in Fig. 18B [107]. Similarly, Castel et al. [27] demonstrated strong P-M coupling by measuring the ISH voltage as a function of the frequency for specific values of applied magnetic fields for a YIG/Pt bilayer mounted on a T-type planar resonator. Furthermore, using the reverse concept, they demonstrated current-induced heating to increase the temperature of the YIG/Pt film, which makes it possible to control the coupling between the magnon and photon modes. This electrical detection of FMR/cavity coupling is not only important as an alternative method of studying magnon-photon coupling but also allows for the direct detection of spin currents in such a coupled system.

5.1.3 Optical detection (Brillouin light scattering)

The optical detection of P-M coupling using BLS was first demonstrated by Klingler et al. [106]. They simultaneously used BLS spectroscopy and microwave absorption measurements to probe both magnonic and photonic excitations in the SRR/YIG system. Fig. 19 shows the experimental setup for the optical detection of P-M coupling.

During the measurement, the YIG/GGG heterostructure is positioned with the GGG side down in the center of the SRR so that the YIG film

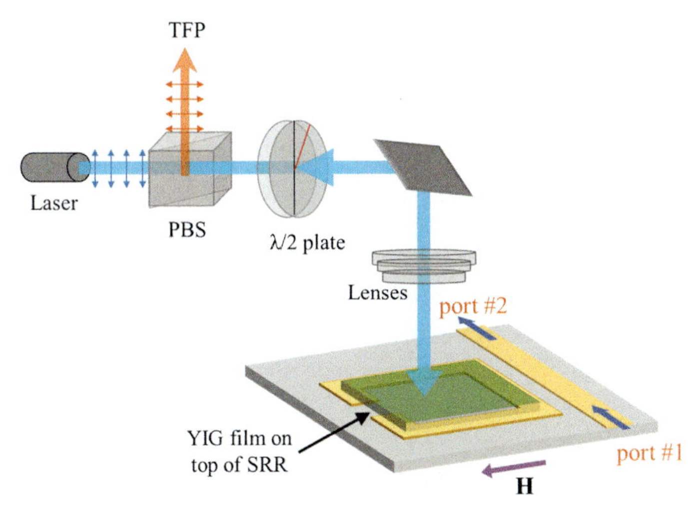

Fig. 19 Schematic illustration of an experimental setup for the optical detection of P-M coupling: A microwave signal is applied to a feeding line that is inductively coupled to the SRR. The YIG film is placed onto the SRR. A polarized laser beam passes through a polarizing beam splitter (PBS) and a $\lambda/2$-plate and then is focused on the surface of the YIG film using a microscope objective lens. The backscattered light passes again through the $\lambda/2$-plate and the PBS, before reaching the Tandem-Fabry-Perot interferometer (TFP). *Adapted from S. Klingler, H. Maier-Flaig, R. Gross, C.-M. Hu, H. Huebl, S. T. B. Goennenwein, et al., Combined Brillouin light scattering and microwave absorption study of magnon-photon coupling in a split-ring resonator/YIG film system, Appl. Phys. Lett. 109 (2016) 072402.*

can be optically accessible from above for the BLS measurements. The laser beam of a wavelength (λ) of 532 nm passes a polarizing beam splitter (PBS) and a $\lambda/2$-plate before it is focused onto the surface of the YIG film using a microscope objective lens. The incident laser photons are inelastically scattered by magnonic excitations in the SRR/YIG system, which results in a shift in the frequency of the inelastically scattered light. The polarization of the inelastically scattered light is rotated by the scattering event according to a certain angle with respect to the incident polarization direction. In contrast, the elastically scattered light retains its incident energy and polarization. The scattered and collected light again passes the $\lambda/2$-plate before it reaches the PBS. The PBS selectively directs the inelastically scattered photons (which earlier undergo a polarization rotation) to a TandemFabry-Perot interferometer (TFP). The $\lambda/2$-plate allows for simultaneous rotation of the polarization of the incident and backscattered light by changing the angle of its fast optical axis relative to the polarization axis of the incoming light. In combination with the PBS, it is possible to analyze the polarization of the backscattered light with respect to the incoming light polarization.

5.2 Anti-crossing effect between photon and magnon modes

One of the advantageous features of studying P-M hybridization is that basic physical signatures are revealed directly in the raw experimental data. An example of such data in a microwave transmission experiment is shown in Fig. 20, taken from Ref. [26]. The ISRR photon mode at a resonance frequency $\omega_p/2\pi = 3.7$ GHz was coupled to an epitaxial YIG/GGG film of 3.7 mm × 3.7 mm × 25 μm. Fig. 20 illustrates the $|S_{21}|$ spectra measured as a function of the microwave frequency $\omega/2\pi$ of oscillating currents flowing in the microstrip line for the indicated different strengths of in-plane magnetic field applied perpendicularly to the microstrip line axis. Only the pure photon mode of the ISSR appeared at the same frequency of 3.7 GHz, having not moved with the field strength, as shown in Fig. 20A. The FMR mode, measured only from the YIG film separately (i.e., without the ISRR), is varied in its frequency position for different field strengths, as shown in Fig. 20B. However, for hybrid ISRR and YIG film, there exist two peaks (see Fig. 20C). One peak (marked by black arrows) is weak in gain and very strongly dependent on the applied field strength. Essentially, it continuously shifts toward the higher-frequency side with increasing *H*, and crosses the other peak position (green arrows), thus

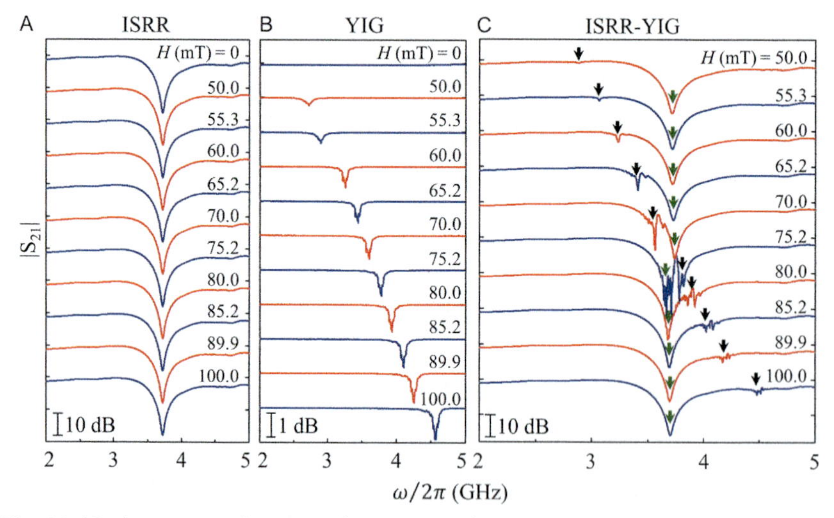

Fig. 20 $|S_{21}|$ spectra as function of microwave frequency of ac currents for indicated different field strengths for (A) ISRR only, (B) YIG film only, and (C) their ISRR-YIG hybrid sample. *A modified version of this figure was originally published in B. Bhoi, B. Kim, J. Kim, Y.-J. Cho, S.-K. Kim, Robust magnon-photon coupling in a planar-geometry hybrid of inverted split-ring resonator and YIG film, Sci. Rep. 7 (2017) 11930, https://doi.org/10.1038/s41598-017-12215-8.*

indicating the FMR mode. The other peak has a relatively high gain and does not much move with increasing H, thus indicating the ISRR mode. It is worth noting that when the lower-gain peak approaches the higher-gain peak, the lower-gain peak gradually increases in gain and attains its highest value just at the moment it crosses the other peak, after which its magnitude gradually decreases again with field strength. The small peak is located on the lower-frequency side before it crosses the higher-gain peak, whereas it is located on the higher-frequency side after it crosses the higher-gain peak. Hereafter, the lower- and higher-frequency peaks are marked as $\omega_-/2\pi$ and $\omega_+/2\pi$, respectively. This evidences that the magnetization-dynamic modes of the YIG film strongly interact with the electrodynamic ISRR mode.

In order to examine the coupling effect shown in Fig. 20, the $|S_{21}|$ spectrum powers on the $(\omega/2\pi) - H$ plane were replotted as indicated in Fig. 21. The $|S_{21}|$ spectra measured separately from the ISSR (Fig. 21A)

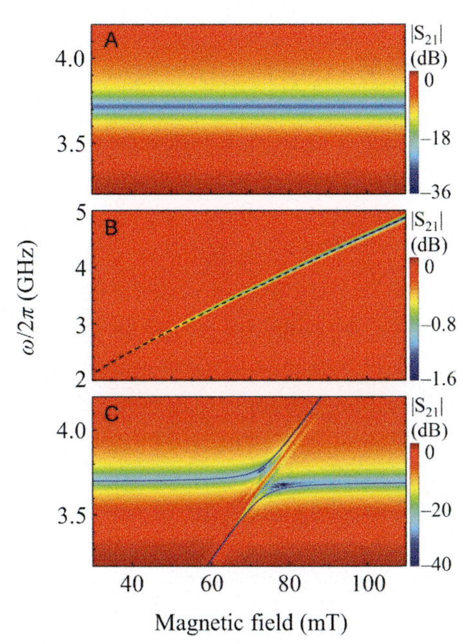

Fig. 21 $|S_{21}|$ power on the plane of microwave frequency and magnetic field ($\omega/2\pi$-H plane) reconstructed from data shown in Fig. 20, for (A) ISRR only, (B) YIG film only, and (C) ISRR-YIG hybrid sample. The dotted line in (B) is the fitting of the Kittel's formula to the data. The solid blue lines in (C) are the results of fitting to the experimental data according to the coupled oscillator model explained in the text. *A modified version of this figure was originally published in B. Bhoi, B. Kim, J. Kim, Y.-J. Cho, S.-K. Kim, Robust magnon-photon coupling in a planar-geometry hybrid of inverted split-ring resonator and YIG film, Sci. Rep. 7 (2017) 11930, https://doi.org/10.1038/s41598-017-12215-8.*

and the YIG film (Fig. 21B) were compared with the results measured for the ISRR-YIG hybrid (Fig. 21C). Far away from the field range of 60–90 mT, the resonant frequencies of the hybrid system are very close to that of the single pure ISRR mode, as shown in Fig. 21A. However, being close to that field region, the slope (in the $(\omega/2\pi) - H$ spectrum in Fig. 21C) becomes similar to the measured FMR frequency-versus-H spectrum of the YIG film shown in Fig. 21B. From Fig. 21C, it is clear that there exists a strong anti-crossing effect between the ISRR and FMR modes. This behavior is also referred to as level repulsion or avoided crossing, as noted in many studies [19–21,25–27,29]. In the anti-crossing region shown in Fig. 21C, fine-featured lines parallel to the Kittel-type FMR mode are observed, which can be attributed to the magnetostatic spin-wave modes. Observation of anti-crossing behavior is the key experimental signature in P-M coupling. This behavior can be the same, irrespective of the detection method.

5.3 Estimation of P-M coupling strength

The coupling strength can be estimated by fitting the experimental data to any of the models described previously. The fitting showed in Fig. 21C is based on a coupled oscillator model that resulted in a coupling strength value of $g/2\pi = 90$ MHz ($k = 0.221$) at 3.7 GHz, where the coupling constant k is defined as $k = [2(g/2\pi)/f_p]^{1/2}$. Comparison of $g/2\pi$ values obtained from different planar and 3-D cavity hybrid systems are presented in Table 1. For a direct comparison of coupling strengths for a variety of different systems (e.g., 3-D or 2-D hybrid structures of different dimensions), spin-number-normalized coupling strength $g/2\pi\sqrt{N}$ (i.e., single spin-photon coupling) for different systems was estimated by considering the spin-density of the effective volume of YIG that can contribute to coupling. It was evident that the value of coupling strengths obtained from the planar-based hybrid systems were higher than those obtained from the 3-D cavity/YIG sphere hybrids.

Although planar hybrid systems exhibit high coupling strengths, linewidth estimation from microwave transmission spectra becomes difficult due to excitation of higher-order spin-wave modes. On the other hand, in the cases of 3-D cavity modes, it is easy to measure both the frequency and linewidth as shown in Fig. 22. The dispersion again highlights the mode hybridization, which is strongest near the crossing point, i.e., $\omega_p = \omega_r$, as being in agreement with the mode composition described by the

Table 1 Comparison of coupling strength and other parameters obtained in different P-M hybrid systems.

System	$\omega/2\pi$ [GHz]	$g/2\pi$ [MHz]	k	$\frac{g}{2\pi\sqrt{N}}$ [Hz]	$k_N \times 10^{-2}$	Reference
YIG film/3-D cavity	10.565	47	0.094	0.031	0.245	[8]
YIG film/3-D cavity	10.847	65	0.109	0.056	0.323	[21]
YIG film/3-D cavity	7.9	23	0.076	0.033	0.292	[25]
YIG sphere/3-D cavity	10.556	31.5	0.077	0.021	0.20	[22]
YIG film/3-D cavity	10.506	80	0.123	0.069	0.364	[10]
YIG film/3-D cavity	9.65	24.2	0.081	0.090	0.434	[24]
YIG film/SRR on stripline	3.2	270	0.411	0.097	0.782	[11]
YIG sphere/SRR on stripline	4.08	65	0.170	0.040	0.445	[23]
YIG cylinder/SRR stripline	4.75	32.85	0.117	0.004	0.137	[29]
ISRR/YIG film	3.7	90	0.221	0.194	1.025	[26]

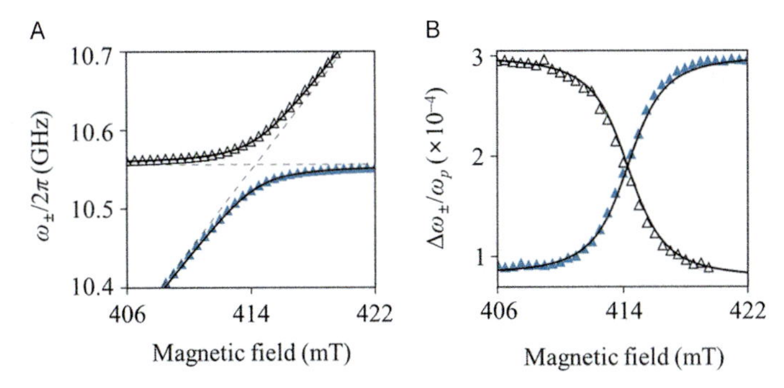

Fig. 22 (A) Dispersion and (B) linewidth profiles for two coupled modes. Symbols and solid curves correspond to experimental data taken from Ref. [5] and the calculations according to Eq. (12b), respectively. The open and closed triangles correspond to the upper and lower branches of the two coupled modes, respectively. *A modified version of these figures were originally published in M. Harder, C.-M. Hu, Cavity spintronics: an early review of recent progress in the study of magnon photon level repulsion, in: Solid State Physics, vol. 69, R.E. Camley, R.L. Stamps (Eds.), Academic Press, Cambridge, 2018, pp. 47–121, https://doi.org/10.1016/bs.ssp.2018.08.001.*

theoretical models of P-M coupling. An important general feature of the damping evolution is that the linewidths of both modes are bounded by $\alpha < \Delta\omega_{\pm}/\omega_p < \beta$, and become equal at the crossing point [5].

5.4 Opposite anti-crossing (level attraction)

The coupling between light and matter is not strictly limited to coherent interactions. If the off-diagonal term of the Hamiltonian of a coupled system (Eqs. 12 and 17) contains an imaginary part, the eigenfrequency of the coupled modes becomes complex and the corresponding real components pull toward each other to meet, which is to say that the hybridized modes coalesce rather than repel [108,109]. This mechanism has been demonstrated in other coupled systems such as quantum optomechanical systems [102,110] and in plasmonic nanostructures due to either near-field or far-field coupling [111–113]. This interaction is referred to as dissipative coupling or level attraction. Very recently, dissipative-coupling-induced level attraction was experimentally discovered in a P-M-coupled system by Harder et al [96], as was, independently by Bhoi et al. [103], opposite anti-crossing.

Harder et al. [96] demonstrated level attraction by moving a YIG sphere in a 3-D Fabry-Perot cavity as shown in Fig. 23. When a magnet in the magnetization precession state falls down in a microwave cavity, the backward action of the induced current in the cavity shall impede the magnetization dynamics so that the magnons shall be coupled with the induced cavity current via the damping-like Lenz effect. This has been referred to as the dissipative P-M coupling that gives rise to level attraction. The microwave transmission S_{21} spectra measured at position A (where the amplitude h of the microwave magnetic field is high) and position B (where h is small) show the coupled modes' level repulsion (Fig. 23C) and level attraction (Fig. 23D), respectively.

On the other hand, Bhoi et al. [103] also demonstrated this effect for a different system of a planar-geometry YIG/ISRR hybrid (case II shown in Fig. 24). Instead of moving the YIG magnet directly, they designed two different ISRR configurations wherein the split-gap position/orientation of the ISRR is placed on the x-axis (case-I) or the y-axis (case-II), as shown in the inset of Fig. 24A. The contrasting dispersions of the coupled modes were observed: normal (case-I) and opposite (case-II) anti-crossing. The opposite anti-crossing was ascribed to the compensation of both intrinsic damping and coupling-induced damping in the magnon modes. This compensation is achievable by controlling the relative strength and phase of the oscillating magnetic fields from the microstrip feeding line and from the

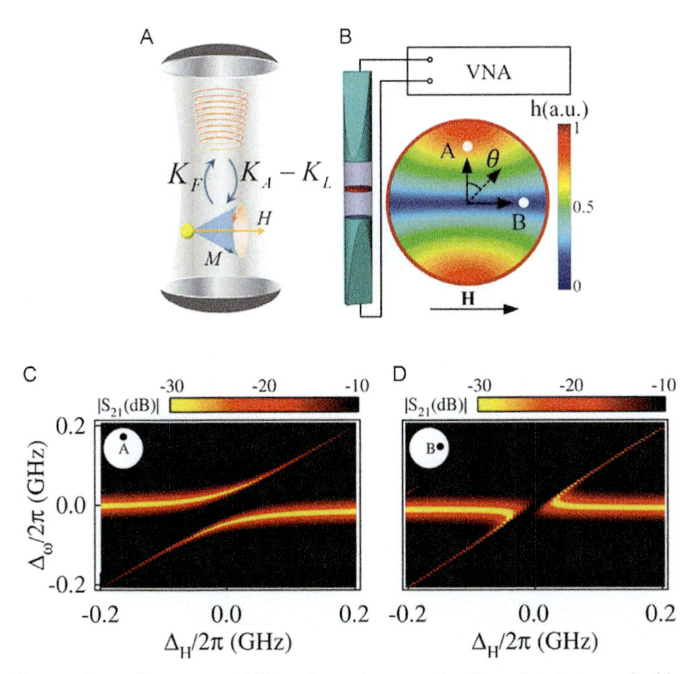

Fig. 23 The motion of a magnet falling down in a conducting pipe is impeded by induced magnetic fields. (A) P-M coupling mechanism including the cavity Lenz effect that impedes the magnetization dynamics. The K_F term stems from Faraday's induction law, which describes the effect of the dynamic magnetization on the rf current. The K_A term comes from Ampere's circuit law, which shows that the current produces an rf magnetic field. The cavity Lenz effect is included in the K_L term, which has the opposite sign to that of the K_A term, since the backward action from the induced rf current impedes the magnetization dynamics. Level repulsion and attraction appear when $K_A - K_L > 0$ and $K_A - K_L < 0$, respectively. These two regimes of P-M coupling are separated by the matching condition of $K_A = K_L$, under which the magnons and photons appear to be decoupled. (B) Experimental setup, with a VNA measuring the microwave transmission through a waveguide loaded with a YIG sphere. The simulated field amplitude h is shown for the TE_{11} mode on the middle plane of the empty waveguide. "A" and "B" in (B) denote the h antinode and node positions for loading of the YIG sphere). The Experimental transmission spectra are shown as a function of $\Delta\omega$ and ΔH for the case of (C) level repulsion and (D) level attraction, respectively, by placing the YIG sphere at the A and B positions, respectively. *Reprinted with permission from Ref. [96], Copyright 2019 American Physical Society. A modified version of this figure was originally published in M. Harder, Y. Yang, B.M. Yao, C.H. Yu, J.W. Rao, Y.S. Gui, R.L. Stamps, C.-M. Hu, Level attraction due to dissipative magnon-photon coupling, Phys. Rev. Lett. 121 (2018) 137203, https://doi.org/10.1103/PhysRevLett.121.137203.*

ISRR's split gap orientation/position with respect to the microstrip line axis [103]. The experimentally observed level attraction/opposite anti-crossing effect in P-M coupling demonstrates the potential and great flexibility of P-M systems for exploration of not-yet-revealed phenomena of light-matter interaction.

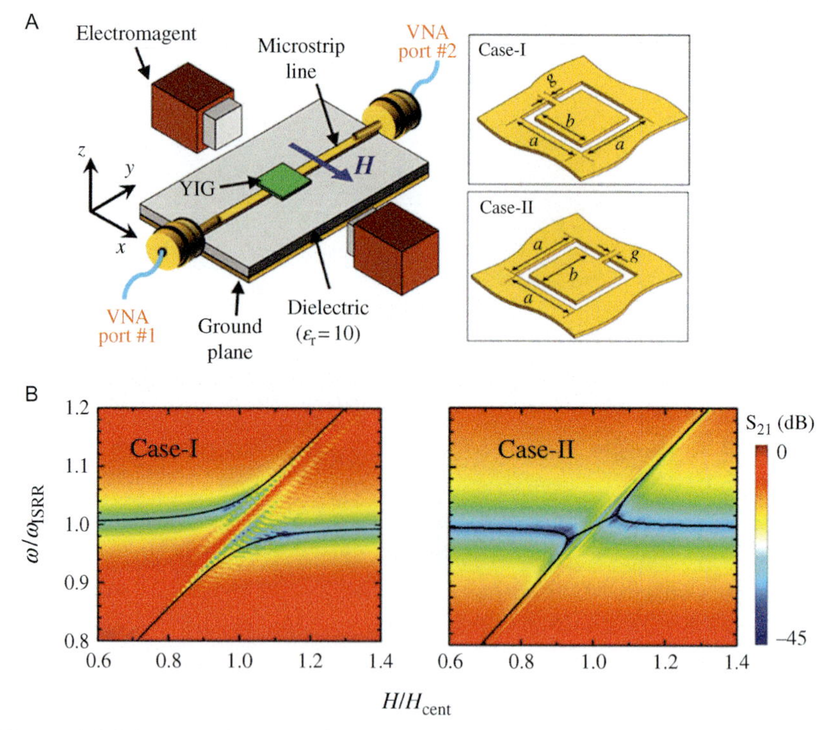

Fig. 24 (A) Schematic drawing of experimental setup for P-M coupling consisting of ISRR and YIG film in the planar geometry. The ISRR is capacitively coupled to a microstrip feeding line. In the experiment, ports 1 and 2 of the feeding line are connected to a VNA, and the static applied magnetic field H is created by an electromagnet applied in the *x*-direction. Insets: dimensions of ISRRs oriented orthogonal (case-I) and parallel (case-II) to microstrip feeding line. (B) Experimentally measured S_{21} power on the plane of normalized microwave angular frequency and magnetic field ($\omega/\omega_p - H/H_{cent}$ plane) of ISRR-YIG hybrid for different orientations of ISRR split-gap with respect to microstrip feeding line: case-I: orthogonal; case-II: parallel. The black solid lines in (B) correspond to the results of the fitting of Eqs. (20a) and (20b) to the higher and lower branches. *A modified version of this figure was originally published in B. Bhoi, B. Kim, S.-H. Jang, J. Kim, J. Yang, Y.-J. Cho, S.-K. Kim, Abnormal anticrossing effect in photon-magnon coupling, Phys. Rev. B 99 (2019) 134426, https://doi.org/10.1103/PhysRevB.99.134426.*

5.5 Exceptional point (EP)

The observable signatures of P-M coupling for level repulsion or level attraction are the presence of a gap in the dispersion spectra or linewidth at the coupling center, respectively, which are directly related to the root of the determinant in Eqs. (12) and (17). If the root (input equation) is exactly zero, there will be no gap in the dispersion of the coupled modes or their linewidth at the crossing point. This point is known as an exceptional point (EP), where the eigenvalues and eigenvectors coalesce [95,97]. Therefore, physically the EP marks the transition between strong coupling, where a gap in the dispersion can be observed, and weak coupling, where no gap can be seen. EPs play an important role in many physical systems [98–102], and the mathematical origin of the P-M EP can be easily understood by examining the hybridized dispersion, as discussed in Section 4.3. Therefore, by controlling both the damping and coupling strength parameters, it is possible to demonstrate the EP point [94].

Experimentally Zhang et al. [97] demonstrated the EP in P-M coupling using a specially designed hybrid system wherein a YIG sphere glued onto a wooden rod is inserted into a 3-D rectangular cavity with two ports (Fig. 25A and B). As per the theory of EP observation, the coupling strength was tuned by adjusting the displacement of the YIG sphere in the x-direction, because the coupling strength is proportional to the magnitude of the microwave magnetic field. Fig. 25C illustrates the measured total output spectrum at different displacements of the YIG sphere. It is clearly seen that when $|x|$ is larger than 1.2mm, there exist two frequency modes (at very low output), corresponding to the two real eigenfrequencies of the coupled modes; when $|x|$ is smaller than 1.2mm, a phase transition occurs and no modes can be found. This system exhibits an EP differentiating the two regimes of unbroken and broken symmetry.

6. Control of dispersion type of P-M coupling

The coupling strength in P-M coupling is the parameter most key to the efficiency of energy exchange between the spin and photonic systems. The coupling strength can be determined by directly measuring the frequency gap (ω_{gap}), the amount of the modes' splitting in frequency, and at the coupling center in normal anti-crossing (i.e., level repulsion). By tuning the geometry parameters, the damping constants of the individual resonators in hybrid systems, and the coupling constant k, the coupling strength varies from a weak coupling regime ($k/\alpha < 1$ and $k/\beta < 1$) to a strong coupling regime ($k/\alpha > 1$ and $k/\beta > 1$) [5,9]. A few successful examples of hybridization controls have been achieved either by directly manipulating the cavity damping [9,21,114] or increasing the size of the YIG sample [8,9]. Furthermore, several experiments on the control of P-M coupling strength by external parameters have also been performed, such as electrical control, [23], control of the direction of the DC magnetic field [26], and thermal control [27]. In the following sections, further relevant details will be discussed.

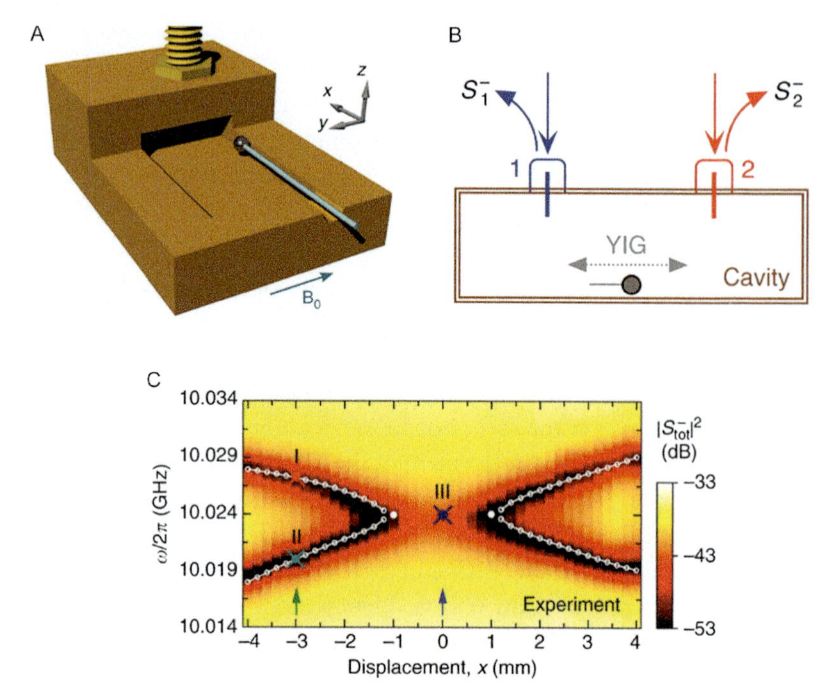

Fig. 25 (A) Sketched structure of the cavity magnon-polariton system, where a YIG sphere glued on a wooden rod is inserted into a 3-D rectangular cavity through a hole of 5 mm diameter in one side of the cavity. The displacement of the YIG sphere can be adjusted in the *x*-direction using a position adjustment stage, and the static magnetic field is applied in the *y*-direction. The cavity has two ports for both measurement and feeding of microwave fields into the cavity. (B) Diagram of the cavity magnon-polariton system with two feedings. The total output spectrum $|S_{tot}|^2$ is the sum of output spectrum $|S_1|^2$ from port 1 and $|S_2|^2$ from port 2. (C) Measured total output spectrum versus the displacement of the YIG sphere in the cavity shows the EP in cavity magnon-polaritons. *A modified version of this figure was originally published in D. Zhang, X.-Q. Luo, Y.-P. Wang, T.-F. Li, J.Q. You, Observation of the exceptional point in cavity magnon-polaritons, Nat. Commun. 8 (2017) 1368. https://doi.org/10.1038/s41467-017-01634-w.*

6.1 Geometry parameters

The net coupling strength can be tuned by the geometry factor Δ_{geom} in the YIG/ISRR, as described by Eqs. (20a) and (20b). In order to examine the effect of Δ_{geom} on the anti-crossing dispersion and the coupling strength in the YIG/ISRR hybrid, Bhoi et al. [103] numerically calculated the $|S_{21}|$ power on the $\omega/\omega_p - H/H_{cent}$ plane by varying both the δ and ϕ values. Fig. 26A shows the calculated $|S_{21}|$ profiles versus ω/ω_p at the center position (H_{cent}). Fig. 26B shows contrasting shapes of anti-crossing dispersion depending on the indicated values of δ and ϕ, instead of the appearance

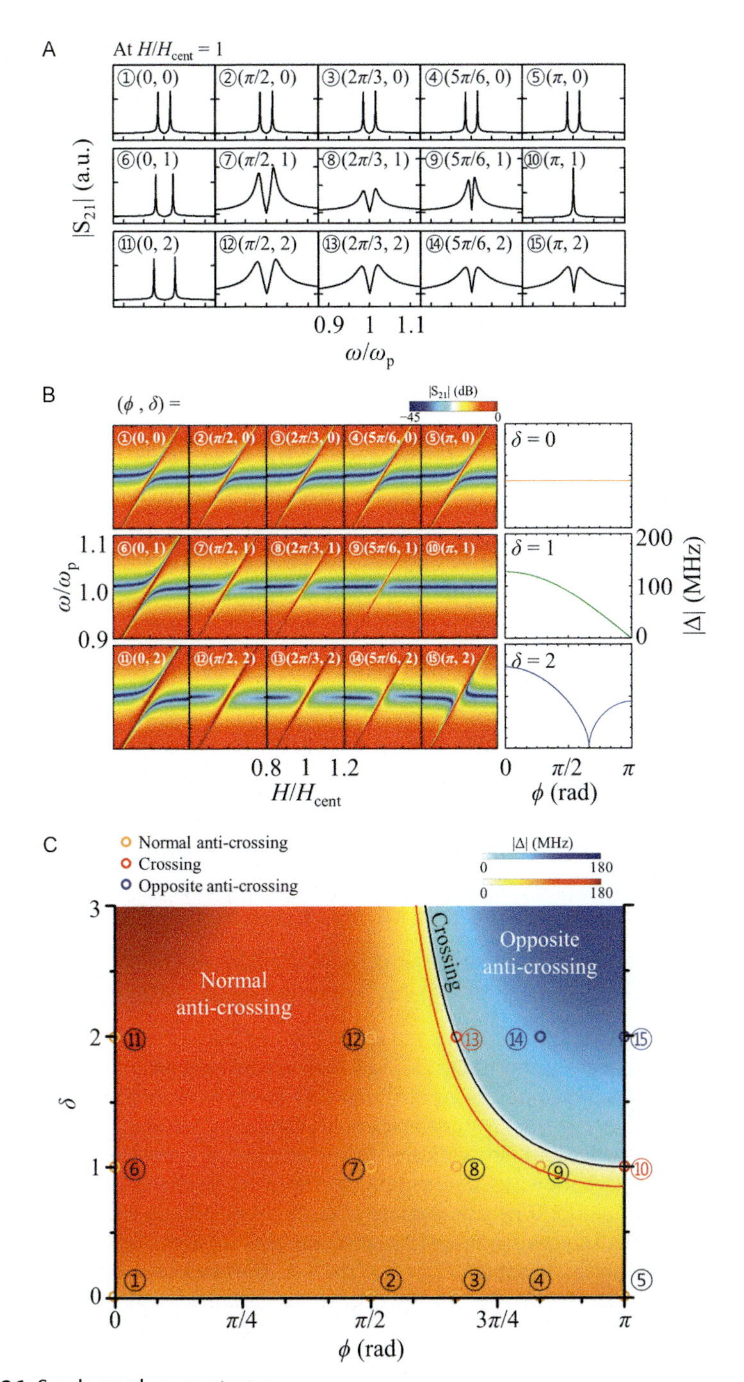

Fig. 26 See legend on next page.

of only the normal and opposite anti-crossing dispersions experimentally found. The shape of dispersion is totally determined by the net coupling strength Δ, which values vary with δ and ϕ, as shown in Fig. 26C.

For example, for $\delta=0$, the net coupling strength is given as $\Delta=\frac{1}{4\pi}\sqrt{2k^2\omega_m\omega_p}=\frac{g}{2\pi}$; thus the variation of ϕ does not affect Δ nor the shape of anti-crossing dispersion; all of the $|S_{21}|$ power contours show the normal type of anti-crossing, as indicated in the first rows of Fig. 26A and B. This condition corresponds to the cases of $|\mathbf{h}_p| \ll |\mathbf{h}_l|$. On the other hand, for the case of $\delta=1$, Eqs. (20a) and (20b) is a function of ϕ; thus, the shape of anti-crossing is remarkably variable with ϕ. However, under a specific condition of $\phi=\pi$, Δ becomes zero, and thus the anti-crossing dispersion completely disappears. For the case of $\delta=1$ and $\phi=0$, $\mathbf{h}_l$ and $\mathbf{h}_p$ are comparable in size and equal in phase, and accordingly, both fields excite the YIG's magnon modes. As ϕ increases from 0 to π, $\mathbf{h}_p$ becomes more out-of-phase with $\mathbf{h}_l$, thereby yielding weaker net coupling strength. For the condition of $\delta=1$, $\phi=\pi$, $\mathbf{h}_p$ and $\mathbf{h}_l$ are exactly out-of-phase, thus yielding a completely zero microwave field, which cannot excite YIG's magnons, as shown by the appearance of only the ISRR's photon mode, without the FMR mode in YIG (⑩ of Fig. 26A and B).

More interestingly, for the case of $\delta=2$, the anti-crossing shape changes from the normal to the opposite one through non-anti-crossing (see the third rows in Fig. 26A and B). The $|\Delta|$ value decreases with ϕ, becoming 0 at $\phi=2\pi/3$ and increasing again with ϕ from $\phi=2\pi/3$ (see the bottom of the right column in Fig. 26B). For the case of $\delta=2$, $\phi=0$, YIG's magnon mode is mainly excited by $\mathbf{h}_p$, and $\mathbf{h}_p$ and $\mathbf{h}_l$ are in-phase, resulting in the normal shape of anti-crossing. At $\phi=2\pi/3$, $|\Delta|$ becomes zero and the anti-crossing disappears. For $\phi>2\pi/3$, $|\Delta|$ increases with ϕ, resulting in

Fig. 26 Analytical calculation of (A) $|S_{21}|$ profiles versus ω/ω_p at the center position (H_{cent}). The y-axis scale for the cases of ①–⑥ and ⑩–⑪ is 10 times larger than that for ⑦–⑨ and ⑫–⑮. (B) The $|S_{21}|$ power spectra on the $\omega/\omega_p - H/H_{cent}$ plane according to both δ and ϕ, the values being indicated by the numbers and positions (open circles) on the phase diagram shown in (C). The right column indicates the $|\Delta|$ as a function of ϕ for each of $\delta=0$, 1, and 2. (C) Phase diagram of various types of anti-crossing dispersion on the $\delta-\phi$ plane. The color indicates the absolute value of net coupling strength $|\Delta|$ noted by the two color bars for $\Delta_{mat}=(\beta-\alpha)^2/2k^2 \sim 0$. The black ($\Delta=0$ for $\Delta_{mat}=0$) and red ($\Delta'=0$ for $\Delta_{mat}=0.215$) lines correspond to the boundaries that distinguish the dispersion types. *A modified version of this figure was originally published in B. Bhoi, B. Kim, S.-H. Jang, J. Kim, J. Yang, Y.-J. Cho, S.-K. Kim, Abnormal anticrossing effect in photon-magnon coupling, Phys. Rev. B 99 (2019) 134426, https://doi.org/10.1103/PhysRevB.99.134426.*

the opposite anti-crossing. With increasing $|\Delta|$, the opposite anti-crossing becomes clearer in its shape. Since $\mathbf{h}_p$ contributes more to the magnon's excitations than does $\mathbf{h}_l$, and since the fields are out-of-phase, the result is the opposite anti-crossing. All of these features clearly indicate that the relative strength and phase of the oscillating magnetic fields generated from both the ISRR's split gap and the microstrip feeding line determine the net coupling strength, consequently resulting in the shape of anti-crossing dispersion in the ISRR-YIG hybrid system.

A phase diagram of anti-crossing dispersion on the plane of ϕ and δ is calculated according to Eqs. (20a) and (20b) for $k=0.03$ and $\omega_p/2\pi=3.7$ GHz, as shown in Fig. 26C. The opposite anti-crossing dispersions (blue region) are separated from the others (the red and yellow regions) by the condition of $\Delta=0$ (i.e., $\delta\cos\phi=-1$), as indicated by the black solid line in Fig. 26C. As noted by the colors in the different regions, as δ increases in the range of $\phi<\pi/2$, the anti-crossing becomes normal with stronger net coupling strength, whereas with increasing δ in the range of $\phi>3\pi/4$ above the marked boundary curve, the anti-crossing becomes the opposite with stronger net coupling strength. For the experimental values of $\beta=2.0\times10^{-2}$, $\alpha=3.2\times10^{-4}$ and $k=0.03$, the condition $\Delta'=0$ is also drawn using Eq. (19) as a red solid line in the phase diagram, as shown in Fig. 26C. The boundary curve is close to that of $\Delta=0$. It is to be noted that the color-bar scale in Fig. 26C represents the net coupling strength only for the case of $\Delta_{mat}=0$.

6.2 Material parameters

To examine how Δ' varies with intrinsic material parameter Δ_{mat}, phase diagrams are plotted for different values of Δ_{mat}, i.e., 0.1, 0.3, 0.5, and 0.9, as shown in Fig. 27. Note that $\Delta'=0$ represents the boundary curves that distinguish the opposite anti-crossing dispersion from the others. The phase diagrams on the planes of ϕ and δ are also modified according to the value of Δ_{mat}. Here, what is most important is the fact that with increasing Δ_{mat}, the opposite anti-crossing region expands toward lower values of both ϕ and δ. As was discussed earlier, when the phase of $\mathbf{h}_p$ becomes deviated to that of $\mathbf{h}_l$, the damping of the photon mode increases while the damping of the magnon mode decreases due to dissipative interaction, thus leading to an opposite anti-crossing dispersion. For larger values of Δ_{mat}, the opposite anti-crossing dispersion occurs for the smaller phase difference between $\mathbf{h}_p$ and $\mathbf{h}_l$ and the lower value of $(\delta=|\mathbf{h}_p|/|\mathbf{h}_{line}|)$. In the case of the ISRR/YIG system in Ref. [103], Δ_{mat} was estimated to be 0.215, and thus the opposite

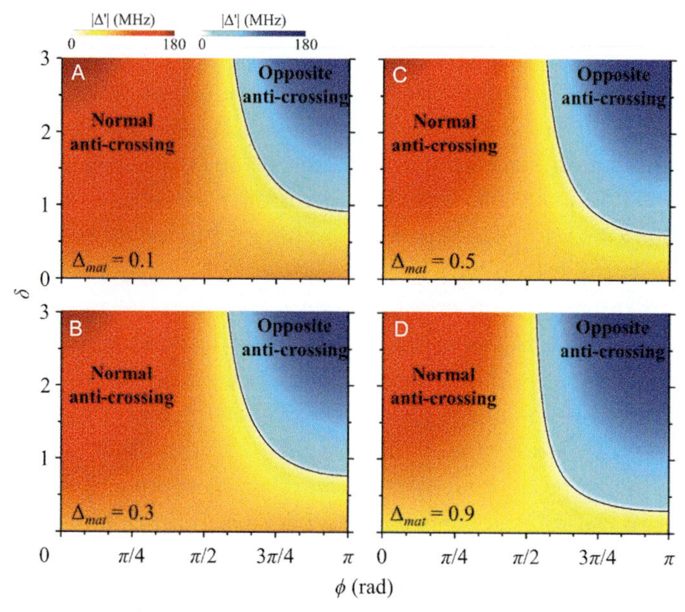

Fig. 27 Numerical calculation of phase diagram of anti-crossing dispersion on $\delta-\phi$ plane. The color indicates the absolute value of net coupling strength $\Delta\prime$ as noted by each of the two color bars. The black lines correspond to the boundaries ($\Delta\prime=0$) between the normal and opposite anti-crossing types for (A) $\Delta_{mat}=0.1$, (B) 0.3, (C) 0.5, and (D) 0.9. *A modified version of this figure was originally published in B. Bhoi, B. Kim, S.-H. Jang, J. Kim, J. Yang, Y.-J. Cho, S.-K. Kim, Abnormal anticrossing effect in photon-magnon coupling, Phys. Rev. B 99 (2019) 134426, https://doi.org/10.1103/PhysRevB.99.134426.*

anti-crossing dispersion for the case-II geometry corresponded to the case of $\Delta_{geom} < 0.215$, while the normal dispersion for the case-I geometry corresponded to the case of $\Delta_{geom} > 0.215$. This material parameter could be changed by changing the damping parameter of the individual systems.

6.3 Size (volume) of YIG and microwave cavity

For a hybrid system of very small damping constants, i.e., α, $\beta \ll 1$, the coupling strength can be directly obtained from the frequency gap $\left(\omega_{gap} = 2g = \sqrt{2k^2\omega_m\omega_p}\right)$ at the anti-crossing center, which can be controlled directly by tuning the value of k, which latter is accomplished by changing the dimensions of the cavity or the YIG. Zhang et al. [9] and Tabuchi et al. [8] suggested that the coupling strength between the Kittel and cavity modes should be expected to be proportional to the square root of the number of the net spins N, i.e., $g/2\pi = (g_0/2\pi)\sqrt{N}$, where g_0 is the strength of coupling of a single Bohr magneton to the cavity, and is given by

$$\frac{g_0}{2\pi} = \frac{\eta}{4\pi}\gamma\sqrt{\frac{\hbar\omega_0\mu_0}{V_c}} \tag{22}$$

where γ is the electron gyromagnetic ratio, μ_0 is the permeability of the vacuum, and V_c corresponds to the volume of the cavity. The coefficient $\eta \leq 1$ describes the spatial overlap and polarization-matching conditions between the microwave field and the magnon mode. The enhancement by the factor of $\sqrt{N}$ is due to magnon excitation in the Kittel mode, i.e., constructive interference of all possible processes in which a cavity photon flips one of the spins in the sphere.

Zhang et al [9] experimentally demonstrated this by increasing the diameter (i.e., volume V) of a YIG sphere, which determines the spin number ($N \propto V$), as shown in Fig. 28A. On the other hand, Tabuchi et al. [8] demonstrated increased coupling strength by simultaneous increase of YIG sphere size with reduction of microwave cavity size, as shown in

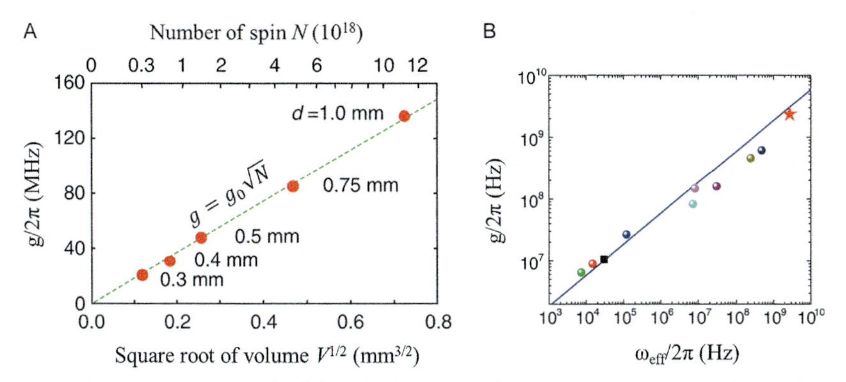

Fig. 28 (A) Coupling strength of the Kittel mode to the microwave cavity mode as a function of the sample diameter. The linear dashed line is plotted to fit the data to the cavity mode according to the single-spin coupling strength $g_0/2\pi$. (B) Coupling strength as a function of modal frequency, ω_{eff}. The solid line and symbols correspond to theoretical prediction and experimental measurements, respectively. *Reprinted with permission from Refs. [8,9], Copyright 2019 American Physical Society. A modified version of this figure was originally published in X. Zhang, C.-L. Zou, L. Jiang, H.X. Tang, Strongly coupled magnons and cavity microwave photons, Phys. Rev. Lett. 113 (15) (2014) 156401, https://doi.org/10.1103/PhysRevLett.113.156401 and Y. Tabuchi, S. Ishino, T. Ishikawa, R. Yamazaki, K. Usami, Y. Nakamura, Hybridizing ferro magnetic magnons and microwave photons in the quantum limit, Phys. Rev. Lett. 113 (2014) 083603, https://doi.org/10.1103/PhysRevLett.113.083603.*

Fig. 28B, where $\frac{g}{2\pi} \propto \frac{\omega_{eff}}{2\pi} = \sqrt{\frac{\omega_c}{2\pi}\frac{V}{V_c}}$. Later, the dependence of coupling strength on the volume of the YIG material was also demonstrated in a planar hybrid system by Castel et al. [105]. This was a more successful hybridization control achieved by manipulating the dimensions of either the cavity or magnetic material directly.

6.4 Damping of microwave cavity

From Eqs. (14) and (19), the frequency gap ω_{gap} can be controlled by tuning the damping rates of spin system α or photon resonator β or both. Tuning the damping of the spin system in a controllable way is challenging. It would be possible either by using different magnetic materials or exploiting the temperature dependence of the damping in certain materials [115]. However, controlling the damping of a microwave cavity is relatively easy. The loss rate of a cavity resonator has two contributions, $\beta_L = \beta_{\text{int}} + \beta_{ext}$, where the intrinsic loss rate, β_{int}, is primarily determined by the cavity conductance by polishing it. On the other hand, the extrinsic loss rate, β_{ext}, is determined by the coupling of external microwaves from the feeding line into the cavity, and is therefore controllable. The quality factor, $Q = 1/2\beta_L$, depends on the loaded cavity loss rate, and thus is controlled by either β_{ext} or β_{int}.

Maier-Flaig et al. [24] studied the influence of the cavity quality factor (Q from 0 to 8000) on P-M coupling spectra by changing the cavity's coupling ratio to the feeding line. Fig. 29A–C show the change in cavity

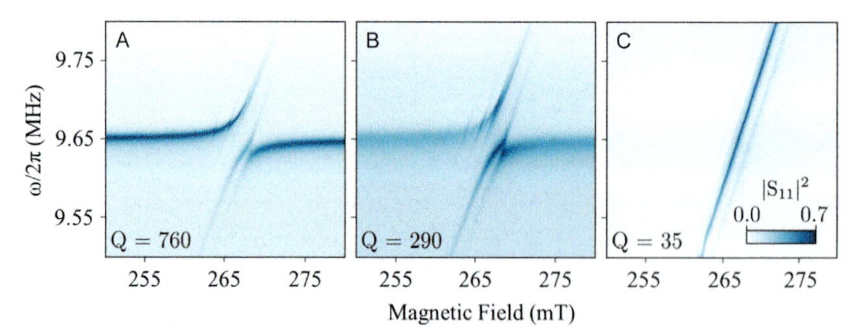

Fig. 29 As the coupling of the cavity to the feeding line (from left to right) increases, the cavity loss rate and the quality factor (Q) decreases. The strong-to-weak coupling transition can be demonstrated by tuning of the cavity quality factor. *Reprinted with permission from Ref. [24], Copyright 2019 American Physical Society. A modified version of this figure was originally published in H. Maier-Flaig, M. Harder, R. Gross, H. Huebl, S.T.B. Goennenwein, Spin pumping in strongly coupled magnon-photon systems, Phys. Rev. B 94 (2016) 054433, https://doi.org/10.1103/Phys-RevB.94.054433.*

reflection spectra [24] for (A) Q=760, (B) Q=260, and (C) Q=35, respectively. With decreasing Q, the cavity mode is broadened with a reduction in the amplitude of the coupled modes, and ω_{gap} decreases, along with, finally, merging of the hybridized peaks at Q=35. This indicates that at extremely low Q, the cavity mode is essentially destroyed, and thus, no hybridization can be observed between the photon and magnon modes. When the cavity mode is narrow, as in Fig. 29A, the bandwidth of microwaves is narrow. As a result, FMR absorption is only observed roughly in a narrow frequency range. However, as the quality factor of the cavity is decreased, the bandwidth of the cavity mode increases, and therefore the FMR resonance can be observed over a much wider frequency. At ultra-low Q, the microwave reflection is essentially flat over the observed frequency range except at the FMR and spin wave resonances. This highlights the importance of the cavity quality factor to the achievement of strong coupling between the excited magnon and photon modes.

6.5 Angular control of P-M coupling

Magnetizations in magnetic materials are excited by a microwave magnetic field generated from a microwave cavity or microstrip line. Since the precession is due to a magnetic torque, $M \times h$, only the components of magnetic fields perpendicular both to the total magnetization M and the static field H will stimulate precession motions efficiently. In this sense, it is also possible to control the coupling strength by manipulating the microwave magnetic field. Bai et al. [21] first studied P-M coupling for two different angles ($\theta' = 0°$ and $90°$) between M and h by placing a YIG/Pt bilayer in the microwave magnetic field of a 3-D cavity. The transmission measurement data are shown in Fig. 30A and B. At $\theta' = 0°$, the coupling gap is small, indicating weak coupling; contrastingly, at $\theta' = 90°$, a large anti-crossing dispersion indicates the presence of strong spin-photon hybridization.

Using the angular control technique, systematic measurement of the angular dependence of coupling strength was carried out in Ref. [31] by varying the angle between M and h (see Fig. 31A). This geometric dependence could be included explicitly in the coupling strength, so that $g/2\pi = (g_0/2\pi)|\sin\theta'|$, where $g_0/2\pi$ is the coupling strength at $\theta' = 90°$. By precisely tuning the angle θ' from $0°$ to $90°$, the coupling strength was controlled over both the strong and weak coupling regimes in the $k/\alpha - k/\beta$ parameter plane, as indicated by the solid line (Fig. 31B). The circles indicate experimental measurements, with different colors corresponding to different angles at which measurements were performed.

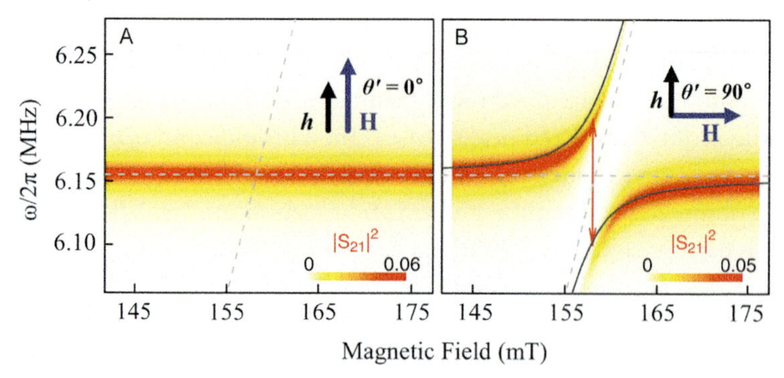

Fig. 30 Frequency-magnetic field dispersion for a YIG film/3-D cavity hybrid system. (A) A small coupling gap is observed when the *H* field is collinear with the local microwave magnetic field *h* ($\theta'=0°$). (B) The coupling gap is largest when the *H* field is perpendicular to the local microwave magnetic field *h* ($\theta'=90°$). Horizontal and vertical dashed lines indicate the uncoupled cavity and FMR dispersions, respectively, while the solid curves in (B) is a fit according to Eqs. (12a) and (12b). *Reprinted with permission from Ref. [31], with Copyright 2019 American Physical Society. A modified version of this figure was originally published in L. Bai, M. Harder, P. Hyde, Z. Zhang, C.-M. Hu, Y.P. Chen, J.Q. Xiao, Cavity mediated manipulation of distant spin currents using a cavity-magnon polariton, Phys. Rev. Lett. 118 (2017) 217201, https://doi.org/10.1103/Phys-RevLett.118.217201.*

6.6 Excitation of spin waves and their coupling with photon modes

In planar-geometry hybrid systems, strongly enhanced signals of higher-order spin-wave modes excited in YIG film were observed due to strong coupling between the YIG's magnons and a photon mode [11,26]. Hybridization is also possible between microwave photons and higher-order spin-wave modes, as demonstrated for standing spin waves (SSW) by different groups [25,29]. As an example, the coupling of a photon mode with different spin-wave modes of MSSW, BVMSW and FVMSW was successfully demonstrated (Fig. 32) according to the direction of externally applied magnetic fields with respect to the spin-wave propagation direction in a hybrid P-M-coupled system consisting of an ISRR and a YIG film [116]. Multiple anti-crossings have been observed along with the largest signal of the anti-crossing between the FMR and photon mode. Due to the hybridization of the multiple-higher-order spin waves with the photon mode, indexing and assignment of multiple higher modes is difficult. However, the strength of the coupling between the spin-wave and photon modes was estimated in some other hybrid systems, for example, by Zhang et al. [25] for a YIG film in a 3-D cavity and by Zhang et al. [29] for a single-crystalline YIG sphere in an SRR under different static magnetic field configurations. The coupling strength was found to be spin-wave-mode-dependent and to decrease as the

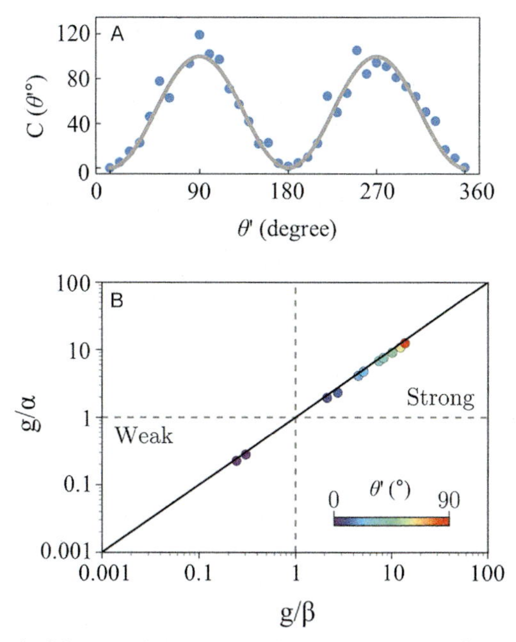

Fig. 31 (A) Control of the coupling strength by tuning the bias field angle. θ indicates the local angle between h and a bias field H at the sample location. The symbols indicate the angular dependent cooperativity ($C \equiv k^2\omega_m/4\alpha\beta\omega_p$) obtained from the experimental data of a YIG/Pt-cavity coupled hybrid system, while the black curve is the result of fitting according to $g/2\pi = (g_0/2\pi)|\sin\theta'|$. (B) The dissipation-normalized coupling strengths always lie along the constant slop, as indicated by the black line. *Panel (A) reprinted with permission from Ref. [31], with Copyright 2019 American Physical Society. Panel (B) reprinted with permission from Ref. [5], with Copyright Clearance Center Elsevier. A modified version of these figures were originally published in L. Bai, M. Harder, P. Hyde, Z. Zhang, C.-M. Hu, Y.P. Chen, J.Q. Xiao, Cavity mediated manipulation of distant spin currents using a cavity-magnon polariton, Phys. Rev. Lett. 118 (2017) 217201, https://doi.org/10.1103/Phys-RevLett.118.217201 and M. Harder and C.-M. Hu, Cavity spintronics: an early review of recent progress in the study of magnon-photon level repulsion, in: R.E. Camley, R.L. Stamps (Eds.), Solid State Physics, vol. 69, Academic Press, Cambridge, 2018, pp. 47–121, Chapter two, https://doi.org/10.1016/bs.ssp.2018.08.001.*

spin-wave mode index increases. The most important result of such work has been the experimental confirmation of the predicted coupling strength dependence $g_n \propto 1/n$, where n is the spin-wave mode index [25,29], which highlights the systematically variable coupling strengths that can be achieved through spin-wave/cavity hybridization. Such experimental results establish a new means of discussing discrete standing wave modes, and represent very useful for magnonics-based devices.

The MSSW mode is more localized at the film surface and exponentially decays with film depth, whereas BVMSWs and FVMSWs are excited

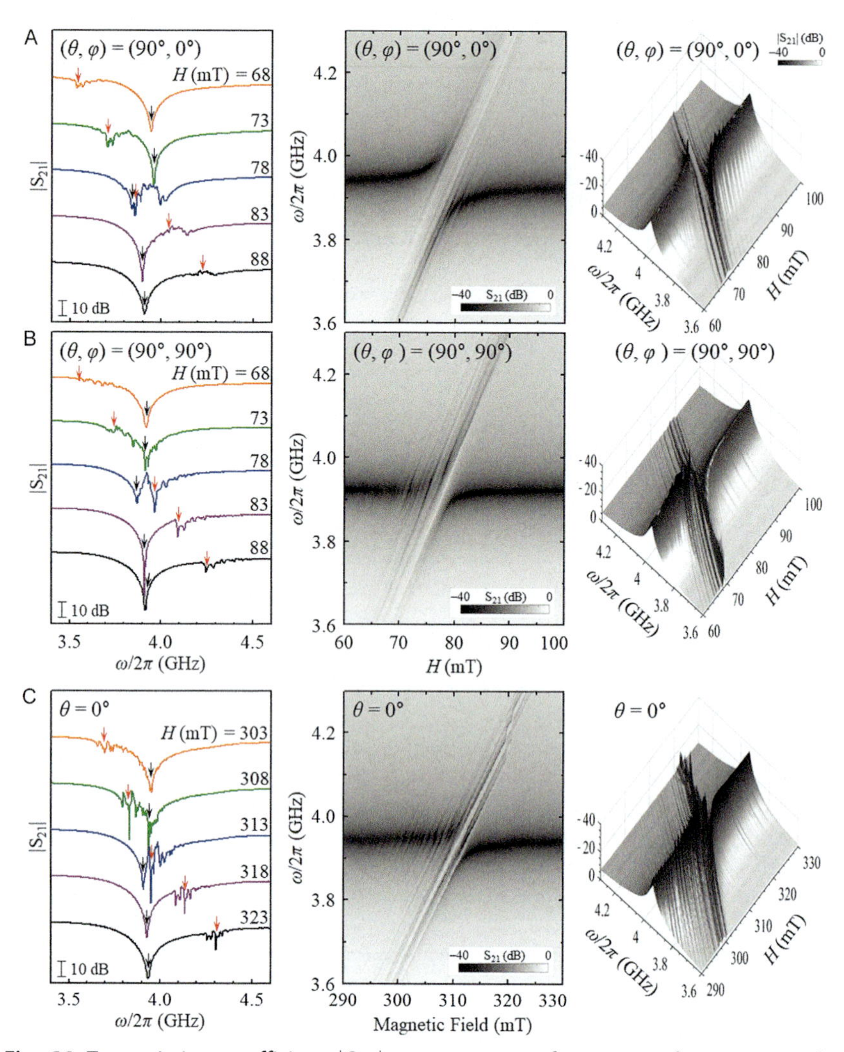

Fig. 32 Transmission coefficient $|S_{21}|$ spectra versus frequency of ac currents for indicated field strengths (first column), and their plane-view (second column) and perspective-view (last column) illustrations. (A–C) correspond to the data for the specific angle sets $(\theta, \varphi) = (90°, 0°)$, $(90°, 90°)$, and (C) $\theta = 0°$, respectively. In the first column, the blue-colored profiles in (A–C) were measured at the corresponding H_{cent} values where the resonance frequency of the FMR mode equated that of the photon mode.

throughout the volume of the film. In this sense, it is also possible to control the coupling strength by manipulating excitation of volume or surface modes. In order to examine the coupling strength for different spin-wave modes, Bhoi et al. [26] performed an experiment using the YIG/ISRR hybrid system in a planar geometry. During the $|S_{21}|$ measurements, both

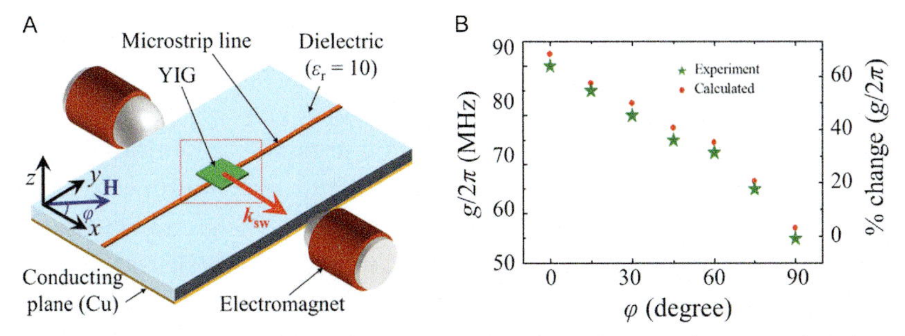

Fig. 33 (A) Schematic drawing of experimental setup and inverted pattern of SRR(ISRR) and YIG film. The ISRR (not shown here) is patterned on the ground plane of the microstrip line. (B) The variation of $g/2\pi$ and the percentage of change in $g/2\pi$ (for both experiment and calculated results) as a function of angle φ. *A modified version of this figure was originally published in B. Bhoi, B. Kim, J. Kim, Y.-J. Cho, S.-K. Kim, Robust magnon-photon coupling in a planar-geometry hybrid of inverted split-ring resonator and YIG film, Sci. Rep. 7 (2017) 11930, https://doi.org/10.1038/s41598-017-12215-8.*

the microwave frequency of oscillating currents flowing in the microstrip line and the static field strength H applied parallel to the film plane were varied for several specific angles φ (every 15° from 0° to 90°) (for the measurement geometry, see Fig. 33A). The excitation of spin-wave modes can be observed from the S_{21} spectra and the anti-crossing dispersion on the f-H plane [26]. The coupling strength monotonically decreases from 90 to 55 MHz with the rotation of φ from 0° to 90°. At $\varphi = 0°$, the BVMSWs are excited, whereas at $\varphi = 90°$, the MSSWs are excited. Since the MSSW mode is localized at the film surface and exponentially decays with film depth, it is less coupled with the ISRR mode than is the BVMSW mode. This leads to the reduction of coupling strength with increasing φ, as shown in Fig. 33B. The collectively enhanced magnon-microwave photon coupling strength therefore leads to the super-strong coupling regime.

6.7 Temperature control

According to the phase correlation model [10,103], the mathematical expression for dispersion spectra is given as

$$S_{21} \propto \frac{\omega^2}{\left(\omega^2 - \omega_p^2 + 2i\beta\omega\omega_p\right) - k^2\omega_m\omega^2(1 + \delta e^{i\phi})(\omega - \omega_r + i\alpha\omega_r)^{-1}}, \quad (23)$$

where the coupling strength $g/2\pi = \sqrt{2k^2\omega_m\omega_p}/4\pi$ and $\omega_m = \gamma\mu_0 M_s$ (Section 4.5; Eqs. (17–20a and 20b). The FMR angular frequency, $\omega_r = \gamma\sqrt{H(H + \mu_0 M_s)}$, depends on the intrinsic material parameters and the applied static magnetic field. The coupling strength is proportional to the square root of the net magnetic moment of the sample, and thus can be varied with the temperature of a magnetic sample.

The temperature control of P-M coupling strength was demonstrated by Maier-Flaig et al. [115] for a system consisting of a gadolinium iron garnet (GdIG) ferrimagnet and a 3-D microwave cavity. GdIG has three magnetic sublattices, where two iron sublattices present at the tetrahedral and octahedral sites are strongly antiferromagnetically coupled and thus results in small net magnetization M_{Fe}. A third sublattice is formed by a Gd^{3+} ion present at dodecahedral (c) sites, coupled weakly to M_{Fe}, and is aligned antiparallel to the net iron magnetization. At room temperature, the net magnetization of GdIG is thus dominated by the magnetization of the iron sublattices, and therefore points along the resultant iron sublattice magnetization M_{Fe}. However, as the magnetization of the Gd sublattice increases with decreasing temperature, the net remanent magnetization of GdIG decreases and then vanishes at a compensation temperature ($T_{comp}=270\,K$). Below T_{comp}, the net magnetization increases again due to the increasing magnetization of the Gd sublattice [68].

The effect of magnetization on the anti-crossing behavior can be observed from the dispersion spectra as shown in Fig. 34A. At 25 K, a strong coupling gives rise to characteristic anti-crossing of the cavity photon and magnon modes, whereas at 110 K, the cavity is only marginally disturbed, as indicative of weak coupling. The scaling of the effective coupling strength g with the magnetic moment (or magnetization) is displayed in Fig. 34B, which shows the variation of g^2 as a function of M_{eff} with its linear fit. According to this linear fit, g becomes zero at $M_{eff}=0$. In this work, it is shown that the coupling strength is scaled with M_{eff}, as $g=g_0\sqrt{M_{eff}}$ with g_0 single spin-single photon coupling strength.

6.8 Electric control

The study of P-M coupling requires bringing their individual resonance modes close to each other. As shown in most studies, a convenient way to accomplish that is to tune the resonance frequency of the magnon mode by externally applied static magnetic fields. However, Kaur et al [23] demonstrated coupling between a cavity and a YIG by control of the resonance of a planar cavity. The resonance frequency of an SRR was continuously controlled while holding the YIG sphere in a constant static magnetic field, by tuning the DC voltage to a varactor diode connected to the SRR, as schematically shown in Fig. 35A. The modes' frequencies estimated from

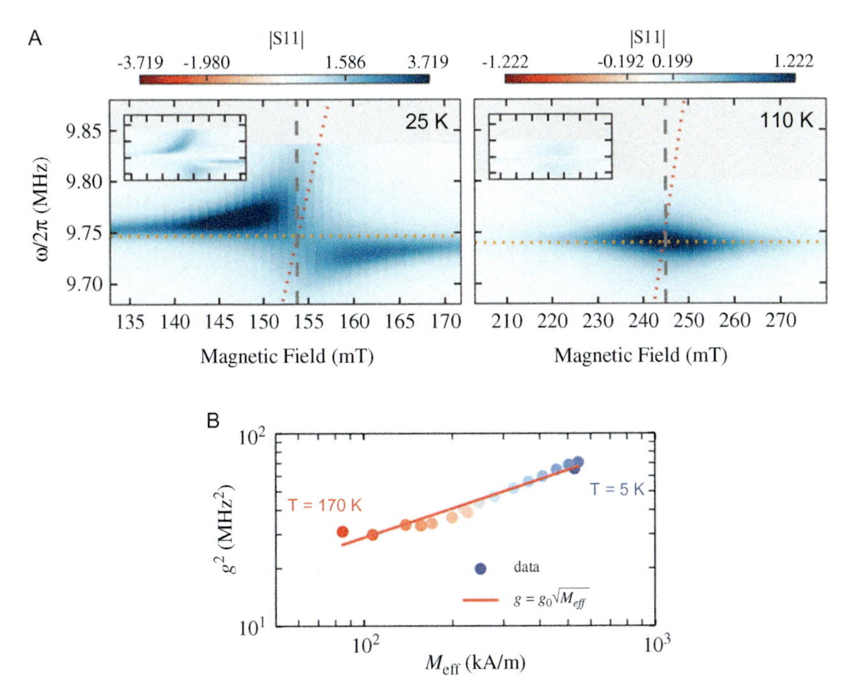

Fig. 34 (A) Derivative of the reflection parameter ($|S_{21}|$) on the plane of the microwave frequency and magnetic field ($\omega/2\pi - H$ plane) for a GdIG film and 3-D cavity coupled hybrid system at two distinct temperatures. The coupling is strong at low temperature and becomes weak at higher temperature. The horizontal orange-color dotted line indicates the resonance frequency of the unperturbed cavity, and the red-color dotted line corresponds to the resonance frequency of the unperturbed spin system. The dashed vertical gray line represents the magnetic field where the unperturbed cavity mode and magnon mode are degenerate (coupling center). (B) Square of coupling strength g^2 as a function of the effective magnetization. The data (symbols) agree well with the expected scaling behavior of $g \propto \sqrt{M_{eff}}$ (see the red line). *Reprinted with permission from Ref. [115], Copyright 2019 by the American Institute of Physics. A modified version of this figure was originally published in H. Maier-Flaig, M. Harder, S. Klingler, Z. Qiu, E. Saitoh, M. Weiler, S. Geprags, R. Gross, S.T.B. Goennenwein, H. Huebl, Tunable magnon-photon coupling in a compensating ferrimagnet—from weak to strong coupling, Appl. Phys. Lett. 110 (2017) 132401, https://doi.org/10.1063/1.4979409.*

the measured S_{21} spectra were plotted as a function of applied voltages, as shown in Fig. 35B. The anti-crossing between the cavity mode and magnon mode due to coupling was then clearly found at around 8.5 V. The experimental results agreed well with the calculations (solid lines) using the coupled harmonic oscillator model, as shown in Fig. 35B. This experimental realization of DC voltage control of coupling in a varactor-loaded cavity not only provides another degree of freedom in the study of coupling physics but also affords the advantage of such a planar structure.

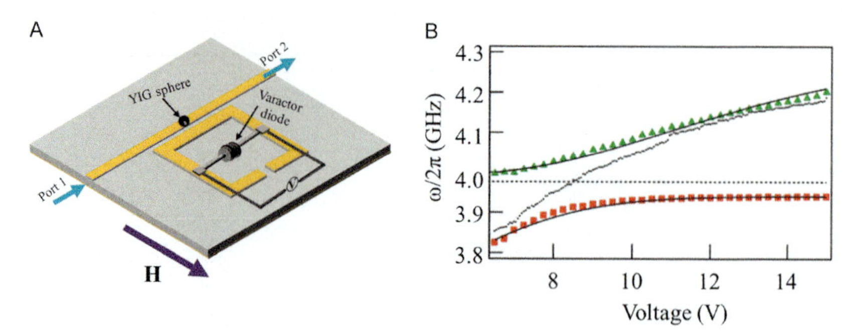

Fig. 35 (A) Schematic drawing of a varactor-loaded SRR used to study the coupling between the cavity and magnon modes. (B) Experimental (symbol) and calculated (solid line) dispersions of coupled modes between ISRR and YIG, as a function of voltage applied to the varactor diode. The dotted black-color lines correspond to each uncoupled mode. *Reprinted with permission from Ref. [23], Copyright 2019 by the American Institute of Physics. A modified version of this figure was originally published in S. Kaur, B.M. Yao, J.W. Rao, Y.S. Gui, C.-M. Hu, Voltage control of cavity magnon polariton, Appl. Phys. Lett. 109 (2016) 032404, https://doi.org/10.1063/1.4959140.*

7. Nonlinear effects in P-M coupling

Nonlinear dynamics in a variety of physical systems differ dramatically from their linear counterparts. Nonlinear properties show significantly modified resonance curves, including the amplitude-dependent shift of the resonance frequency from its "natural" value and the distortion of the resonance line shape (Fig. 36A) to produce a foldover effect [117,118]. This nonlinearity is related to the Kerr effect of magnons [119,120]. Recently such nonlinear effects have been demonstrated in P-M-coupled hybrid systems [93,119,120]. These microwave-power-induced nonlinearities are directly related to the strength of the microwave magnetic field that drives magnetization dynamics in magnetic films [117,121,122]. These nonlinear effects impact not only the physical understanding of a system's dynamics but also have important technological implications in modern electronics and advanced optical devices [123,124].

7.1 Magnon-Kerr effect in P-M coupling

Distortion of the FMR absorption profile occurs when the driving microwave power exceeds a threshold value, which causes nonlinear effects as shown in Fig. 36 [117,121,122]. Such effects can also be observed in a

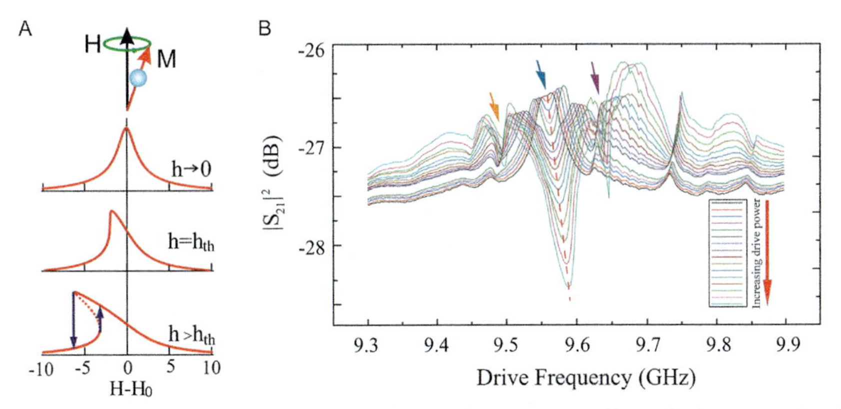

Fig. 36 (A) Schematic representation of FMR absorption profile with the strength of microwave magnetic fields. (B) Transmission spectrum of the cavity measured as a function of the drive frequency by successively increasing the driving power. The probe field is fixed at 10.1035 GHz. The blue arrow indicates the response of the Kittel mode, whereas the orange and purple arrows indicate two magnetostatic modes. *Panel (A) adapted from Y.S. Gui, A. Wirthmann, N. Mecking, C.M. Hu, Direct measurement of nonlinear ferromagnetic damping via the intrinsic foldover effect, Phys. Rev. B 80 (2009) 060402. Reprinted with permission from Ref. [119], Copyright 2019 by the American Physical Society. A modified version of this figure was originally published in Y. P. Wang, G. Q. Zhang, D. Zhang, X.Q. Luo, W. Xiong, S.P. Wang, T.F. Li, C.M. Hu, and J.Q. You, Magnon Kerr effect in a strongly coupled cavity-magnon system, Phys. Rev. B 94 (2016) 224410, https://doi.org/10.1103/PhysRevB.94.224410.*

strongly coupled P-M hybrid system. When the magnetic material is pumped to generate considerable magnons, the Kerr effect yields a perceptible shift of the cavity's central frequency and more appreciable shifts of the magnon modes. This was first experimentally demonstrated by Wang et al [119] at cryogenic temperatures using a small YIG sphere strongly coupled to the photons in a specially designed 3-D cavity with three ports. The two ports are used for transmission spectra measurement, where a third port is connected to a loop antenna in the vicinity of the YIG sphere to efficiently drive the magnon mode.

Fig. 36B shows the measured transmission spectra by tuning of the frequency of the driving field where the frequency of the probe field is fixed at the central frequency of 10.1035 GHz of the cavity containing the YIG sphere at a bias magnetic field of 346.8 mT. The dips indicated by the blue, orange and purple arrows correspond to the Kittel mode and two different magnetostatic modes, respectively. With increasing driving power, the main dip becomes successively deeper, and simultaneously shifts rightward. This reveals that the Kittel mode has a blueshift with the increase in the driving power. The responses of magnetostatic modes are similar. This study paves the way to the exploration of further nonlinear effects (i.e., bi-stability) in the cavity P-M-coupled system.

7.2 Bi-stability of P-M coupling

The appearance of bi-stability in FMR spectra has been widely studied since the early 1980s [121,122]. The bi-stable behaviors in P-M coupling were demonstrated as a sharp frequency switching of the coupled modes (also called cavity magnon polaritons, CMP) [93,120]. For example, Fig. 37B measure the frequency shift Δ_{LP} of the lower-branch CMPs versus the driving power P_d for different values of the drive-field frequency detuning, where the [100] crystallographic axis of the YIG sphere is aligned parallel to the static magnetic field. Here, the angular frequency ω_{LP} of the lower-branch CMPs is tuned to be at the anticrossing point A in Fig. 37A and the drive-field frequency detuning $\delta_{LP} \equiv \omega_{LP} - \omega_d$ is relative to the lower-branch CMPs, where ω_d is the angular frequency of the driving field. The observation of a hysteresis loop clearly reveals the emergence of CMP bi-stability in the cavity magnonics system. This hysteresis loop is counter-clockwise when considering the increasing and decreasing directions of the driving power. Its area reduces with the decrease in the frequency detuning $|\delta_{LP}|$. Fig. 37C shows the frequency shift Δ_{LP} of the same lower branch of the coupled mode versus the frequency detuning δ_{LP} for different values of P_d. When the increasing and decreasing directions of δ_{LP} are considered, a counter-clockwise hysteresis loop is also clearly shown, the area of which decreases with the decrease in P_d.

The frequency shifts of the lower-branch CMPs as a function of both the driving field frequency detuning δ_{LP} and the drive power P_d can be obtained by solving

$$\left[(\Delta_{LP} + \delta_{LP})^2 + \left(\frac{\gamma_{LP}}{2}\right)^2\right]\Delta_{LP} - cP_d = 0, \tag{24}$$

where γ_{LP} is the damping rate of the lower-branch CMP, and c is a coefficient characterizing the coupling strength between the drive field and the lower-branch CMPs (for details on the derivation, see Ref. [93]). The two stable solutions of Δ_{LP} in Eq. (24) correspond to two states of the system with large and small numbers of polaritons in the lower branch, which fit well with the experimental results shown in Fig. 37B and C. These bi-stable behaviors are also remarkably dependent on the direction of the crystallographic axes of the YIG sphere along which the static magnetic field

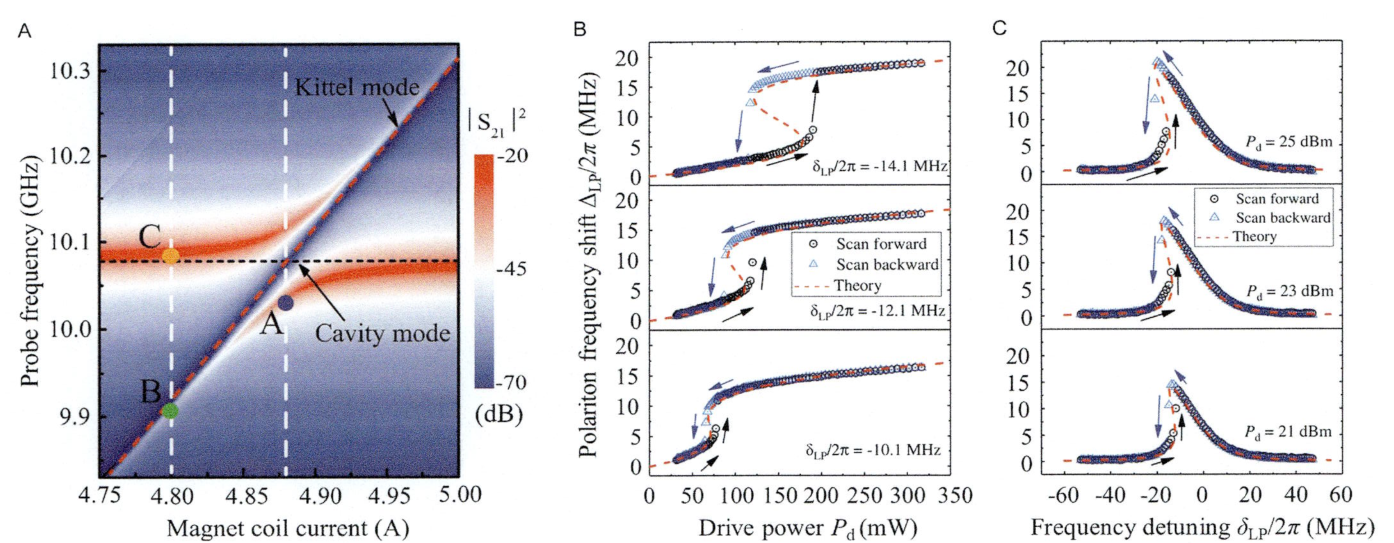

Fig. 37 (A) Transmission power of a P-M-coupled system, measured versus the magnet coil current (i.e., the static magnetic field) and the frequency of the probe field. The two vertical dashed lines indicate, respectively, the resonance and the very-off resonance at which bi-stability was measured. (B) Frequency shift Δ_{LP} as a function of driving power P_d for different frequency detuning δ_{LP}. (C) Frequency shift Δ_{LP} versus frequency detuning δ_{LP} for different values of P_d. The black circle (blue triangle) dots are the forward (backward)-scanning results. The red dashed curves are the theoretical results obtained using Eq. (24). *Reprinted with permission from Ref. [93], Copyright 2019 by the American Physical Society. A modified version of this figure was originally published in Y.-P. Wang, G.-Q. Zhang, D. Zhang, T.-F. Li, C.-M. Hu, J.Q. You, Bistability of cavity magnon polaritons, Phys. Rev. Lett. 120 (2018) 057202, https://doi.org/10.1103/PhysRevLett.120.057202.*

is applied. For the [100,109] directions, clockwise and counter-clockwise hysteresis loops are observed in the bi-stable behaviors due to the Kerr nonlinearity [119,120]. The simultaneous bi-stability of both magnons and cavity photons at very off-resonance points B and C shown in Fig. 37A was also demonstrated by applying the drive field only to the lower branch, where the optical bi-stability is achieved via the magnetic bi-stability.

Hyde et al. [125] proposed an approach to directly measure the nonlinear foldover effect for cavity magnon polaritons by placing a YIG sphere into a Fabry-Perot-like microwave cavity. Here, a high-power microwave generator drives the cavity subsystem, and the fields produced by the cavity resonance mode are then used to excite resonance in the YIG. At high input power, this excites the coupled CMP system to nonlinear regimes, the dynamics of which can then be detected with a spectrum analyzer. This implementation allows use of the same frequency for the driving and probing fields and performance of an in-tune two-port measurement. The resulting bi-stability features can exhibit clockwise, counter-clockwise, and butterfly-like hysteresis loops and are solely dependent on the relative weight of the magnon-like and photon-like components. Analytically it was demonstrated that the coupling strength and damping of the cavity should be carefully designed in order to produce foldover effects for photon-like CMPs. Fig. 38 shows the critical microwave power required

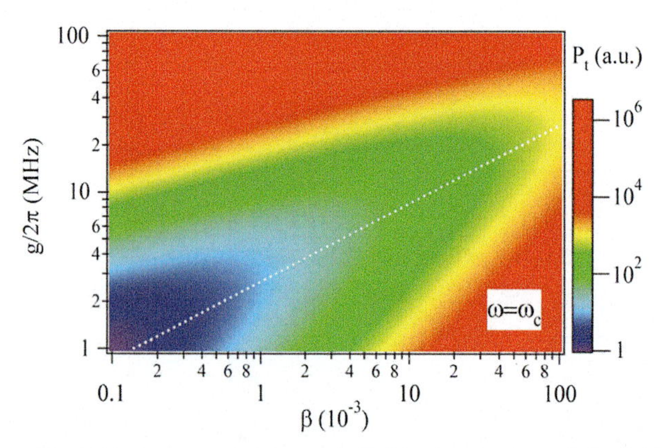

Fig. 38 The critical microwave power required for observation of the foldover effect, calculated as a function of the coupling strength g and the damping parameter β of the cavity. The dotted line indicates the lower limit of P_t versus β. *Reprinted with permission from Ref. [125], Copyright 2019 by the American Physical Society. A modified version of this figure was originally published in P. Hyde, B.M. Yao, Y.S. Gui, G-Q. Zhang, J.Q. You, and C.-M. Hu, Direct measurement of foldover in cavity magnon-polariton systems, Phys. Rev. B 98 (2018) 174423, https://doi.org/10.1103/PhysRevB.98.174423.*

for an observation of the foldover effect as a function of the coupling strength g and the damping parameter β of the cavity.

Although the photon subsystem has no nonlinear components of its own, these photon-like foldover effects are produced through light-matter interactions with a nonlinear magnetic subsystem (i.e., YIG). These observations indicate that CMPs can serve as a bridge or transducer between optical and magnetic bi-stabilities, introducing a new approach for use of one effect to manipulate and control the other.

8. Prospective applications of P-M coupling

Innovative condensed matter physicists are often inspired by atomic physics. The successful demonstration of strong coupling between the magnon modes and microwave-frequency photon modes attracted immediate and broad interest from groups working on spintronics, optics, and micro-mechanical systems. This section discusses newly emerging subfields based on the concept of P-M couplings, including cavity spintronics [5,10], cavity magnomechanics [15], and cavity optomagnonics [13].

8.1 Cavity spintronics

Spin currents are the foundation of spintronics as an information carrier for next-generation information-processing devices. Therefore, the generation (excitation), manipulation, and detection of spin currents have attracted much attention over the past two decades. The coupled P-M systems are of great relevance for spintronic applications. A spin pumping experiment was performed by Bai et al. [10] on an electrodynamically coupled P-M system, which revealed distinct coherent features that could potentially be used to manipulate spin currents via light-matter interaction. Based on the intriguing physics of coherent magnon-photon coupling, a new research field, namely, "cavity spintronics," is emerging, which connects some of the most exciting modern physics, such as quantum information and quantum optics, with one of the oldest researches and sciences.

This was achieved by designing a special microwave cavity as schematically shown in Fig. 39. This set-up enables both microwave transmission measurement of the cavity and microwave-induced voltage measurements in a ferromagnet. By inserting a bilayer of YIG/Pt into the cavity, microwave voltages generated in Pt were measured via the spin currents stimulated by the FMR in the YIG film. This enables studies of the impact of P-M coupling on spin current transport.

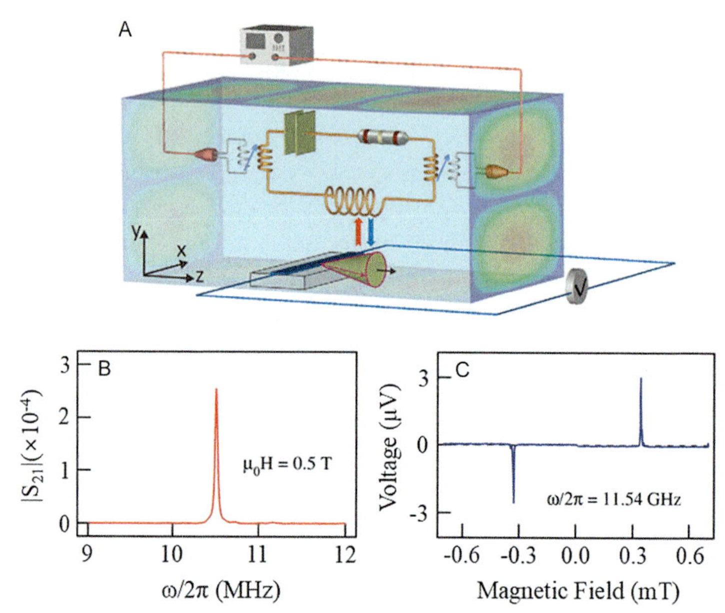

Fig. 39 (A) Sketch of the experimental setup for the study of P-M coupling. The setup was designed to enable both transmission measurements of the cavity and the electrical detection of FMR in samples loaded in the cavity. The artificial LCR circuit models the microwave current carried by the cavity mode, which couples with the magnons in the magnetic material. (B) A typical cavity mode measured by transmission spectrum. (C) The uncoupled FMR electrically detected by measuring the photo-voltage due to spin pumping. *Reprinted with permission from Ref. [10], Copyright 2019 by the American Physical Society. A modified version of this figure was originally published in L. Bai, M. Harder, Y.P. Chen, X. Fan, J.Q. Xiao, and C.-M. Hu, Spin Pumping in Electrodynamically Coupled Magnon-Photon Systems, Phys. Rev. Lett. 114 (2015) 227201, https://doi.org/10.1103/PhysRevLett.114.227201.*

Furthermore, a manipulation of spin currents over a long distance of several centimeter was demonstrated by inserting two YIG/Pt bilayer films in a microwave cavity as shown in Fig. 40. The cavity photons act as a bridge to carry information from the first sample, denoted as YIG-1, to the second sample, denoted as YIG-2. As indicated in the figure, the rotation of the cavity lid enables tuning of the angle θ' between the local microwave and the static magnetic field at the YIG-1 location. When θ' was tuned from 0° to 90°, the voltage, and hence the spin current generated in YIG-1, increased as the field torque on the magnetization increased, and additionally, a simultaneous increase in the spin current in the YIG-2 was observed owing to photon-mediated interaction between the two YIG

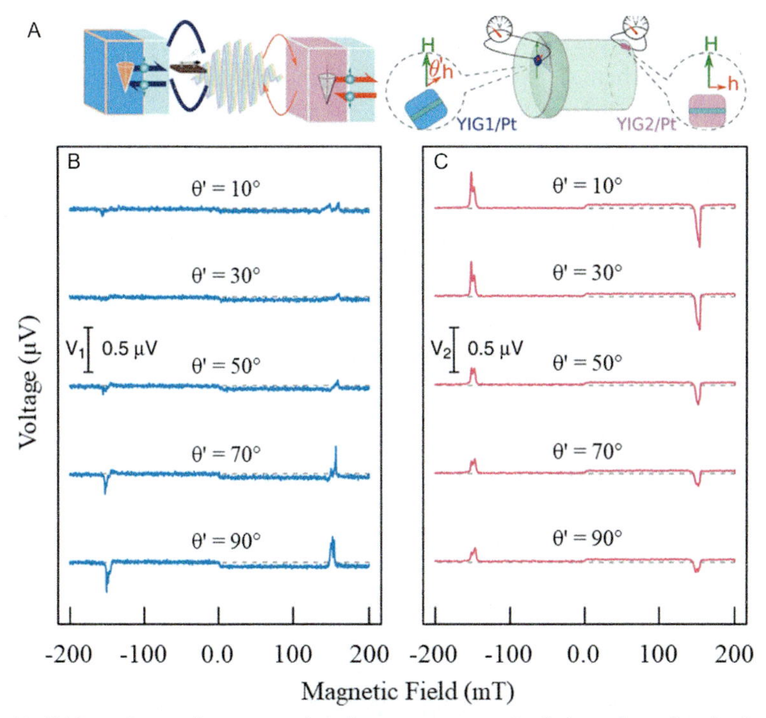

Fig. 40 (A) Experimental setup used to demonstrate manipulation of non-local spin current. The general idea is to use cavity photons, which are coupled to both spin systems, as a bridge to carry information from YIG-1 to YIG-2. Therefore, by locally tuning YIG-1, an influence on YIG-2 can be observed. Experimentally this idea was realized by placing YIG-1 on the lid of a cylindrical microwave cavity which was rotated within a static magnetic field, H. YIG-2, at the bottom of the cavity, had a fixed orientation with respect to H. (B) The YIG-1 voltage signal depends directly on the angle θ', which controls the cooperativity of YIG-1, while (C) the voltage on YIG-2 is also tuned by θ'. *Reprinted with permission from Ref. [31], Copyright 2019 by the American Physical Society. A modified version of this figure was originally published in L. Bai, M. Harder, P. Hyde, Z. Zhang, C.-M. Hu, Y.P. Chen, J.Q. Xiao, Cavity mediated manipulation of distant spin currents using a cavity-magnon-polariton, Phys. Rev. Lett. 118 (2017) 217201, https://doi.org/10.1103/Phys-RevLett.118.217201.*

films [31]. By controlling the coupling strength of only one magnetic system, a simultaneous change in the spin current of another magnetic system is observed, which is well separated and not directly tuned. Such long-distance manipulation of spin currents and its dependence on the strength of P-M coupling opens a new avenue to the generation and manipulation of spin currents in the development of cavity spintronics.

8.2 Cavity magnomechanics

The use of mechanical degrees of freedom in ferromagnetic crystals is a new field in P-M-coupled hybrid quantum systems [6,15]. Cavity magnomechanics has emerged from the combination of a cavity photon and coupling between phonon and magnon modes through a magnetostrictive interaction. Notably, deformation modes in a ferromagnetic crystal, which is to say, phonon modes intrinsic to the sample that couple to magnetostatic modes through magnetostrictive forces, can be used as mechanical modes (Fig. 41). Therefore, coherent signal conversions among these three different information carriers (photons, phonons, magnons) can be realized in a single hybrid device.

Cavity magnomechanics was first demonstrated at room temperature with a 250 μm diameter YIG sphere having a phonon frequency $\omega_d/2\pi = 11.42$ MHz inside a 3-D microwave cavity [15]. Mechanical modes at larger frequencies could be possible using smaller-sized ferromagnetic

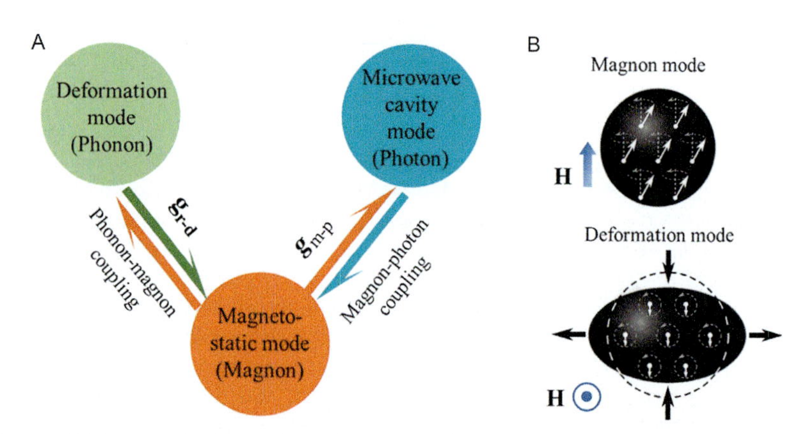

Fig. 41 (A) An intuitive illustration of magnomechanical coupling, showing the coupling between a deformation mode (phonon), a magnetostatic mode (magnon), and a microwave cavity mode (photon) of corresponding resonance frequencies ω_d, ω_r and ω_p, respectively. The magnetostrictive force leads to phonon-magnon coupling between deformation and magnetostatic modes of strength g_{r-d}, which can be enhanced beyond the damping of the mechanical mode by driving the magnetostatic mode. A magnetic dipole interaction leads to P-M coupling of strength g_{m-p}, which couples the magnetostatic and microwave cavity modes. (B) Schematic representation of uniform magnon excitation in the YIG sphere (top). The bottom figure illustrates how the dynamic magnetization of magnons (vertical black arrows) causes the deformation (compression in the direction perpendicular to the magnetic field direction) of the YIG sphere (and vice versa), which rotates at the magnon frequency. *Adapted from X. Zhang, C.-L. Zou, L. Jiang, H.X. Tang, Cavity magnomechanics, Sci. Adv. 2 (2016) e1501286.*

mechanical oscillators [126]. In the experiment of Ref. [15], in order to greatly reduce the clamping losses of the deformation (phonon) modes, the YIG sphere is glued to an optical fiber, enabling a mechanical linewidth of $\gamma_d/2\pi = 150$ Hz. In this system, the magnomechanical coupling strength $g_{r-d}/2\pi \sim 10$ MHz is much smaller than every linewidth in the P-M-phonon-coupled hybrid system. However, as the magnetostrictive interaction is of the radiation pressure type, the coupling strength can be parametrically enhanced to approximately 30 kHz by strongly pumping the magnetostatic mode [15]. This enhancement leads to a coupling strength larger than the linewidth γ_d of the deformation mode, but still smaller than the MHz linewidth of the magnetostatic modes in the YIG spheres. Despite this limitation, a rich diversity of phenomena has been observed in this cavity P-M-phonon-coupled hybrid system [15]. Notably, under the so-called triple resonance condition, the frequency of the mechanical mode matches the magnetic dipole coupling strength between the resonant magnetostatic and microwave cavity modes, as demonstrated in Ref. [15].

8.3 Cavity optomagnonics

In cavity optomagnonics, magneto-optical effects allow coupling between the magnetostatic modes in ferromagnetic crystals and optical cavity modes (THz) in combination with microwave cavity (MHz-GHz) modes [13]. By exploiting such a hybrid system formed by the Kittel mode and the microwave cavity mode, Hisatomi et al. [124] demonstrated bi-directional coherent conversions between the microwave photon and optical photon, where the microwave field is coupled to the hybrid system through the cavity mode, while the THz optical field addresses the hybrid system through the Kittel mode via the Faraday and inverse Faraday effects.

As schematically illustrated in Fig. 42A, a 0.75 mm-diameter YIG sphere is embedded in a 3-D microwave cavity to form a strongly coupled hybrid system between the Kittel mode and the cavity mode. The YIG sphere is illuminated by a 1550 nm continuous-wave laser. After passing the sample, the polarization of the transmitted beam laser oscillates at the frequency of the induced magnetization oscillation by the magneto-optical Faraday effect, thus producing the two optical sidebands around the laser frequency, which can be detected by a heterodyne measurement with a high-speed photodiode. On the other hand, the creation of microwave photons by light (optical photons) is established using two phase-coherent laser fields generated from a monochromatic laser simultaneously impinged on the YIG

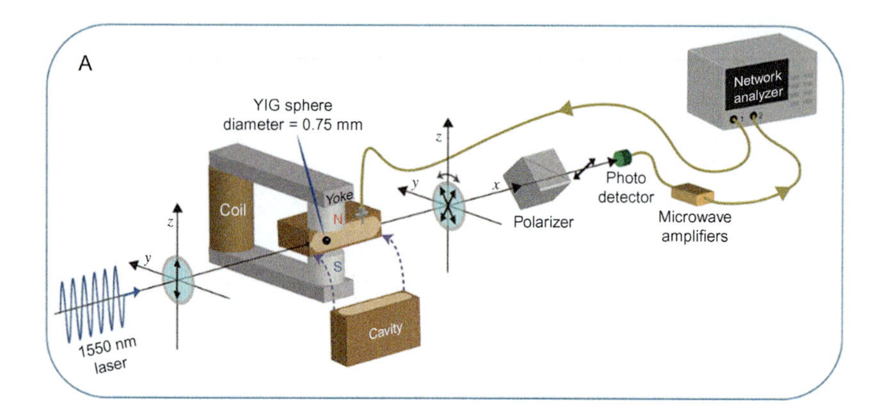

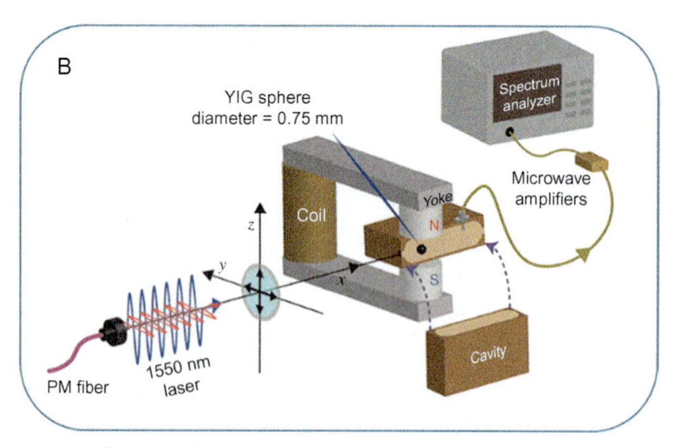

Fig. 42 Set-up of experimental measurements. (A) Conversion of microwave-frequency signal (photons) to optical signal. A YIG sphere is placed in a microwave cavity to form a strongly coupled hybrid system between the Kittel mode and the microwave cavity mode, which is characterized by measuring the microwave reflection coefficient from the system using VNA. A static magnetic field is applied to the YIG sample with permanent magnets. Under the microwave drive, when the laser is impinged on the YIG sample, the polarization of the carrier laser produces the optical sideband field. The beat signal between the carrier and the sideband field is measured using a polarizer and a fast photodetector, is amplified with two low-noise microwave amplifiers, and is fed into the vector network analyzer. (B) Conversion of optical signal to microwave-frequency signal. Two phase-coherent laser fields generated from a monochromatic laser are simultaneously impinged on the YIG sample to induce the inverse Faraday effect. The created magnons predominantly decay to the microwave cavity, and the coupled-out microwave signal from the cavity is amplified and fed into a spectrum analyzer. *Reprinted with permission from Ref. [124], Copyright 2019 by the American Physical Society. This figure was originally published in R. Hisatomi, A. Osada, Y. Tabuchi, T. Ishikawa, A. Noguchi, R. Yamazaki, K. Usami, and Y. Nakamura, Bidirectional conversion between microwave and light via ferromagnetic magnons, Phys. Rev. B 93 (2016) 174427, https://doi.org/10.1103/PhysRevB.93.174427.*

sample to induce the inverse Faraday effect. The created magnons relax in the microwave cavity, resulting in a microwave signal measured with a vector network analyzer (Fig. 42B). The phase coherence is preserved in the conversion process, resulting in a photon conversion efficiency of $\sim 10^{-10}$. The prospect of efficient bi-directional conversion of microwaves and optical photons is appealing for quantum limited microwave amplifiers [127] and quantum telecommunications [128]. This kind of an inaugural research calls for further development to increase the strength of optomagnonic interaction, especially for the purposes of experiments in optical cavities.

9. Summary and future directions

Understanding the interactions of excited modes in hybrid quantum systems such as coupling between magnon and photon modes are key to building large-scale many-body quantum systems for quantum information technology. In this context, the coupling of magnon modes collectively excited in ferromagnets or ferrimagnets to elementary excitations of electromagnetic waves (photons) is mostly necessary in the development of quantum information technologies, because such a hybrid system provides long coherence time, fast operation, and scalability.

We herein present an overview of the current status of P-M coupling, especially devices that integrate spin systems with microwave resonators. Considerable progress in the study of P-M coupling has been made recently in both theory and experimentation. Even though the basic ideas for coupling between different systems (e.g., relying on natural coupling to electric or magnetic fields) are quite simple, there are a variety of different approaches for integrating different physical systems in their hybrids. Moreover, careful consideration of the intrinsic dynamic properties of individual systems should be made before selecting the one that is best suited for achieving desired types and strengths of coupling. Experimental measurements of P-M coupling have only just begun to demonstrate coherent and dispersive coupling between different physical systems. However, with the help of advances in fabrication and design, we expect to have promising and advanced results within a few years. This review provides a useful introduction and different perspectives on P-M coupling for beginners in this emerging research field.

We now conclude this review with a discussion of the possible future directions of research in the area of P-M coupling. One of the interesting applications of P-M coupling is the realization of manipulations of local and non-local spin currents in a hybrid system of normal anticrossing dispersion [10,31]. It would be exciting to investigate the effect of dispersion type on the generation of spin currents produced in a YIG/Pt bilayer coupled to a variety of photon cavities.

In principle, a P-M hybrid system with EPs is potentially applicable for topological energy transfer, quantum sensing, and nonreciprocal photon transmission [102,129]. The experimental observations of level attraction, abnormal anti-crossing and EPs are interesting for their further demonstration of nonreciprocal energy transfer and relevance to the development of related one-way devices.

All of the relevant studies completed to date are restricted to coupling in single or narrow-band operating frequencies. Meanwhile, recent demands for solving data traffic jams in communications have stimulated the development of P-M hybrid systems for operation in broadband frequency ranges [46,130]. Therefore, the design and measurements of a reliable, low-cost and less-complex simple hybrid structure are in high demand for development of future broadband communication devices using controllable strong P-M coupling.

So far, the coupling of microwave photons with magnons in ferro- and ferrimagnets has been the focus, though magnetization dynamics in antiferromagnets (AFMs) have been much studied also, owing to their better stability against external field perturbations and negligible cross-talking with neighboring spins due to the absence of stray fields. The AFM dynamics also are on the order of THz frequency, much faster than GHz frequency in typical ferromagnets. However, only a few theoretical studies on P-M coupling in AFMs have been reported [131]. Antiferromagnetic insulators with low damping are promising candidates for realization of strong P-M coupling and THz information processing device applications.

Magnetostrictive materials (e.g., FeCo or CoPd) combined with piezoelectric materials (e.g., PZT) are also interesting with respect to the study of coupling between photons and magnons, specifically as they can manipulate the anisotropy of magnetic thin films using both magnetoelastic coupling and voltage-driven piezoelectric effect. Thus, multiferromic materials/photon-cavity hybrid systems are also useful to the study of voltage-controlled P-M coupling.

Optomagnonic coupling in an optical frequency range requires new magnetic materials such as bismuth-doped YIG [132,133] or magneto-optical bilayers [134]. Ideal materials for this coupling would require a small magnon linewidth, strong magneto-optical effects, low optical absorption, and high optical quality factors. The magnetostatic volume mode could also be tuned by changing the shape of the optical resonator. For example, as recently demonstrated in Ref. [33], samples' geometry such as a disk shape would allow the exploration of optomagnonic coupling by exciting vortex modes in an optical resonator.

Nonlinear dynamics in magnons is an important issue in the area of magnetization dynamics. The coupling of photon modes with non-linear magnon modes would present novel opportunities for the development of integrated on-chip photonics and also spintronics devices. As recently demonstrated in Ref. [135], fractals of nonlinear spin-wave modes rely on strong dispersion and arise spontaneously in a quasi-1-D magnonic crystal with the increase of microwave power. This newly found fractal is also interesting with respect to its application to P-M coupling for implementation in telecommunications and multimode signal processing.

Last but not least, the research field of hybrid quantum systems is still evolving with new-emergent ideas. It is quite likely that in a few years, the designs of samples to be studied and practical devices will be significantly more sophisticated than the prototypes studied in recent years. Thus, many more exciting P-M coupling behaviors and their further applications would be demonstrated, though this research field is small relative to magnetization dynamics and related spintronics.

Acknowledgments

The authors are thankful to Bosung Kim and Yongsub Kim for their valuable assistance. This work was supported by the Basic Science Research Program through the National Research Foundation of Korea (NRF) funded by the Ministry of Science, ICT and Future Planning (Grant No. NRF-2018R1A2A1A05078913).

References

[1] P. Forn-Díaz, L. Lamata, E. Rico, J. Kono, E. Solano, Ultrastrong coupling regimes of light-matter interaction, Rev. Mod. Phys. 91 (2019) 025005.
[2] M. Wallquist, K. Hammerer, P. Rabl, M. Lukin, P. Zoller, Hybrid quantum devices and quantum engineering, Phys. Scr. T137 (2009) 014001.
[3] Z.-L. Xiang, S. Ashhab, J.Q. You, F. Nori, Hybrid quantum circuits: superconducting circuits interacting with other quantum systems, Rev. Mod. Phys. 85 (2013) 623–653.
[4] A. Frisk Kockum, A. Miranowicz, S. De Liberato, S. Savasta, F. Nori, Ultrastrong coupling between light and matter, Nat. Rev. Phys. 1 (2019) 19–40.
[5] M. Harder, C.-M. Hu, Cavity spintronics: an early review of recent progress in the study of magnon–photon level repulsion, in: R.E. Camley, R.L. Stamps (Eds.), Solid State Physics, vol. 69, Academic Press, 2018, pp. 47–121 (Chapter two).
[6] D. Lachance-Quirion, Y. Tabuchi, A. Gloppe, K. Usami, Y. Nakamura, Hybrid quantum systems based on magnonics, Appl. Phys. Express 12 (2019) 070101.

[7] H. Huebl, C.W. Zollitsch, J. Lotze, F. Hocke, M. Greifenstein, A. Marx, et al., High cooperativity in coupled microwave resonator ferrimagnetic insulator hybrids, Phys. Rev. Lett. 111 (2013) 127003.
[8] Y. Tabuchi, S. Ishino, T. Ishikawa, R. Yamazaki, K. Usami, Y. Nakamura, Hybridizing ferromagnetic magnons and microwave photons in the quantum limit, Phys. Rev. Lett. 113 (2014) 083603.
[9] X. Zhang, C.-L. Zou, L. Jiang, H.X. Tang, Strongly coupled magnons and cavity microwave photons, Phys. Rev. Lett. 113 (2014) 156401.
[10] L. Bai, M. Harder, Y.P. Chen, X. Fan, J.Q. Xiao, C.M. Hu, Spin pumping in electrodynamically coupled magnon-photon systems, Phys. Rev. Lett. 114 (2015) 227201.
[11] B. Bhoi, T. Cliff, I.S. Maksymov, M. Kostylev, R. Aiyar, N. Venkataramani, et al., Study of photon–magnon coupling in a YIG-film split-ring resonant system, J. Appl. Phys. 116 (2014) 243906.
[12] I.S. Maksymov, Perspective: strong microwave photon-magnon coupling in multiresonant dielectric antennas, J. Appl. Phys. 124 (2018) 150901.
[13] A. Osada, R. Hisatomi, A. Noguchi, Y. Tabuchi, R. Yamazaki, K. Usami, et al., Cavity optomagnonics with spin-orbit coupled photons, Phys. Rev. Lett. 116 (2016) 223601.
[14] D. Zhang, X.-M. Wang, T.-F. Li, X.-Q. Luo, W. Wu, F. Nori, et al., Cavity quantum electrodynamics with ferromagnetic magnons in a small yttrium-iron-garnet sphere, Npj Quantum Inf. 1 (2015) 15014.
[15] X. Zhang, C.-L. Zou, L. Jiang, H.X. Tang, Cavity magnomechanics, Sci. Adv. 2 (2016) e1501286.
[16] D.D. Stancil, A. Prabhakar, Spin Waves: Theory and Applications, Springer US, 2009.
[17] A.A. Serga, A.V. Chumak, B. Hillebrands, YIG magnonics, J. Phys. D Appl. Phys. 43 (2010) 264002.
[18] S. Haroche, J.M. Raimond, P. Oxford University, Exploring the Quantum: Atoms, Cavities, and Photons, OUP Oxford, 2006.
[19] M. Goryachev, W.G. Farr, D.L. Creedon, Y. Fan, M. Kostylev, M.E. Tobar, High-cooperativity cavity QED with magnons at microwave frequencies, Phys. Rev. Appl. 2 (2014) 054002.
[20] Y. Cao, P. Yan, H. Huebl, S.T.B. Goennenwein, G.E.W. Bauer, Exchange magnon-polaritons in microwave cavities, Phys. Rev. B 91 (2015) 094423.
[21] L. Bai, K. Blanchette, M. Harder, Y.P. Chen, X. Fan, J.Q. Xiao, et al., Control of the magnon–photon coupling, IEEE Trans. Magn. 52 (2016) 1–7.
[22] M. Harder, L. Bai, C. Match, J. Sirker, C. Hu, Study of the cavity-magnon-polariton transmission line shape, Sci. China Phys. Mech. Astron. 59 (2016) 117511.
[23] S. Kaur, B.M. Yao, J.W. Rao, Y.S. Gui, C.M. Hu, Voltage control of cavity magnon polariton, Appl. Phys. Lett. 109 (2016) 032404.
[24] H. Maier-Flaig, M. Harder, R. Gross, H. Huebl, S.T.B. Goennenwein, Spin pumping in strongly coupled magnon-photon systems, Phys. Rev. B 94 (2016) 054433.
[25] X. Zhang, C. Zou, L. Jiang, H.X. Tang, Superstrong coupling of thin film magnetostatic waves with microwave cavity, J. Appl. Phys. 119 (2016) 023905.
[26] B. Bhoi, B. Kim, J. Kim, Y.-J. Cho, S.-K. Kim, Robust magnon-photon coupling in a planar-geometry hybrid of inverted split-ring resonator and YIG film, Sci. Rep. 7 (2017) 11930.
[27] V. Castel, R. Jeunehomme, J. Ben Youssef, N. Vukadinovic, A. Manchec, F.K. Dejene, et al., Thermal control of the magnon-photon coupling in a notch filter coupled to a yttrium iron garnet/platinum system, Phys. Rev. B 96 (2017) 064407.
[28] B. Yao, Y.S. Gui, J.W. Rao, S. Kaur, X.S. Chen, W. Lu, et al., Cooperative polariton dynamics in feedback-coupled cavities, Nat. Commun. 8 (2017) 1437.

[29] D. Zhang, W. Song, G. Chai, Spin-wave magnon-polaritons in a split-ring resonator/single-crystalline YIG system, J. Phys. D Appl. Phys. 50 (2017) 205003.
[30] Z.J. Tay, W.T. Soh, C.K. Ong, Observation of electromagnetically induced transparency and absorption in yttrium iron garnet loaded split ring resonator, J. Magn. Magn. Mater. 451 (2018) 235–242.
[31] L. Bai, M. Harder, P. Hyde, Z. Zhang, C.-M. Hu, Y.P. Chen, et al., Cavity mediated manipulation of distant spin currents using a cavity-magnon-polariton, Phys. Rev. Lett. 118 (2017) 217201.
[32] Y. Tabuchi, S. Ishino, A. Noguchi, T. Ishikawa, R. Yamazaki, K. Usami, et al., Coherent coupling between a ferromagnetic magnon and a superconducting qubit, Science 349 (2015) 405.
[33] K. Schultheiss, R. Verba, F. Wehrmann, K. Wagner, L. Körber, T. Hula, et al., Excitation of whispering gallery magnons in a magnetic vortex, Phys. Rev. Lett. 122 (2019) 097202.
[34] X. Zhang, C.-L. Zou, N. Zhu, F. Marquardt, L. Jiang, H.X. Tang, Magnon dark modes and gradient memory, Nat. Commun. 6 (2015) 8914.
[35] C. Cohen-Tannoudji, J. Dupont-Roc, G. Grynberg, Photons and Atoms: Introduction to Quantum Electrodynamics, Wiley-Interscience, New York, 1997, pp. 1–3. 1997/03/28.
[36] R. Loudon, The Quantum Theory of Light, Oxford University Press, 2000.
[37] D.T. Haar, The Old Quantum Theory, Elsevier Science, 2016.
[38] A.H. Compton, A quantum theory of the scattering of X-rays by light elements, Phys. Rev. 21 (1923) 483–502.
[39] G.N. Lewis, The conservation of photons, Nature 118 (1926) 874–875.
[40] M. Dirac Paul Adrien, D. Bohr Niels Henrik, The quantum theory of the emission and absorption of radiation, Proc. R. Soc. Lond. Ser. A 114 (1927) 243–265.
[41] D.M. Pozar, Microwave Engineering, third ed., Wiley India Pvt. Limited, 2009.
[42] G.R. Eaton, S.S. Eaton, D.P. Barr, R.T. Weber, Quantitative EPR, Springer, Vienna, 2010.
[43] J.B. Pendry, A.J. Holden, D.J. Robbins, W.J. Stewart, Magnetism from conductors and enhanced nonlinear phenomena, IEEE Trans. Microwave Theory Tech. 47 (1999) 2075–2084.
[44] F. Bloch, Zur theorie des ferromagnetismus, Z. Phys. 61 (1930) 206–219.
[45] C. Kittel, On the theory of ferromagnetic resonance absorption, Phys. Rev. 73 (1948) 155–161.
[46] I.S. Maksymov, M. Kostylev, Broadband stripline ferromagnetic resonance spectroscopy of ferromagnetic films, multilayers and nanostructures, Phys. E. 69 (2015) 253–293.
[47] J.H. Van Vleck, Concerning the theory of ferromagnetic resonance absorption, Phys. Rev. 78 (1950) 266–274.
[48] C. Herring, C. Kittel, On the theory of spin waves in ferromagnetic media, Phys. Rev. 81 (1951) 869–880.
[49] J.R. Eshbach, R.W. Damon, Surface magnetostatic modes and surface spin waves, Phys. Rev. 118 (1960) 1208–1210.
[50] R.W. Damon, J.R. Eshbach, Magnetostatic modes of a ferromagnet slab, J. Phys. Chem. Solid 19 (1961) 308–320.
[51] M.G. Cottam, Linear and Nonlinear Spin Waves in Magnetic Films and Superlattices, World Scientific Publishing Company, 1994.
[52] G. Srinivasan, A.N. Slavin, High Frequency Processes in Magnetic Materials, World Scientific, 1995.
[53] A.G. Gurevich, G.A. Melkov, Magnetization Oscillations and Waves, Taylor & Francis, 1996.

[54] B.A. Kalinikos, A.N. Slavin, Theory of dipole-exchange spin wave spectrum for ferromagnetic films with mixed exchange boundary conditions, J. Phys. C Solid State Phys. 19 (1986) 7013–7033.
[55] B. Heinrich, Z. Frait, Temperature dependence of the FMR linewidth of iron single-crystal platelets, Phys. Status Solidi B 16 (1966) K11–K14.
[56] C.E. Patton, Linewidth and relaxation processes for the main resonance in the spin-wave spectra of Ni–Fe alloy films, J. Appl. Phys. 39 (1968) 3060–3068.
[57] V.E. Demidov, S.O. Demokritov, K. Rott, P. Krzysteczko, G. Reiss, Linear and nonlinear spin-wave dynamics in macro- and microscopic magnetic confined structures, J. Phys. D Appl. Phys. 41 (2008) 164012.
[58] A.V. Chumak, P. Pirro, A.A. Serga, M.P. Kostylev, R.L. Stamps, H. Schultheiss, et al., Spin-wave propagation in a microstructured magnonic crystal, Appl. Phys. Lett. 95 (2009) 262508.
[59] F. Bertaut, F. Forrat, Sur les déformations dans les pérovskites à base de terres rares et d'éléments de transition trivalents, J. Phys. Radium 17 (1956) 129–131.
[60] S. Geller, M.A. Gilleo, Structure and ferrimagnetism of yttrium and rare-earth-iron garnets, Acta Crystallogr. 10 (1957) 239.
[61] S. Geller, M.A. Gilleo, The crystal structure and ferrimagnetism of yttrium-iron garnet, Y3Fe2(FeO4)3, J. Phys. Chem. Solid 3 (1957) 30–36.
[62] M.A. Gilleo, S. Geller, Magnetic and crystallographic properties of substituted yttrium-iron garnet, $3Y_2O_3{\cdot}xM_2O_3{\cdot}(5-x)Fe_2O_3$, Phys. Rev. 110 (1958) 73–78.
[63] B. Lax, K.J. Button, Microwave Ferrites and Ferrimagnetics, Literary Licensing, LLC, 2012.
[64] H.A. Algra, P. Hansen, Temperature dependence of the saturation magnetization of ion-implanted YIG films, Appl. Phys. A 29 (1982) 83–86.
[65] M. Sparks, Ferromagnetic-Relaxation Theory, McGraw-Hill, 1964.
[66] Z. Celinski, B. Heinrich, Ferromagnetic resonance linewidth of Fe ultrathin films grown on a bcc Cu substrate, J. Appl. Phys. 70 (1991) 5935–5937.
[67] Y.-Y. Song, S. Kalarickal, C.E. Patton, Optimized pulsed laser deposited barium ferrite thin films with narrow ferromagnetic resonance linewidths, J. Appl. Phys. 94 (2003) 5103–5110.
[68] G. Winkler, Magnetic Garnets, Vieweg, 1981.
[69] E.T. Jaynes, F.W. Cummings, Comparison of quantum and semiclassical radiation theories with application to the beam maser, Proc. IEEE 51 (1963) 89–109.
[70] M. Tavis, F.W. Cummings, Exact solution for an *N*-molecule-radiation-field Hamiltonian, Phys. Rev. 170 (1968) 379–384.
[71] C. Ciuti, I. Carusotto, Input-output theory of cavities in the ultrastrong coupling regime: the case of time-independent cavity parameters, Phys. Rev. A 74 (2006) 033811.
[72] J. Plumridge, E. Clarke, R. Murray, C. Phillips, Ultra-strong coupling effects with quantum metamaterials, Solid State Commun. 146 (2008) 406–408.
[73] T. Niemczyk, F. Deppe, H. Huebl, E.P. Menzel, F. Hocke, M.J. Schwarz, et al., Circuit quantum electrodynamics in the ultrastrong-coupling regime, Nat. Phys. 6 (2010) 772.
[74] M. Gessner, M. Ramm, T. Pruttivarasin, A. Buchleitner, H.P. Breuer, H. Häffner, Local detection of quantum correlations with a single trapped ion, Nat. Phys. 10 (2013) 105.
[75] A. Wallraff, D.I. Schuster, A. Blais, L. Frunzio, R.S. Huang, J. Majer, et al., Strong coupling of a single photon to a superconducting qubit using circuit quantum electrodynamics, Nature 431 (2004) 162–167.
[76] J.P. Reithmaier, G. Sęk, A. Löffler, C. Hofmann, S. Kuhn, S. Reitzenstein, et al., Strong coupling in a single quantum dot–semiconductor microcavity system, Nature 432 (2004) 197–200.

[77] Y. Sato, Y. Tanaka, J. Upham, Y. Takahashi, T. Asano, S. Noda, Strong coupling between distant photonic nanocavities and its dynamic control, Nat. Photonics 6 (2011) 56.
[78] K. Henschel, J. Majer, J. Schmiedmayer, H. Ritsch, Cavity QED with an ultracold ensemble on a chip: prospects for strong magnetic coupling at finite temperatures, Phys. Rev. A 82 (2010) 033810.
[79] R. Amsüss, C. Koller, T. Nöbauer, S. Putz, S. Rotter, K. Sandner, et al., Cavity QED with magnetically coupled collective spin states, Phys. Rev. Lett. 107 (2011) 060502.
[80] T. Gaebel, M. Domhan, I. Popa, C. Wittmann, P. Neumann, F. Jelezko, et al., Room-temperature coherent coupling of single spins in diamond, Nat. Phys. 2 (2006) 408–413.
[81] P. Bushev, A.K. Feofanov, H. Rotzinger, I. Protopopov, J.H. Cole, C.M. Wilson, et al., Ultralow-power spectroscopy of a rare-earth spin ensemble using a superconducting resonator, Phys. Rev. B 84 (2011) 060501.
[82] F. Flamini, N. Spagnolo, F. Sciarrino, Photonic quantum information processing: a review, Rep. Prog. Phys. 82 (2018) 016001.
[83] A. Imamoğlu, Cavity QED based on collective magnetic dipole coupling: spin ensembles as hybrid two-level systems, Phys. Rev. Lett. 102 (2009) 083602.
[84] Y. Kubo, F.R. Ong, P. Bertet, D. Vion, V. Jacques, D. Zheng, et al., Strong coupling of a spin ensemble to a superconducting resonator, Phys. Rev. Lett. 105 (2010) 140502.
[85] H. Wu, R.E. George, J.H. Wesenberg, K. Mølmer, D.I. Schuster, R.J. Schoelkopf, et al., Storage of multiple coherent microwave excitations in an electron spin ensemble, Phys. Rev. Lett. 105 (2010) 140503.
[86] X. Zhu, S. Saito, A. Kemp, K. Kakuyanagi, S.-i. Karimoto, H. Nakano, et al., Coherent coupling of a superconducting flux qubit to an electron spin ensemble in diamond, Nature 478 (2011) 221.
[87] H. Yu, O. d'Allivy Kelly, V. Cros, R. Bernard, P. Bortolotti, A. Anane, et al., Magnetic thin-film insulator with ultra-low spin wave damping for coherent nanomagnonics, Sci. Rep. 4 (2014) 6848.
[88] A. Krysztofik, H. Głowiński, P. Kuświk, S. Ziętek, L.E. Coy, J.N. Rychły, et al., Characterization of spin wave propagation in (111) YIG thin films with large anisotropy, J. Phys. D Appl. Phys. 50 (2017) 235004.
[89] Ö.O. Soykal, M.E. Flatté, Size dependence of strong coupling between nanomagnets and photonic cavities, Phys. Rev. B 82 (2010) 104413.
[90] Ö.O. Soykal, M.E. Flatté, Strong field interactions between a nanomagnet and a photonic cavity, Phys. Rev. Lett. 104 (2010) 077202.
[91] S.A. Gregory, L.C. Maple, G.B.G. Stenning, T. Hesjedal, G. van der Laan, G.J. Bowden, Angular control of a hybrid magnetic metamolecule using anisotropic FeCo, Phys. Rev. Appl. 4 (2015) 054015.
[92] S.A. Gregory, G.B.G. Stenning, G.J. Bowden, N.I. Zheludev, P.A.J. de Groot, Giant magnetic modulation of a planar, hybrid metamolecule resonance, New J. Phys. 16 (2014) 063002.
[93] Y.-P. Wang, G.-Q. Zhang, D. Zhang, T.-F. Li, C.M. Hu, J.Q. You, Bistability of cavity magnon polaritons, Phys. Rev. Lett. 120 (2018) 057202.
[94] M. Harder, L. Bai, P. Hyde, C.-M. Hu, Topological properties of a coupled spin-photon system induced by damping, Phys. Rev. B 95 (2017) 214411.
[95] M. Müller, I. Rotter, Exceptional points in open quantum systems, J. Phys. A Math. Theor. 41 (2008) 244018.
[96] M. Harder, Y. Yang, B.M. Yao, C.H. Yu, J.W. Rao, Y.S. Gui, et al., Level attraction due to dissipative magnon-photon coupling, Phys. Rev. Lett. 121 (2018) 137203.
[97] D. Zhang, X.-Q. Luo, Y.-P. Wang, T.-F. Li, J.Q. You, Observation of the exceptional point in cavity magnon-polaritons, Nat. Commun. 8 (2017) 1368.

[98] M. Philipp, P.v. Brentano, G. Pascovici, A. Richter, Frequency and width crossing of two interacting resonances in a microwave cavity, Phys. Rev. E 62 (2000) 1922–1926.
[99] C. Dembowski, B. Dietz, H.D. Gräf, H.L. Harney, A. Heine, W.D. Heiss, et al., Observation of a chiral state in a microwave cavity, Phys. Rev. Lett. 90 (2003) 034101.
[100] B. Dietz, T. Friedrich, J. Metz, M. Miski-Oglu, A. Richter, F. Schäfer, et al., Rabi oscillations at exceptional points in microwave billiards, Phys. Rev. E 75 (2007) 027201.
[101] C. Hahn, Y. Choi, J.W. Yoon, S.H. Song, C.H. Oh, P. Berini, Observation of exceptional points in reconfigurable non-Hermitian vector-field holographic lattices, Nat. Commun. 7 (2016) 12201.
[102] H. Xu, D. Mason, L. Jiang, J.G.E. Harris, Topological energy transfer in an optomechanical system with exceptional points, Nature 537 (2016) 80.
[103] B. Bhoi, B. Kim, S.-H. Jang, J. Kim, J. Yang, Y.-J. Cho, et al., Abnormal anticrossing effect in photon-magnon coupling, Phys. Rev. B 99 (2019) 134426.
[104] V.L. Grigoryan, K. Shen, K. Xia, Synchronized spin-photon coupling in a microwave cavity, Phys. Rev. B 98 (2018) 024406.
[105] V. Castel, A. Manchec, J.B. Youssef, Control of magnon-photon coupling strength in a planar resonator/yttrium-iron-garnet thin-film configuration, IEEE Magn. Lett. 8 (2017) 1–5.
[106] S. Klingler, H. Maier-Flaig, R. Gross, C.-M. Hu, H. Huebl, S.T.B. Goennenwein, et al., Combined Brillouin light scattering and microwave absorption study of magnon-photon coupling in a split-ring resonator/YIG film system, Appl. Phys. Lett. 109 (2016) 072402.
[107] Z.J. Tay, W.T. Soh, C.K. Ong, Strong excitation of surface and bulk spin waves in yttrium iron garnet placed in a split ring resonator, J. Phys. D Appl. Phys. 51 (2018) 065003.
[108] A.P. Seyranian, O.N. Kirillov, A.A. Mailybaev, Coupling of eigenvalues of complex matrices at diabolic and exceptional points, J. Phys. A Math. Gen. 38 (2005) 1723–1740.
[109] N.R. Bernier, E.G. Dalla Torre, E. Demler, Unstable avoided crossing in coupled spinor condensates, Phys. Rev. Lett. 113 (2014) 065303.
[110] N.R. Bernier, L.D. Tóth, A.K. Feofanov, T.J. Kippenberg, Level attraction in a microwave optomechanical circuit, Phys. Rev. A 98 (2018) 023841.
[111] J. Wiersig, Formation of long-lived, scarlike modes near avoided resonance crossings in optical microcavities, Phys. Rev. Lett. 97 (2006) 253901.
[112] Q.H. Song, H. Cao, Improving optical confinement in nanostructures via external mode coupling, Phys. Rev. Lett. 105 (2010) 053902.
[113] S.-H.G. Chang, C.-Y. Sun, Avoided resonance crossing and non-reciprocal nearly perfect absorption in plasmonic nanodisks with near-field and far-field couplings, Opt. Express 24 (2016) 16822–16834.
[114] P. Hyde, L. Bai, M. Harder, C. Dyck, C.-M. Hu, Linking magnon-cavity strong coupling to magnon-polaritons through effective permeability, Phys. Rev. B 95 (2017) 094416.
[115] H. Maier-Flaig, M. Harder, S. Klingler, Z. Qiu, E. Saitoh, M. Weiler, et al., Tunable magnon-photon coupling in a compensating ferrimagnet—from weak to strong coupling, Appl. Phys. Lett. 110 (2017) 132401.
[116] B. Kim, B. Bhoi, and S-K. Kim, "Spin-wave excitation and critical angles in a hybrid photon-magnon-coupled system" J. Appl. Phys. (2019) (accepted).
[117] Y.S. Gui, A. Wirthmann, N. Mecking, C.M. Hu, Direct measurement of nonlinear ferromagnetic damping via the intrinsic foldover effect, Phys. Rev. B 80 (2009) 060402.

[118] Y.K. Fetisov, C.E. Patton, V.T. Synogach, Nonlinear ferromagnetic resonance and foldover in yttrium iron garnet thin films-inadequacy of the classical model, IEEE Trans. Magn. 35 (1999) 4511–4521.
[119] Y.-P. Wang, G.-Q. Zhang, D. Zhang, X.-Q. Luo, W. Xiong, S.-P. Wang, et al., Magnon Kerr effect in a strongly coupled cavity-magnon system, Phys. Rev. B 94 (2016) 224410.
[120] G. Zhang, Y. Wang, J. You, Theory of the magnon Kerr effect in cavity magnonics, Sci. China Phys. Mech. Astron. 62 (2019) 987511.
[121] Y.T. Zhang, C.E. Patton, G. Srinivasan, The second-order spin-wave instability threshold in single-crystal yttrium-iron-garnet films under perpendicular pumping, J. Appl. Phys. 63 (1988) 5433–5438.
[122] M. Chen, C.E. Patton, G. Srinivasan, Y.T. Zhang, Ferromagnetic resonance foldover and spin-wave instability in single-crystal YIG films, IEEE Trans. Magn. 25 (1989) 3485–3487.
[123] Y. Gui, L. Bai, C. Hu, The physics of spin rectification and its application, Sci. China Phys. Mech. Astron. 56 (2013) 124–141.
[124] R. Hisatomi, A. Osada, Y. Tabuchi, T. Ishikawa, A. Noguchi, R. Yamazaki, et al., Bidirectional conversion between microwave and light via ferromagnetic magnons, Phys. Rev. B 93 (2016) 174427.
[125] P. Hyde, B.M. Yao, Y.S. Gui, G.-Q. Zhang, J.Q. You, C.M. Hu, Direct measurement of foldover in cavity magnon-polariton systems, Phys. Rev. B 98 (2018) 174423.
[126] Y.-J. Seo, K. Harii, R. Takahashi, H. Chudo, K. Oyanagi, Z. Qiu, et al., Fabrication and magnetic control of Y3Fe5O12 cantilevers, Appl. Phys. Lett. 110 (2017) 132409.
[127] S. Barzanjeh, S. Guha, C. Weedbrook, D. Vitali, J.H. Shapiro, S. Pirandola, Microwave quantum illumination, Phys. Rev. Lett. 114 (2015) 080503.
[128] H.J. Kimble, The quantum internet, Nature 453 (2008) 1023.
[129] K. Fang, J. Luo, A. Metelmann, M.H. Matheny, F. Marquardt, A.A. Clerk, et al., Generalized non-reciprocity in an optomechanical circuit via synthetic magnetism and reservoir engineering, Nat. Phys. 13 (2017) 465.
[130] F. Martin, Artificial Transmission Lines for RF and Microwave Applications, Wiley, 2015.
[131] H.Y. Yuan, X.R. Wang, Magnon-photon coupling in antiferromagnets, Appl. Phys. Lett. 110 (2017) 082403.
[132] D. Lacklison, G. Scott, H. Ralph, J. Page, Garnets with high magnetooptic figures of merit in the visible region, IEEE Trans. Magn. 9 (1973) 457–460.
[133] S. Parchenko, A. Stupakiewicz, I. Yoshimine, T. Satoh, A. Maziewski, Wide frequencies range of spin excitations in a rare-earth Bi-doped iron garnet with a giant Faraday rotation, Appl. Phys. Lett. 103 (2013) 172402.
[134] A. Stu akiewicz, Magnetization statics and ultrafast photoinduced dynamics in Co/garnet heterostructures (Chapter 9), in: M. Khan (Ed.), Magnetic Materials, BoD—Books on Demand, 2016, pp. 195–222.
[135] D. Richardson, B.A. Kalinikos, L.D. Carr, M. Wu, Spontaneous exact spin-wave fractals in magnonic crystals, Phys. Rev. Lett. 121 (2018) 107204.

CHAPTER TWO

The influence of the internal domain wall structure on spin wave band structure in periodic magnetic stripe domain patterns

Pawel Gruszecki[a,*], Chandrima Banerjee[b], Michal Mruczkiewicz[c], Olav Hellwig[d,e], Anjan Barman[f], Maciej Krawczyk[a]

[a]Faculty of Physics, Adam Mickiewicz University in Poznan, ul. Uniwersytetu Poznańskiego, Poznań, Poland
[b]School of Physics, Trinity College Dublin, College Green, Dublin, Ireland
[c]Institute of Electrical Engineering, Slovak Academy of Sciences, Bratislava, Slovakia
[d]Institute of Physics, Chemnitz University of Technology, Chemnitz, Germany
[e]Institute of Ion Beam Physics and Materials Research, Helmholtz-Zentrum Dresden-Rossendorf, Dresden, Germany
[f]Department of Condensed Matter Physics and Material Sciences, S. N. Bose National Centre for Basic Sciences, Kolkata, India
*Corresponding author: e-mail address: pawel.gruszecki@amu.edu.pl

Contents

Solid State Physics, Volume 70
ISSN 0081-1947
https://doi.org/10.1016/bs.ssp.2019.09.003

1. Introduction

In 1947, the invention of the transistor [1] ignited a series of radical and phenomenological changes in the world of communication and technology. The ability of on-chip digitization has fueled the information-age, whose fundamental sources of economic wealth are knowledge and communication rather than natural resources and physical labor as in earlier times. In present days, if we look around, people are seen communicating with others by cell phone, working on laptops, or performing their household duties using the World Wide Web. There are several other applications in devices as diverse as video cameras, jumbo jets, modern automobiles, manufacturing equipment, *etc.* Overall, the explosive growth and development of high-performance electronic devices and integrated circuits have transformed the foundation of the world economy and have an increasing impact on all areas from everyday life to the most advanced science, technology, and business.

The evolution of semiconductor-based technology has always been boosted by the ever increasing data processing capability, together with astonishingly high integration density and faster speed. Combined with advances in optical communication and magnetic materials, these advantages have enabled easy and affordable transmission, processing and storage of a massive amount of information. Nevertheless, according to the International Technology Roadmap for Semiconductors, the semiconductor-based technology will face serious obstacles hindering downscaling and power reduction in the near future, since both the miniaturization of the single-element size and the operational speed will reach their ultimate limits [2]. Another fundamental drawback, inherent in electronics, is the generation of waste heat during switching (since it is associated with a translational motion of electrons), which is responsible for an increase in the power consumption of electronic devices. Consequently, the challenge is not only to develop new particle-less technologies to accommodate increasing amounts of information but also how to organize information in the new media. A possible way out is to move toward alternative integrated circuits. An example is all-optical devices, which employ photons, the quanta of electromagnetic waves as the information carrier. Researchers have opened a new direction in the field of nanophotonics: the photonic crystals [3–5]. Photonic crystals are a class of artificial optical materials with the periodically modulated refractive index. In a semiconductor, the periodic lattice of atoms causes electrons to have energy bands and energy bandgaps; in photonic crystals, the periodic dielectric constant causes electromagnetic waves to have frequency bands and frequency bandgaps, the latter often being called photonic bandgaps. If the frequency of light (or photon) falls in the gap, the photonic crystals will reflect the light totally irrespective of its angle of incidence. The properties of the bandgaps can be reliably controlled and manipulated by changing the physical parameters of the structure, such as dielectric constant of the used materials or lattice parameters. In other words, photonic crystal structures suggest high spectral selectivity that has led to various applications from the photonic crystal to optical nanodevices, such as photonic waveguides [6–10] and integrated circuits [11–13].

In the meantime, the areas of nanomagnetism and magnetization dynamics [14], a popular sub-area of spintronics [15], have become a subject of growing interest, owing to their promising applications in nanoscale signal processing and information transfer devices [16–18] such as filters [19], delay lines [20], transistor [21], and storage elements [22,23]. The term spintronics originally was an acronym for SPIN Transport electronics and this field encompasses the transmission, processing, and storage of information based on the magnetization state of a system. Historically, the concept of associating the spin degree of freedom with electric signals (*i.e.*, electronic charge) came with the experimental observation of the giant magnetoresistance (GMR) effect in Fe/Cr multilayered systems by Fert and Grünberg [24,25], for which they were awarded the Nobel prize in 2007. The phenomenon is based on the fact that for a spin-polarized current, *i.e.*, a current which is not carried by an equal amount of up- and down-spins, flowing from one ferromagnetic layer into another, the resistance depends on the relative orientation of the two ferromagnetic layers. This was a major breakthrough in the technology for the improvement of magnetic hard disk drives and led to the possibility of novel non-volatile magnetic memories, like Magnetoresistive Random Access Memory (MRAM) [16]. Later on, the discoveries of Spin Transfer Torque (STT) [26,27], which allows for current assisted magnetization switching and (Inverse) Spin Hall Effect ((I)SHE) [28–30], provided new ways to create and detect spin currents, while the Spin Pumping Effect [31,32], the Spin Seebeck Effect (SSE) [33] and the Dzyaloshinskii-Moriya Interaction (DMI) [34–36] have added further interesting prospects to the field of spintronics.

Besides using electrical current, the transfer of magnetic moment or spin can also occur with the help of spin waves [37,38] or magnons (the quanta of spin waves with spin-1 obeying Bose-Einstein statistics) in magnetic materials giving rise to the nascent scientific field of so-called magnonics. From a classical point of view, the spin wave represents a phase-coherent precession of microscopic vectors of magnetization of the magnetic medium. The concept of the spin wave was first introduced by Bloch in 1930 [39] and presently is considered as a potential data carrier for computing devices, providing the elemental base for Joule-heat-free transfer of spin information over macroscopic distances. In that sense, the spin waves are equivalent to the electromagnetic waves in photonics, however, having wavelengths that are a few orders of magnitude shorter than that of electromagnetic waves at the same frequency. Magnonics thus inherently foster the miniaturization of microwave devices toward the nanoscale. Ultimate control and functionality of spin wave-based logic are expected from the so-called magnonic crystal (the magnetic counterpart of a photonic crystal) [40–46], which is formed by periodic magnetic nanostructures. In such structures allowed excitations form frequency bands, separated by partial or full bandgaps. The properties can be manipulated via a wide range of parameters, including the choice of magnetic material, the size, and shape of the sample, the orientation, and magnitude of the applied bias magnetic field. A striking example is given by the recently discovered magnetic lattices created by skyrmions [47,48] and artificial spin ice [49,50], which stimulated abundant new physics and many new applications. An interesting property of some of the magnonic crystals is that they can arise spontaneously. Due to the interaction between spins, the magnetization configuration can form periodic patterns that minimize the total energy of the system. These spontaneous non-collinear alignments allow generating a nanosize pattern below the lithography limit. In a magnonic crystal, spin waves of different frequencies can be used for parallel logic operations where information or data is coded into magnon phase or density. On the other hand, the intermediate gaps are important for creating spin-wave filters [19] and transistor [51], which can be tailored by applying electric fields to construct spin wave logic gates [52]. Further advantages of magnonic devices lie in the fact, that perhaps more than any other kind of wave (for example electromagnetic or acoustic), spin waves display a diversity of dispersion characteristics. For example, the dispersion of dipole-dominated spin waves in thin ferromagnetic films strongly depends on their propagation direction with respect to an in-plane magnetic field, giving rise to Damon-Eshbach modes for k-vector perpendicular to the magnetic field and backward volume modes for k-vector parallel to the magnetic field.

Despite the fact that the concept of magnonics provides unprecedented controllability in terms of ultrafast information transmission and processing, its growth is limited by the challenges like complex fabrication process, a requirement of a high magnetic field and geometrical confinements limiting the spin wave propagation. This has prompted the study of spin waves in stable topological magnetic structures such as magnetic domain walls, vortices, and skyrmions. A domain wall, mathematically represented as a topological soliton, is a gradual reorientation of neighboring magnetic moments typically across a few (up to several dozen) nanometers widths which occurs at the boundary of two magnetic domains. Depending on the interplay among exchange, anisotropy and dipolar energy of the system, the domain wall can be of different types and can have a nontrivial impact on the spin-wave dynamics. For example, the Néel domain wall formed in an in-plane magnetized waveguide promotes channeling of the spin wave through it [53]. On the other hand, the spin waves propagating across the domain wall either get scattered or undergoes a phase shift, depending on their wavelength and the nature of the domain wall [54–56]. A particularly fascinating scenario emerges when the symmetry in the exchange interaction is broken. In this case, the domain wall prefers a chirality which invokes a nonreciprocity in the properties of counter-propagating spin waves [57,58]. Overall, the study of spin-wave dynamics in the presence of domain walls is still flourishing which may open a route for low power magnonic computing.

The framework of this chapter consists of a review on the study of spin-wave dynamics in the presence of periodical sequences of domain walls. The rest of this chapter will be divided into five sections. In Section 2, we will briefly introduce different magnetic interactions, relevant types of domain walls and the basics of spin-wave dynamics. Section 3 will be dedicated to a review of spin-wave characteristics when propagating along and across a single domain wall. The role of domain wall type on the spin-wave propagation is explained. Section 4 will deal with the scenario when spin waves propagate in a periodic arrangement of domain walls, emphasizing its effect on the spin-wave band structure. Finally, we will summarize and discuss the future prospects of spin-wave manipulation using domain walls (Section 5).

2. Theoretical background

The fundamental origin of the intrinsic magnetic properties of materials lies in its atomic or molecular structure as well as in its electronic arrangement. The spin of the electrons, together with the number of unpaired electrons and the interactions between their spin and orbital momenta determine the magnetic response, both at the microscopic and macroscopic level. Based on that, materials can be classified as diamagnet, paramagnet, ferromagnet, ferrimagnet, or antiferromagnet. Formation of magnetic domains is a characteristic of the materials having spontaneous magnetic ordering. The ferromagnetic transition metals (like iron, cobalt, and nickel) are characterized by partially filled internal electron shell corresponding to the *3d* levels. The *3d* band is very narrow and hence the kinetic energy of *3d* electrons is small while the density of states is very high. These facts lead to the intricate interaction between the *3d* electrons of neighboring atoms (known as exchange interaction), which is responsible for the atomic or molecular spin alignment of the material. According to the Weiss theory [59], this exchange interaction has to compete with the dipolar interaction caused by the free magnetic poles on the surface (demagnetizing field), which for ferromagnets results in a break up into small domains in order to minimize the total free energy of the system.

The individual moments of a ferromagnetic material collectively contribute to a overall volume magnetization of the system, which is maximum when the moments are fully aligned under a large enough external magnetic field. If the magnetization is excited away from the external field direction the coupled moments precess about the external field direction, which is called ferromagnetic resonance (FMR) in case of uniform precession (all the moments have the same frequency and phase) [60]. When the precession is not uniform, each individual dipole precesses slightly out-of-phase with its nearest neighbor and a propagating wave, known as spin wave [39], with its energy and momentum related by characteristic dispersion relations are produced. The parameter that describes the rate of energy loss of precessing magnetization, is called magnetic damping, which can be intrinsic (*e.g.*, due to spin-orbit coupling) or extrinsic (*e.g.*, due to two-magnon scattering) depending upon its origin. Another exotic property observed in ferromagnetic materials is magnetic anisotropy that is the dependence of magnetic energy on the direction of the spins with respect to the crystal lattice, or, in patterned structures, with respect to the geometrical axes.

This section starts off with an overview of the energies that play a role in the dynamics of a ferromagnetic system. Next, we describe the magnetization dynamics followed by the illustration of different aspects of spin waves.

2.1 Magnetic energies

The static and dynamic properties of magnetic materials depend on the relative contributions of the different magnetic energy terms, such as Zeeman energy, magnetostatic self-energy (also known as demagnetizing energy), exchange energy, and magnetic anisotropy energy. In equilibrium, the system is in a local minimum of the total free energy. This section presents a brief outline of some of the vital energy factors.

2.1.1 Zeeman energy

The Zeeman energy describes the interaction of the magnetization **M** with an external magnetic field $\mathbf{H}_0$. The energy is given by:

$$E_{\text{Zee}} = -\mu_0 \int_V \mathbf{M} \cdot \mathbf{H}_0 dV, \tag{1}$$

where V is the volume occupied by the magnetic structure and $d\tau$ the volume element. This equation states that in the presence of the magnetic field, the magnetization tends to align along H_0 to minimize the energy.

2.1.2 Exchange energy

The exchange interaction, which is quantum-mechanical in nature, is responsible for the long-range magnetic order in ferromagnets [61]. Besides ferromagnets, this interaction is pronounced in ferrimagnets and antiferromagnets as well, where the exchange interaction between the neighboring magnetic ions will force the individual moments into parallel or antiparallel alignment with their neighbors. It actually stems from the Coulomb interaction energy and Pauli Exclusion Principle, it can be phenomenologically described by the Heisenberg exchange Hamiltonian, given by

$$H_{\text{ex}} = -2J \sum_{i \neq j} \mathbf{S}_i \cdot \mathbf{S}_j, \tag{2}$$

where $\mathbf{S}_i$ is the spin operator of the i-th atom and J is known as the exchange integral. In the continuum model, assuming exchange interaction only between nearest neighbors in the cubic lattice, the above equation can be written as:

$$E_{\text{ex}} = A_{\text{ex}} \int (\nabla \mathbf{m})^2 dV, \tag{3}$$

where the reduced magnetization $\mathbf{m} = \mathbf{M}/M_S$ is a continuous vector and A_{ex} is called the exchange stiffness constant for the simple cubic lattice of spins takes the form:

$$A_{\text{ex}} = \frac{2JS^2}{a}, \tag{4}$$

where a is the lattice constant.

The aforementioned exchange is also known as direct exchange interaction, where the electrons of magnetic atoms interact with its nearest neighbors. Apart from that, the exchange can also occur in indirect ways, which couples moments over relatively larger distances. For example, Ruderman–Kittel–Kasuya–Yosida (RKKY) exchange, where the metallic ions are coupled via itinerant electrons, super-exchange, where the exchange is mediated via different nonmagnetic ions, anisotropic exchange interaction (also known as Dzyaloshinskii-Moriya interaction) [62], where the spin-orbit interaction plays a major role and often leads to canting of adjacent spins by a small angle.

2.1.3 Magnetostatic energy

The interaction between two magnetic dipoles is common for any type of dipoles, irrespective of whether it is in a ferromagnetic system or not. The corresponding energy term for interacting magnetic dipole moments $\boldsymbol{\mu}_1$ and $\boldsymbol{\mu}_2$ reads as

$$E_{\text{dipole}} = \frac{\mu_0}{4\pi r^3}\left[\boldsymbol{\mu}_1 \cdot \boldsymbol{\mu}_2 - \frac{3}{r^2}(\boldsymbol{\mu}_1 \cdot \mathbf{r})(\boldsymbol{\mu}_2 \cdot \mathbf{r})\right], \tag{5}$$

where μ_0 is the permeability of vacuum and $\mathbf{r}$ is the vector connecting the two dipoles. The above equation tells, that the energy decreases with the third order of their distance. In the case of ferromagnets, the magnetostatic energy plays an important role in a number of phenomena such as the formation of domains, demagnetizing field, and spin waves in the long-wavelength regime.

A magnetic system with finite boundaries exhibits poles at its surfaces which lead to a stray field outside the sample. This effect also gives rise to the demagnetizing field inside the sample. The energy corresponding to the stray field is given by

$$E_{\text{dem}} = -\frac{1}{2}\int \mu_0 \mathbf{M} \cdot \mathbf{H}_{\text{dem}} dV. \tag{6}$$

In case of a homogeneously magnetized ellipsoid, the demagnetizing field is uniform and can be written as

$$\mathbf{H}_{\text{dem}} = -\mathbf{N}\mathbf{M}, \tag{7}$$

where $\mathbf{N}$ is called the demagnetizing tensor. Thus, the stray field energy (demagnetizing energy) reduces to

$$E_{\text{dem}} = \frac{1}{2}\mu_0 \int \mathbf{M}\mathbf{N} dV, \tag{8}$$

and finally

$$E_{dem} = \frac{1}{2} V \mu_0 \mathbf{MNM}, \tag{9}$$

where V is the volume of the sample. For an arbitrarily shaped element, **N** is dependent on the shape and geometry of the element and forms a complex function of position. In the case where anisotropy doesn't play a significant role, the preferred orientation of magnetization is the one for which E_{dem} is minimum.

2.1.4 Magnetic anisotropy

Magnetic anisotropy [63] is the directional dependence of the material's magnetic moment. Because of anisotropy, there are preferred orientations of the magnetic moment in space (known as easy magnetization axes). Deviation from these directions imposes an additional energy penalty on the system, the so-called anisotropy energy. Magnetic anisotropy originates from various sources, such as:

(i) *Magnetocrystalline anisotropy*: the magnetocrystalline anisotropy is an intrinsic property of the material, which is caused by the spin-orbit interaction of the electrons. The spatial arrangement of the electron orbitals is strongly linked to the crystallographic structure. Consequently, when the electrons interact via their spins they force the latter to align along well-defined crystallographic axes. The symmetry of the lattice structure is responsible for the "foldedness" of the magnetic anisotropy. For instance, in cubic systems, the energy density due to crystal anisotropy reads

$$E_{ani} = K_0 + K_1 \left(\alpha_x^2 \alpha_y^2 + \alpha_y^2 \alpha_z^2 + \alpha_z^2 \alpha_x^2 \right) + K_2 \alpha_x^2 \alpha_y^2 \alpha_z^2, \tag{10}$$

where α_i are the directional cosines of the normalized magnetization **m** with respect to the Cartesian axes of the lattice. K_i's are the magnetocrystalline anisotropy constants, *i.e.*, K_0, K_1, and K_2 are the crystalline anisotropy constants of zero, first and second order, respectively. On the other hand, for crystals having uniaxial anisotropy, the energy density is given by

$$E_{ani} = K^V \alpha_x^2, \tag{11}$$

where K^V is the uniaxial volume magnetocrystalline anisotropy constant.

(ii) *Shape anisotropy*: Another common origin of magnetic anisotropy belongs to the anisotropic shape of the magnetic element. It is basically the anisotropic dipolar interaction of free magnetic poles (stray and demagnetization fields) that tends to magnetize magnetic elements with magnetic moments directed parallelly to the surfaces (to minimize magnetostatic energy). The shape anisotropy causes *e.g.*, vortex state of magnetic disks and preferable magnetization of magnetic stripes along their axis. The value of the shape anisotropy can be calculated solving Eq. (9) assuming demagnetizing tensor **N** related to the shape of the considered magnetic element. For instance, for a thin film (with the diagonal components of the demagnetizing tensor $N_x = N_y = 0$, $N_z = 1$ and the z-axis directed across film's thickness) the resulting shape anisotropy is equal $K^{\mathrm{V}}_{\mathrm{shape}} = \mu_0 M_{\mathrm{S}}^2/2$.

(iii) *Surface and interface anisotropy*: The broken symmetry at surfaces and interfaces of magnetic thin films and multilayers often induces some anisotropy in the system. This results in the effective anisotropy (K^{eff}) constant to be divided into two parts:

$$K^{\mathrm{eff}} = K^{\mathrm{V}} + \frac{2K^{\mathrm{S}}}{d}, \tag{12}$$

with K^{S} being the surface anisotropy. The latter exhibits an inverse dependence on the thickness d of the system and often prefers an out of plane magnetization of the sample. The competition between volume and surface anisotropy gives rise to a dependence of the magnetization on the film thickness. Below a critical thickness d_{c} (given by $d_{\mathrm{c}} = -2K^{\mathrm{S}}/K^{\mathrm{V}}$), the magnetization favors an out of plane orientation and *vice versa*.

In the case of some multilayered systems (like Co/Pd and Co/Pt multilayers), the observed interfacial anisotropy is rooted in the interfacial hybridization of the electronic band structure [64,65]. These hybridizations are sensitive to the local interface structure and affect the spin-orbit coupling strength, which in turn changes the magnetocrystalline anisotropy energy.

2.2 Magnetic domain wall

The interplay between the aforementioned magnetic energies governs the magnetic response of the system including the hysteresis and the equilibrium magnetic structure such as domains. In absence of the saturating magnetic field, the system minimizes the dipolar energy by dividing itself into many small regions of uniform magnetization, called magnetic domains. The neighboring domains have different directions of magnetization, which are separated by domain walls, a region where the magnetization gradually rotates from the direction in one domain to that in the next domain. The domain wall width (Δ) depends on the relative strength between the exchange energy (A_{ex}) and the effective anisotropy (K^{eff}):

$$\Delta = \sqrt{A_{ex}/K^{eff}}. \quad (13)$$

The exchange interaction prefers a wider domain wall so that the angle between neighboring spins is small. The anisotropy, in contrast, favors a minimization of the number of the magnetic moments not aligned along the easy axis.

Depending on the interactions present in the system and the film thickness the different types of domain walls can form. These types of domain walls can be distinguished based on the way the magnetization varies inside the domain wall. In the case of magnetic films, two main kinds of domain walls are observed, *i.e.*, the Bloch wall and the Néel wall. When the magnetization gradually rotates along the direction parallel to the wall, it is called Néel wall. For a Bloch domain wall, the magnetic moments rotate about the axis perpendicular to the wall. In a magnetic thin film with an in-plane easy axis, a Néel wall is favored as the formation of a Bloch domain wall will create magnetic charges at the sample surface. On the other hand, for samples with out of plane anisotropy, a magnetization rotation perpendicular to the domain wall would cost less dipolar energy as compared to the rotation along the domain wall. Hence, a Bloch wall will be favored. Depending on the film thickness, anisotropy, *etc.*, other kinds of domain walls may appear, which are mixtures of Néel and Bloch-type structures, for example, cross tie [66] and twisted "corkscrew"-type domain walls characteristic for bubbe films that consists of a Bloch-type wall in the film's center with two Néel caps at the film's surfaces [67,68]. In the presence of Dzyaloshinskii-Moriya interaction (an antisymmetric exchange interaction) [62], a tilt is induced between two neighboring spins of the system. Due to this, a chiral Néel domain wall is favored even in the presence of perpendicular anisotropy [69]. Fig. 1 shows the domain wall configuration (upper panel) and the corresponding magnetization profile across it (lower panel) for Néel wall, Bloch wall and Néel wall in presence of interfacial DMI.

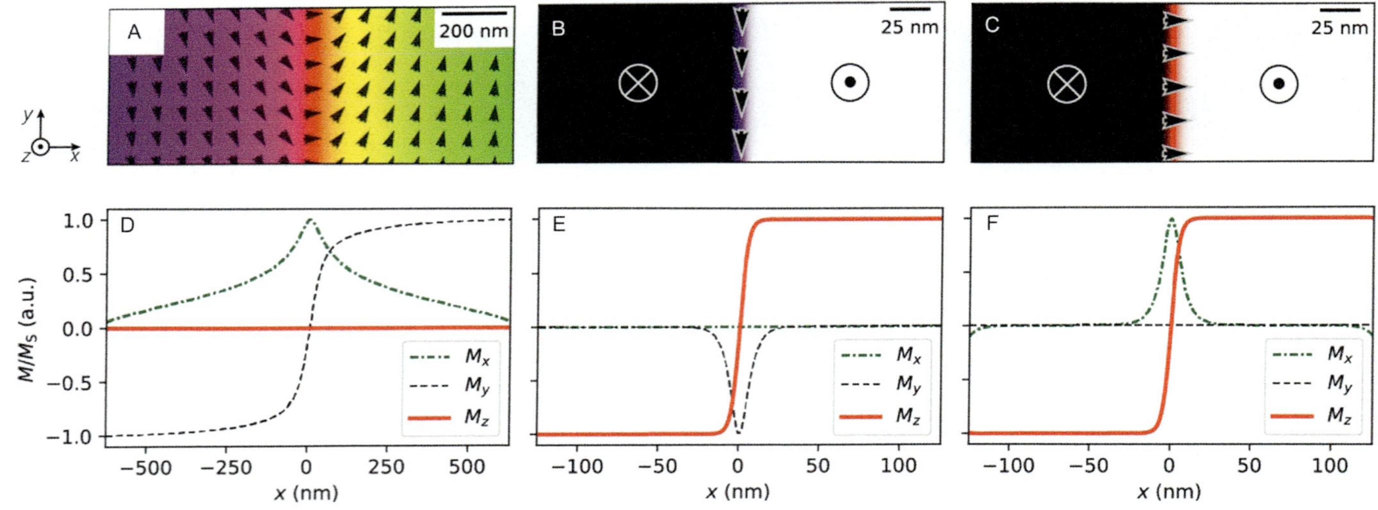

Fig. 1 Schematic of the domain wall configuration of (A) a Néel wall in an in-plane magnetized film (calculated for parameters: $M_S = 800$ kA/m, $A_{ex} = 13$ pJ/m, $d = 10$ nm), (B) a Bloch wall in an out-of-plane magnetized film (calculated for parameters: $M_S = 1$ MA/m, $K_u = 1$ MJ/m$_3$, thickness $d = 1$ nm and $A_{ex} = 10$ pJ/m, $D = 0$), (C) a Néel wall in an out-of-plane magnetized film possessing Dzyaloshinsky-Moriya interaction [the same parameters as in (B) but with $D = 1$ mJ/m^2], and the corresponding magnetization profiles (d, e, f, respectively).

2.3 Magnetization dynamics and spin waves

If the magnetization is not in equilibrium, *i.e.*, if **M** is not parallel to the internal magnetic field, the magnetic moments of a magnetic material experience a torque which induces a precessional motion about the direction of the external field. In addition, the moments try to align themselves along the external field to minimize the Zeeman energy. Effectively, they execute a damped spiral motion about the field direction which is referred to as magnetization dynamics (see Fig. 2). The net behavior of dynamic magnetization is phenomenologically illustrated by the Landau-Lifshitz and Gilbert equation of motion. The Landau-Lifshitz-Gilbert (LLG) equation is a torque equation, which was first introduced by Lev Landau and Evgeny Lifshitz in 1935 as the Landau-Lifshitz (LL) equation [70]. Later, Gilbert modified it by inserting a Gilbert damping term [71].

When a magnetic moment $\boldsymbol{\mu}_{\mathrm{m}}$ is placed in an effective magnetic field $\mathbf{H}_{\mathrm{eff}}$, it experiences a torque given as:

$$\tau = \boldsymbol{\mu}_{\mathrm{m}} \times \mu_0 \mathbf{H}_{\mathrm{eff}} \tag{14}$$

In a semiclassical approach, the magnetic moment $\boldsymbol{\mu}_{\mathrm{m}}$ can be related to the angular momentum $\mathbf{L}$ of electrons as follows:

$$\boldsymbol{\mu}_{\mathrm{m}} = -\gamma \mathbf{L}, \tag{15}$$

where $\gamma = g\mu_B/\hbar$ is the gyromagnetic ratio, g is the Landé-factor ($g \approx 2$), μ_B is the Bohr magnetron and $\hbar$ is the reduced Planck constant. By applying the

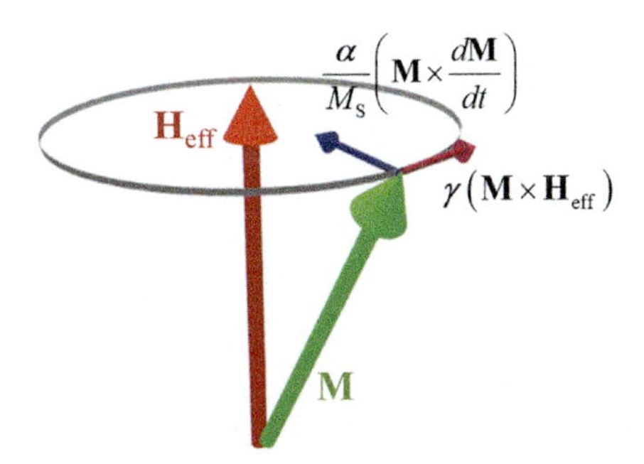

Fig. 2 The precessional motion of the magnetization around the effective magnetic field direction.

momentum theorem one can express Eq. (14) as the rate of change of angular momentum $\mathbf{L}$:

$$\frac{d\mathbf{L}}{dt} = \boldsymbol{\mu}_{\mathrm{m}} \times (\mu_0 \mathbf{H}_{\mathrm{eff}}). \tag{16}$$

Using Eq. (15) the above expression reduces to

$$\frac{d\boldsymbol{\mu}_{\mathrm{m}}}{dt} = \mu_0 \gamma (\boldsymbol{\mu}_{\mathrm{m}} \times \mathbf{H}_{\mathrm{eff}}). \tag{17}$$

Here, the effective magnetic field $\mathbf{H}_{\mathrm{eff}}$ is a sum of all external and internal magnetic fields:

$$\mathbf{H}_{\mathrm{eff}} = \mathbf{H}_0 + \mathbf{h}(t) + \mathbf{H}_{\mathrm{ex}} + \mathbf{H}_{\mathrm{dem}} + \mathbf{H}_{\mathrm{ani}}, \tag{18}$$

where $\mathbf{H}_0$ is the applied bias magnetic field, $\mathbf{h}(t)$ is the dynamic component, $\mathbf{H}_{\mathrm{ex}}$ is the exchange field and $\mathbf{H}_{\mathrm{dem}}$ represents the demagnetization field created by the dipolar interaction of magnetic surface and volume charges. The field $\mathbf{H}_{\mathrm{ani}}$ includes all kinds of anisotropic fields described above. All these magnetic fields can be calculated as the first variational derivatives of the corresponding energies with respect to the magnetization vector, for instance for the exchange energy $\mathbf{H}_{\mathrm{ex}} = -\mu_0^{-1} \delta E_{\mathrm{ex}} / \delta \mathbf{M}$.

In the continuum limit, the atomic magnetic moment can be replaced by the macroscopic magnetization $\mathbf{M}$ resulting in the equation of motion, *i.e.*, Landau-Lifshitz (LL) equation:

$$\frac{d\mathbf{M}}{dt} = \mu_0 \gamma (\mathbf{M} \times \mathbf{H}_{\mathrm{eff}}). \tag{19}$$

Physically, the above equation features a continuum precession which means that the system is non-dissipative. To avoid this impractical situation, Landau and Lifshitz proposed the damping term proportional to: $-\mathbf{M} \times (\mathbf{M} \times \mathbf{H}_{\mathrm{eff}})$. Later, Gilbert introduced another damping term into the LL equation resulting in the so-called Landau-Lifshitz-Gilbert (LLG) equation as:

$$\frac{d\mathbf{M}}{dt} = \mu_0 \gamma (\mathbf{M} \times \mathbf{H}_{\mathrm{eff}}) + \frac{\alpha}{M_{\mathrm{S}}} \left(\mathbf{M} \times \frac{d\mathbf{M}}{dt} \right), \tag{20}$$

where α is the dimensionless Gilbert damping parameter. An interesting aspect of this parameter is its viscous nature, *i.e.*, an increase in the rotation of magnetization $d\mathbf{M}/dt$ increases the damping of the system.

Fig. 2 schematically illustrates the interplay between different torques acting on the magnetization vector. The precessional torque (the first term on the right-hand side of Eq. 20) acts tangentially to the circle traced by the tip of the magnetization vector, while the damping torque (the second term) acts radially to align the magnetization along the effective magnetic field. The mechanism of damping primarily arises from the spin-orbit interaction [72] and therefore the damping parameter α is a material property. Other than spin-orbit interaction, different external channels, namely, magnon-electron scattering [73], eddy currents [74], spin pumping [75], multi-magnon scattering [76,77], *etc.*, may also take part in the relaxation of the magnetization precession amplitude.

In 1930, the idea of spin waves was introduced by Bloch [39] in order to describe the reduction of the spontaneous magnetization in a ferromagnet. At absolute zero temperature, all the atomic magnetic moments of a ferromagnet are aligned in the same direction (as shown in Fig. 3A), which is the ground state of the system. As the temperature rises, the thermal agitation perturbs the spontaneous magnetization causing a deviation of the spins from the aligned direction. This phase of the disturbance propagates through the system in the form of a wave as depicted in Fig. 3B and C.

In case of uniform precession in presence of a magnetic field, the spins of the system precess in phase about the field direction, *i.e.*, the wavelength (wavevector) of the spin wave is nearly infinity (zero). By solving the LLG equation (Eq. 19) the frequency of uniform precession (known as Ferromagnetic

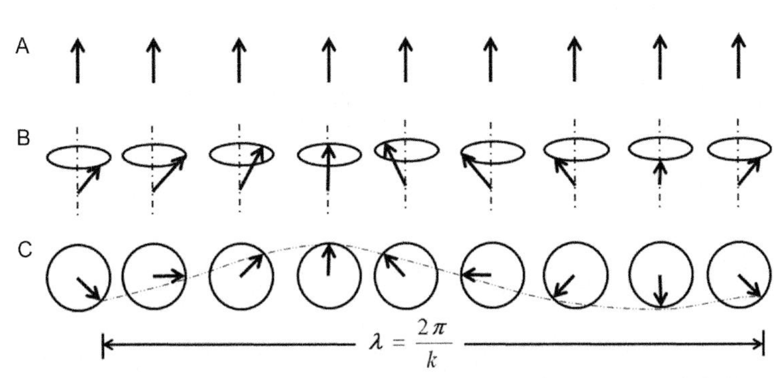

Fig. 3 Semiclassical representation of a spin wave in a ferromagnet: (A) the ground state, (B) a spin wave of precessing magnetic moments and (C) the spin wave (viewed from above) showing a complete wavelength.

Resonance [FMR] frequency) can be obtained. The corresponding solutions were calculated by Kittel [60] and are given by,

$$f_{\text{FMR}} = \frac{\mu_0 \gamma}{2\pi} \sqrt{[H_0 + (N_x - N_z)M_S]\left[H_0 + \left(N_y - N_z\right)M_S\right]}, \tag{21}$$

where N_x, N_y, and N_z are the diagonal components of the demagnetizing tensor in the x, y and z directions. For instance, for in-plane magnetized thin films $N_x = 1$ and $N_y = N_z = 0$, whereas for out-of-plane magnetized films with the external magnetic field aligned along the z-axis we get $N_x = N_y = 0$ and $N_z = 1$.

On the other hand, the non-uniform precession is signified as the propagation of phase shift across the spin system and hence the mutual interactions between the moments play a crucial role in determining the wavelength, frequency and overall dispersion behavior of the spin wave. In the long-wavelength regime, the phase difference between consecutive spins is rather small, the spin wave energy is primarily dominated by dipolar energy and the spin waves are referred to as dipolar-dominated or magnetostatic spin waves. In contrast, the short-wavelength spin waves are governed by exchange interaction and known as exchange spin waves.

The characteristics of magnetostatic spin waves propagating in thin ferromagnetic films were first reported by Damon and Eshbach [78] in 1961. The dipolar interaction being anisotropic, the frequency, amplitude and propagation properties of magnetostatic modes strongly depend on the geometry of their propagation direction with respect to the applied field and the film plane. When both the applied field and the spin wave wavevector lie in the film plane and are perpendicular (parallel) to each other, a magnetostatic surface mode (magnetostatic backward volume mode) appears. On the contrary, if the magnetic field is applied out of the film plane and the spin wave propagates parallel to the surface, it is known as the magnetostatic forward volume mode. The frequency vs wavevector dispersion characteristics of these modes is presented in Fig. 4. The magnetostatic surface wave (MSSW) mode, also known as Damon and Eshbach mode (DE), shows a positive dispersion starting from the Kittel mode at zero wave vectors. This mode is further characterized by the localization of spin wave amplitude at the top or bottom surface. Spin waves propagating in opposite directions have localization at opposite surfaces, *i.e.*, they exhibit a nonreciprocal behavior. Even that the nonreciprocal property is seen only in the amplitude localization, not in dispersion. However, introducing inhomogeneity in magnetic properties

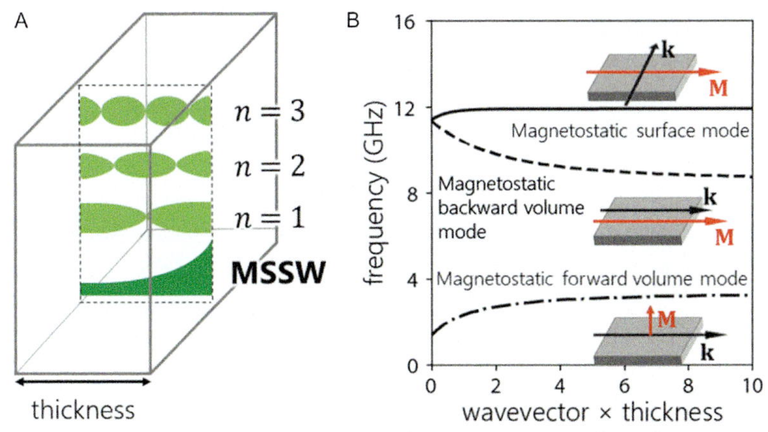

Fig. 4 Schematics of (A) perpendicular standing spin wave mode and magnetostatic surface spin wave mode (MSSW) amplitudes are shown for a ferromagnetic thin film. (B) The dispersion relations for three different types of the magnetostatic spin waves modes calculated for $M_S = 0.2$ MA/m and $\mu_0 H_0 = 0.3$ T.

and breaking reflectional symmetry about the midplane can induce nonreciprocity of the dispersion, as well. As a result, oppositely propagating spin waves with the same frequency will have different wavelength. In absence of any exchange interactions and magnetic anisotropy, the dispersion relation of the DE mode is given by [79]

$$f_{DE} = \frac{\gamma \mu_0}{2\pi} \left[H_0 (H_0 + M_S) + \frac{M_S^2}{4} \left(1 - \exp\left(-2k_{\parallel} d\right)\right) \right]^{\frac{1}{2}}, \tag{22}$$

where $k_{\parallel}$ is the in-plane component of the wavevector and d is the thickness of the film.

In the case of magnetostatic backward volume waves (MSBVW), the spin wave amplitude prevails throughout the thickness of the sample and the dispersion features a negative slope (group and phase velocities are in opposite directions). The corresponding dispersion relation is given by [80].

$$f_{MBVSW} = \frac{\gamma \mu_0}{2\pi} \left[H_0 \left(H_0 + M_S \frac{1 - \exp\left(-k_{\parallel} d\right)}{k_{\parallel} d} \right) \right]^{\frac{1}{2}}. \tag{23}$$

2.3.1 Exchange spin waves

When the wavevector of spin wave is increased (wavelength is decreased), the exchange interaction comes into play. Basically, the exchange interaction becomes dominant when the spin wave wavelength is of the order of the exchange length which is given by.

$$l_{\mathrm{ex}} = \sqrt{\frac{2A_{\mathrm{ex}}}{\mu_0 M_{\mathrm{S}}^2}}. \tag{24}$$

It is worth to note, that the exchange dominated spin wave mode does not depend on the relative orientation of the wavevector and the magnetic field. The spin wave dispersion relation with dipolar-exchange interactions taken into account in thin in-plane magnetized ferromagnetic films is given by [80].

$$f = \frac{\gamma\mu_0}{2\pi}\left[\left\{H_0 + M_{\mathrm{S}} l_{\mathrm{ex}}^2 k_{\|}{}^2\right\} \times \left\{H_0 + M_{\mathrm{S}} l_{\mathrm{ex}}^2 k_{\|}{}^2 + M_{\mathrm{S}}\left(1 - P\cos^2\varphi + \frac{M_{\mathrm{S}} P(1-P)}{H_0 + M_{\mathrm{S}} l_{\mathrm{ex}}^2 k_{\|}{}^2}\sin^2\varphi\right)\right\}\right]^{\frac{1}{2}}, \tag{25}$$

where $P = 1 - [1 - \exp(-kd)]/(kd)$, and φ is the angle between the applied in-plane field H_0 and $k_{\|}$.

In addition to $k_{\|}$ it is also possible to excite spin waves with the non-zero component of the wavevector perpendicular to the film thickness necessary to form perpendicular standing spin waves (Fig. 4A). For a film of thickness d, the wavevector of a perpendicular standing spin wave mode can be written as

$$k_{\perp} = \frac{n\pi}{d}. \tag{26}$$

The dynamic magnetization profile is sinusoidal for free magnetization on the surfaces and depends on the quantization number n:

$$m_n(z) = a_n . \cos\left[k_{\perp,n}\left(z + \frac{d}{2}\right)\right] \text{ for } -\frac{d}{2} < z < \frac{d}{2}, \tag{27}$$

where z denotes the coordinate along the film thickness. The above equation actually describes a standing mode consisting of two counter-propagating waves with quantized wave vectors. Without considering the contribution of the in-plane wavevector, the frequency for the perpendicular standing spin wave mode for the in-plane magnetized film is given by [81]

$$f = \frac{\gamma\mu_0}{2\pi}\left[\left(H_{\mathrm{eff}} + \frac{2A_{\mathrm{ex}}}{\mu_0 M_{\mathrm{S}}} k_{\perp}{}^2\right) \times \left(H_{\mathrm{eff}} + \frac{2A_{\mathrm{ex}}}{\mu_0 M_{\mathrm{S}}} k_{\perp}{}^2 + M_{\mathrm{eff}}\right)\right]^{\frac{1}{2}}. \tag{28}$$

2.3.2 Confined spin wave modes in magnetic structures

In the discussion above, we considered the properties of spin waves in an infinite thin film. In the case of a confined structure, the spectrum of spin wave modes is modified by the boundary conditions imposed by the lateral dimensions [82]. In particular, the spin waves may find a propagation channel [83] or form standing waves [84] in the distribution of "potential wells" defined by the geometry of the structure. The standing wave often leads to a localized mode or quantized mode (with multiple quantization numbers) when the feature dimensions are of the order of the wavelength of the spin wave. Overall, the number of spin wave modes within a given frequency range increases with increasing size of the structure, whose properties are strongly dependent on the system geometry as well as on the orientation of the magnetic field.

In a magnonic crystal, this confinement occurs periodically. The resulting inhomogeneity of the internal magnetic field can then be taken into account in the form of the superposition of plane waves and the Bloch theorem can be used, similar to the crystal field of an electronic crystal. The calculation of the spin wave dispersion, therefore, yields the magnonic band structure, consisting of a number of allowed and forbidden magnonic bands [85]. The corresponding Brillouin zones are determined by the crystal structure, which provides the flexibility to mold the spin wave properties by the changing geometry or confinement of the system. For instance, Fig. 5

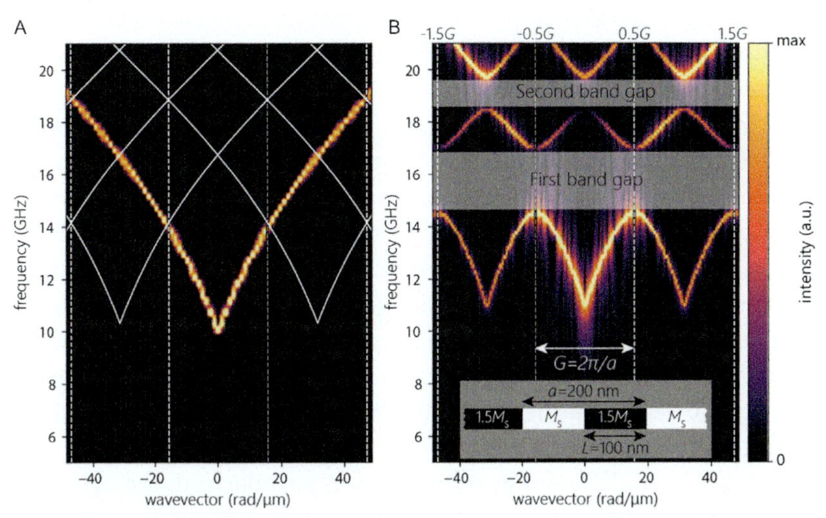

Fig. 5 Spin wave dispersion in a ferromagnetic film in DE geometry (A) and in a bi-component magnonic crystal (B). The white lines in the background of (A) are analytically derived dispersions and shifted by $\pm nG$ (empty lattice model), where G is the reciprocal lattice vector, as shown in (B) and n is integer number.

presents a comparison between the spin wave dispersion in a uniform ferromagnetic film in DE geometry and that in a one-dimensional bi-component magnonic crystal (Fig. 5A and B, respectively). In the former case, the magnetic parameters used are saturation magnetization $M_S = 1$ MA/m and exchange constant $A_{ex} = 10$ pJ/m. The magnonic crystal consists of a periodic arrangement of two kinds of magnetic stripes (the lattice parameters are shown in the inset of Fig. 5B) where M_S of the second material is $M_{S2} = 1.5$ MA/m, all other parameters are kept constant. The spin wave dispersion, as shown by the colormap, depicts a single spin wave mode for a thin film, which turns into three spin wave bands in the magnonic crystal, separated by bandgaps at the Brillouin zone boundaries (dashed vertical lines) and centers.

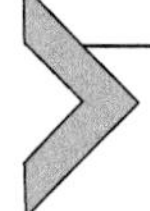

3. Spin wave propagation in the presence of a single domain wall

3.1 Spin wave propagation along a 180° Néel-type domain wall for in-plane magnetized films

First, let us consider a long, ferromagnetic stripe with a 180° Néel-type domain wall separating two antiparallel in-plane magnetic domains with magnetic moment parallel to the stripe's edges, see the visualization in Fig. 6A. The calculations are done for a 10 μm wide, infinitely long, and 10 nm thick Py stripe ($M_S = 800$ kA/m, $A_{ex} = 13$ pJ/m) with assumed periodic boundary conditions along the sides perpendicular to the domain wall. In this configuration, the magnetic moments within the Néel-type domain wall gradually rotate in the film's plane as one moves between two antiparallel domains (along the x-axis). This magnetization non-collinearity induces volume magnetic charges, *i.e.*, the density of magnetic charges ($-\rho_M = \nabla \cdot \mathbf{M}$) in that region is non-zero. In consequence, the strong demagnetizing field of value $\mu_0 H_{demag}(x = 0) = -27$ mT in the center of the domain wall is directed antiparallel to the magnetization orientation (see Fig. 6C). Noteworthy, that the magnetization's non-collinearity creates also a non-uniformity in the exchange field, however, much weaker than the demagnetizing field, see the exchange field profile in Fig. 6D and the resulting static effective magnetic field ($\mathbf{H}_{eff} = \mathbf{H}_{dem} + \mathbf{H}_{ex}$) in Fig. 6B.

As it has been discussed in the subsection introduction, the decrease of the effective magnetic field (as in Fig. 6B) shifts down the bottom of the spin wave spectrum. In other words, one may expect to find spin wave modes residing in the domain wall that acts as a potential well. In order to visualize

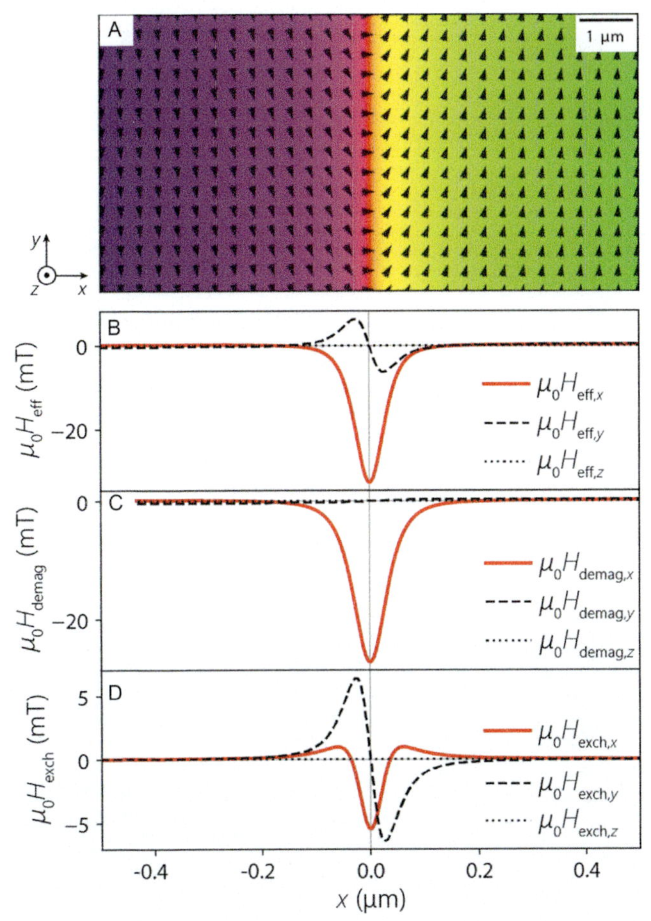

Fig. 6 (A) Simulated static magnetic configuration for a 10 μm wide and 10 nm thick Py stripe. (B) The static effective magnetic field, (C) the demagnetizing field, and (D) the exchange field plotted across the domain wall (along the *x*-axis from (A)).

that, a set of micromagnetic simulations performed using MuMax3 [86] for spin waves excited at frequencies of 1 GHz, 2 GHz, and 3 GHz by a source placed in the center of the domain wall have been performed. The results are presented in Fig. 7, where the location of the spin wave source is indicated by the white star. It is visible that for 1 GHz spin wave modes are localized inside the domain wall (Fig. 7A), whereas the modes of higher frequencies radiate spin waves outward the domain wall. It is especially visible for the frequency of 3 GHz in Fig. 7C (and higher frequencies that are not depicted). Note, that the dispersion plotted along the center of the domain

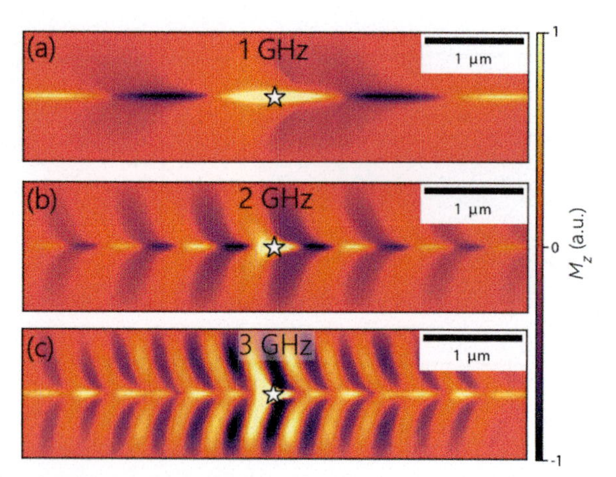

Fig. 7 Simulated spin wave amplitude maps for spin waves excited at frequencies of (A) 1 GHz, (B) 2 GHz, and (C) 3 GHz by a point source located in the domain wall center (indicated by the white star).

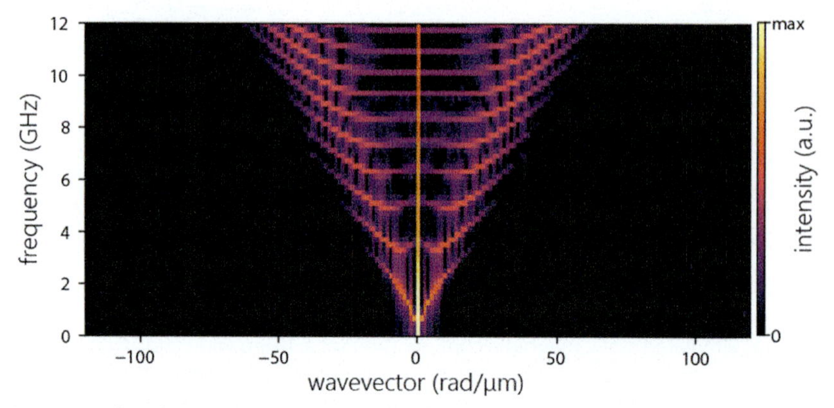

Fig. 8 Simulated spin wave dispersion for spin waves propagating along a Néel-type domain wall.

wall (Fig. 8) is reciprocal and the wavelengths of spin waves propagating in the opposite directions measured at the center of the domain wall do not differ (Fig. 7). However, one may see that the symmetry with respect to the mirroring y-axis (perpendicular to the domain wall) is broken, the "V-shaped" unidirectional phase fronts emerge. This bending of the wavefronts is present in the transition region where the magnetic moments are in-plane rotating by 90°.

The dispersion relation is inherently anisotropic and depends on the magnetization direction. It suggests that the emerging "V-shaped" wavefronts

[53,87,88] are rather a consequence of the domain wall non-collinearity combined with the anisotropy of spin wave dispersion.

Moreover, the analysis of the spin wave dispersion relation of waves in the domain wall computed by micromagnetic simulations shows a relatively dense spin wave spectrum (10 modes below 12 GHz, see Fig. 8). It is due to the wide distance at which spin texture rotates, resulting in a relatively shallow potential well. In result, the frequencies of bulk-like spin waves are only slightly higher than frequencies of domain wall localized spin waves.

The experimental realization of 180° Néel-type domain walls as a waveguide has been studied by Wagner et al. [53]. There, spin waves have been excited by a microstrip antenna placed on top of the stripe across its width (perpendicularly to 180° Néel-type domain wall) and spin wave dynamics have been measured using micro Brillouin light scattering confirming spin wave localization in the domain wall. Moreover, they have shown that applying an in-plane magnetic field the position of the domain wall can be dynamically shifted. This topic has been further studied by the same group in Ref. [88] where micromagnetic simulations have been employed to study spin wave reciprocity in the presence of the Néel domain wall in a rectangular quasi-infinite permalloy thin film. The subject of spin wave confinement in Néel-type domain walls (180° and also 90°) for in-plane magnetized thin films are covered in the Ph.D. dissertation of Wagner [89].

3.2 Spin wave propagation along domain walls in out-of-plane magnetized thin films: Bloch-type and DMI-induced Néel-type domain walls

Films with strong perpendicular anisotropy are an interesting class of materials especially from the data storage point of view [90]. They can host non-collinear stable states in remanence, *e.g.*, very narrow domain walls separating "up" and "down" domains. Typically, these domains are separated by Bloch-type domain walls with the magnetic moments rotating about the normal of the domain wall, a configuration that is preferential in terms of dipole energy [91]. However, in materials with strong interfacial DMI these domain walls can change their structure since the DMI energy contribution prefer perpendicular canting. The magnetic moments within the domain wall can then rotate perpendicularly to the domain wall axis creating DMI-induced Néel-type domain walls, where the magnetic moments rotate about a line that is orthogonal to the normal of the domain wall and the direction of the tilt between neighboring spins is determined by the sign of the DMI, *i.e.*, they are chiral.

In the following paragraphs, the basic properties of spin waves propagating within both types of domain walls in films with strong perpendicular anisotropy will be discussed. But first, let us discuss spin wave dispersion of the bulk-like spin wave modes residing in "up" and "down" domains. For exchange dominated spin waves, it takes the following form [92]

$$f = \frac{1}{2\pi}\left(\omega_k + \frac{2\gamma K^{\mathrm{eff}}}{M_{\mathrm{S}}}\right), \tag{29}$$

where $\omega_k = 2\gamma A_{\mathrm{ex}} k_x{}^2 / M_{\mathrm{S}}$ the quadratic exchange part and $K^{\mathrm{eff}} = K^{\mathrm{V}} - \mu_0 M_{\mathrm{S}}{}^2/2$ is the effective perpendicular anisotropy constant. An exemplary dispersion obtained for parameters: $M_{\mathrm{S}} = 1$ MA/m, $K_{\mathrm{u}} = 1\,\mathrm{MJ/m^3}$, thickness $d = 1$ nm and $A_{\mathrm{ex}} = 10$ pJ/m is displayed in Fig. 9A by the solid parabolic line. It is visible, that this dispersion is isotropic and the wavelength doesn't depend on the direction of propagation. The bottom of the spin wave spectrum for this mode depends linearly on the effective perpendicular

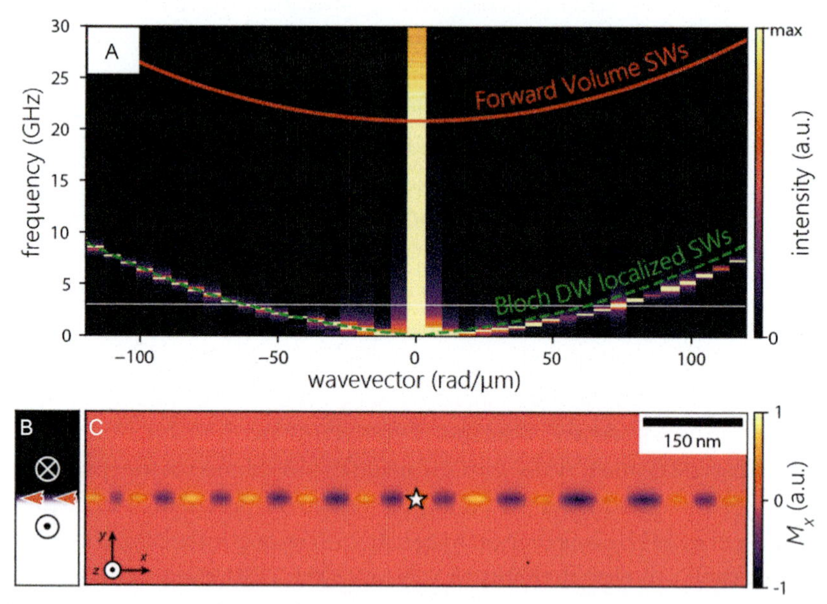

Fig. 9 (A) Dispersion relation of spin waves propagating in the Bloch-type domain wall (colormap in the background) computed by means of micromagnetic simulations. The dispersion of the forward volume spin waves is marked by the solid line. (B) The static magnetic configuration used in dynamic simulations with the Bloch-type domain wall separating two magnetic domains (down and up). (C) The profile of spin waves excited at frequency 3 GHz. The white star indicates the position of point source used for spin wave excitation.

anisotropy constant, *i.e.*, $f(k=0)=2\gamma K^{\mathrm{eff}}/(2\pi M_S)$. In our case, the resonant frequency in the absence of the external field is above 20 GHz.

The dispersion for the spin waves confined within Bloch-type domain wall is [92]

$$f_B=\frac{1}{2\pi}\sqrt{\omega_k(\omega_k+\omega_\perp)}, \tag{30}$$

where $\omega_\perp=2\gamma K_\perp/M_S$ and $K_\perp=\mu_0 N_y {M_S}^2/2$ is a transverse anisotropy taking into account the role of dipolar interactions due to volume charges at the center of the domain wall, $N_y\approx d/(d+\pi\Delta)$represents an effective demagnetization constant across the walls, and $\Delta=\sqrt{A_{\mathrm{ex}}/K^{\mathrm{eff}}}$ is the characteristic wall width parameter.

The dispersion relation obtained for spin waves residing within the Bloch-type domain wall computed by micromagnetic simulations and using Eq. (30) for the same magnetic parameters is displayed in Fig. 9A. It is well visible, that the dispersion is gapless (starts at $f=0$). These oscillations propagating along the domain wall can be treated as flexure oscillations of the domain wall shape (so-called Winter's magnons) [66,93,94].

An exemplary mode excited at frequency 3 GHz excited in the center of the domain wall (indicated by white star) is plotted in Fig. 9C, whereas the profile of domain wall is presented in Fig. 9B (and also in Fig. 2B with the rotated coordinate system). It is visible that spin waves are localized at the very center of the domain wall.

The introduction of strong interfacial DMI changes the structure of domain wall, *i.e.*, the up and down domain walls can be separated by Néel-type domain walls with magnetic moments rotating in the plane perpendicular to the domain wall. The dispersion relation for the spin waves propagating along DMI-induced Néel-type domain walls is [92]:

$$f_N=\frac{1}{2\pi}\sqrt{\omega_k\left(\omega_k-\omega_\perp+\frac{\omega_{D,k}}{k_x\Delta}\right)}+\mathrm{sgn}(k_x)\omega_{D,k}, \tag{31}$$

where $\omega_{D,k}=\pi\gamma D k_x/2M_S$. An exemplary dispersion relation for the same magnetic parameters as above and added non-zero interfacial DMI $D=1\,\mathrm{mJ/m^2}$ is displayed in Fig. 10A, where solid line presents dispersion obtained for the bulk spin waves. Analytical (Eq. 31) and simulated dispersion relation for spin waves propagating along the DMI-induced Néel-type domain wall are presented by the dashed line and the colormap in the background, respectively. It is visible that the introduction of non-zero DMI

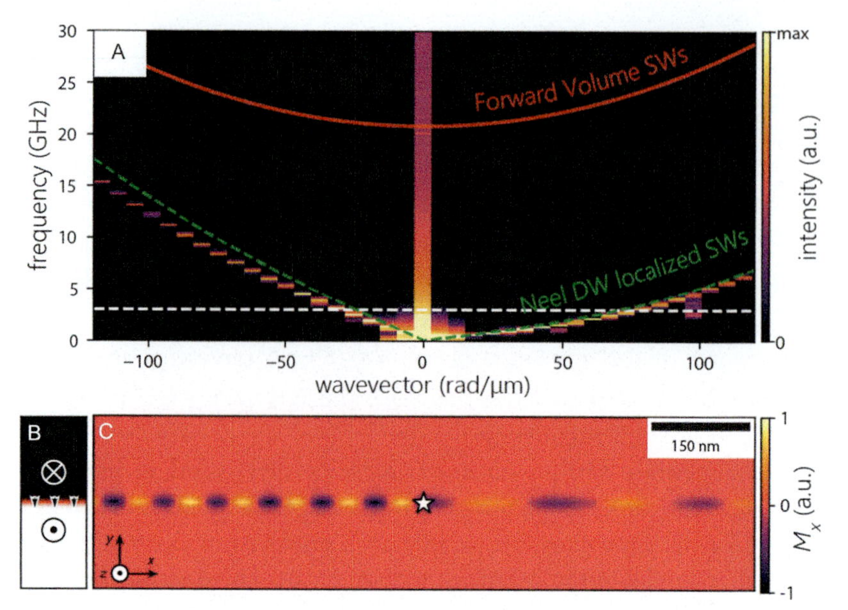

Fig. 10 Dispersion relation of spin waves propagating in the Néel-type domain wall (colormap in the background) computed by means of micromagnetic simulations. The dispersion of the forward volume spin waves is marked by the solid line. (B) The static magnetic configuration used in dynamic simulations with the Néel-type domain wall separating two magnetic domains (down and up). (C) The profile of spin waves excited at frequency 3 GHz. The white star indicates the position of point source used for spin wave excitation.

results in nonreciprocity of waves confined in the domain wall, whereas the dispersion of bulk spin waves is reciprocal. It is also presented in Fig. 10C, where an exemplary mode of spin waves excited at frequency 3 GHz in the region of the domain wall center (indicated by the white star) is displayed. It is visible that spin waves are well confined within the domain wall and the wavelength of waves propagating leftward differs with respect to waves propagating rightward which confirms the spin wave nonreciprocity visible in the dispersion relation. The profile and the location of the domain wall are shown in Fig. 10B (more detailed domain wall profile is also shown in Fig. 3C).

The topic of spin wave propagation along these types of domain walls separating "up" and "down" domains have been studied in Refs. [92,95] where, *e.g.*, nonreciprocal properties of spin waves channeling in domain walls and propagation along DMI-induced Néel-type domain walls confined in curved track have been presented.

3.3 Spin wave propagation across the domain wall

Spin waves propagating across a domain wall experience a scattering potential. Interestingly, the potential related to static one-dimensional Bloch-type domain wall can be described as the Pöschl-Teller potential that is reflectionless and leads only to a phase shift of the transmitted waves [96–98]. The situation is different in case of other types of domain walls, *e.g.*, Néel-type domain walls [99], where part of the incident spin wave amplitude is reflected. Interestingly, the oscillations of both Bloch [100] and Néel [101,102] domain walls also leads to the spin wave excitation. In the non-linear regime, however, even for Bloch-type domain walls, the internal domain wall oscillation modes can interact with spin waves resulting in inelastic scattering of spin waves [103,104]. Furthermore, the obliquely incident spin waves at the domain wall can be bent in or reflected from the region where magnetization rotates [105].

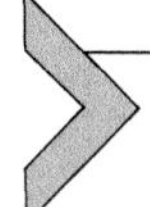

4. Spin wave dynamics in Co/Pd multilayers magnetized in the form of a periodic stripe domain pattern

4.1 Spin wave dynamics in periodic up and down stripe domain pattern separated by Bloch-type and DMI-stabilized Néel-type domain walls

The topic of spin wave propagation in periodic stripe domain structures has been theoretically studied by Borys et al. [98]. In this paper spin wave propagation through an ultra-thin ferromagnet with strong perpendicular magnetocrystalline anisotropy magnetized in form of a periodic pattern of up and down stripe domains separated by Bloch and Néel-type domain walls has been considered. Different types of domain walls were stabilized depending on the strength of the DMI, *i.e.*, in the case of absent DMI, the up and dawn domains are separated by the Bloch domain walls, whereas for strong DMI Néel-type domain walls (also referred to as Dzyaloshinskii domain walls) are favored.

One of the interesting findings is the fact that the magnon dispersion in a periodic pattern of "up" and "down" domains separated by one-dimensional static Bloch-type domain walls does not have bandgaps. It is related to the reflectionless Pöschl-Teller potential that characterizes Bloch-type domain walls. Therefore, it doesn't produce reflections for the spin waves propagating through the domain wall and band gap opening doesn't occur. However, it is not the case for the Néel-type domain walls present in the system with strong DMI. DMI-stabilized Néel-type domain walls which reflect spin waves, and

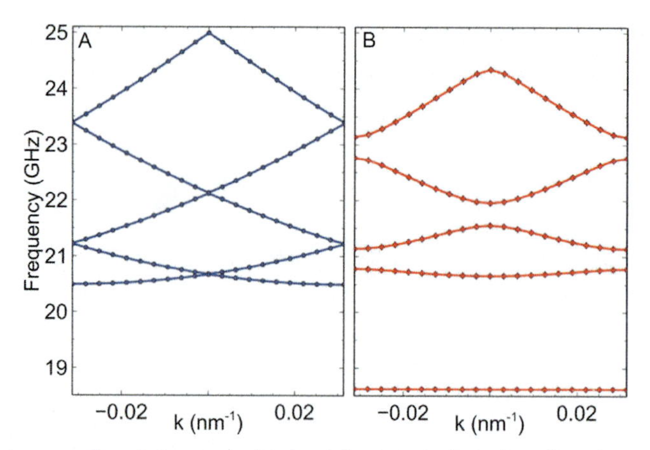

Fig. 11 Spin wave band structure obtained for a periodic stripe domain pattern with (A) Bloch-type domain walls and (B) DMI-stabilized Néel-type domain walls. *Reprinted Figure with permission from P. Borys, F. Garcia-Sanchez, J.-V. Kim, R.L. Stamps, Spin-wave eigenmodes of Dzyaloshinskii domain walls, Adv. Electron. Mater. 2 (2016) 1500202. Copyright (2015) by John Wiley and Sons.*

the stronger DMI the wider bandgap. The exemplary dispersion relations obtained analytically for systems of the same magnetic parameters with a different value of DMI, $D=0$ and $D=1.56\,\text{mJ/m}^2$ are presented in Fig. 11A and B, respectively. It is well visible that the introduction of DMI and change of the domain wall type results in bandgap opening at the edges of the Brillouin zone. Note, that analytical calculations from Ref. [98] do not take into account the dynamic dipolar interaction between subsequent domain walls that can influence spin wave band structure.

4.2 Spin wave dynamics in periodic stripe domain pattern in Co/Pd multilayers

Borys et al. [98] have studied spin wave dynamics in ultra-thin ferromagnets where domain walls are uniform across the thickness. The dynamics of spin spiral states stabilized by strong DMI was investigated in Refs. [106–108]. In these cases, the domain walls can be classified to either Bloch or Néel-type.

However, periodic patterns of stripe domains can be also stabilized in more complex systems, *e.g.*, multilayers possessing perpendicular magnetocrystalline anisotropy caused by the interface anisotropy (induced by the broken translational symmetry at the Co/Pd interfaces) combined with d-d hybridization at the ferromagnet/nonmagnetic interfaces. There, the dipole contribution to the total energy is significant and emerging of a stripe domain walls structure is much more complex across the system thickness than in the case of both Bloch and Néel-type domain walls [91]. Significant stray fields arise due to the greater overall thickness (larger than domain wall width) and perpendicular magnetocrystalline anisotropy causing spins to align perpendicular to the surface. Therefore the wall is strongly influenced by the surface, resulting in a twisted corkscrew structure of the domain wall with a Bloch-type wall in the film's center with two Néel caps at the film's surfaces [109,110].

The static and dynamic magnetic properties in Co/Pd multilayers with stripe domains by using magnetic force microscopy (MFM), vibrating sample magnetometry (VSM) and Brillouin light scattering (BLS) spectroscopy have been studied in Ref. [68]. In this paper, the influence of the stable spin configuration on the spin wave spectrum has been analyzed both experimentally and theoretically. For that, the spin waves were measured in saturated as well as in the remanent state (*i.e.*, in presence of stripe domains) and their wavevector dispersions were analyzed and compared. Finally, micromagnetic simulations were employed to calculate spin wave dispersion and mode profiles of the spin waves.

4.2.1 Experimental details

DC magnetron sputtering was used to deposit the multilayered sample which has the following structure: Ta(15)/Pd(30)/$[Co(10)/Pd(7)]_{25}$/Pd(20), where the numbers are the thicknesses in Å. For deposition, a multi-source confocal sputter up geometry was used with a source/substrate distance of 4–6 in. During the deposition, the substrate was rotated for uniformity at about 3 Hz and was kept at room temperature using a low Argon pressure of 3 mTorr. Co was sputtered at 250 W with a deposition rate of 0.7 Å/s and Pd was sputtered at 100 W with a deposition rate of $\sim$2 Å/s. Given the low Ar pressure used during deposition, the film is relatively smooth, with sharply defined interfaces [111].

The aligned stripe domain structure forms by applying in-plane ac demagnetizing field starting from saturation. In other words, the applied in-plane field is switched back and forth between positive and negative polarity starting above the saturation field strength down to about a few decimals of millitesla, where the field amplitude is reduced about 1% in each step. Note that, the Co/Pd multilayer is isotropic for an in-plane direction and the alignment of the stripe domains is along the applied field direction. The images of the distribution of up and down domains at remanence were taken using MFM. VSM was employed at room temperature to obtain the magnetization curves (out of plane and in-plane). To further investigate the dynamic magnetization characteristics, spin wave dispersions (frequency vs wavevector) were recorded using BLS technique in the 180° backscattered geometry. This technique is based on the inelastic scattering of a continuous light wave from thermal magnons. During the interaction between photons and magnons, the energy as well as the angular momentum are conserved, hence the information of the spin waves can be extracted by analyzing the scattered beam. In the experiment, a monochromatic laser light of wavelength $\lambda = 532\,\text{nm}$ from a solid-state laser was focused on the sample surface using an achromatic doublet which collects the backscattered beam as well. From the conservation of momentum, the magnitude of the in-plane transferred wave-vector **k** can be selected by changing the incidence angle of light θ according to: $k = (4\pi/\lambda)\sin\theta$. The direction of magnon wave-vector **k** lies along the intersection of the scattering plane with the film surface. The scattered beam then passes through an analyzer which is aligned in a cross position with respect to the incident beam, in order to minimize the contribution from other vibrations, like phonons. Subsequently, the frequencies of the scattered light are analyzed using a Sandercock-type six-pass tandem Fabry-Perot interferometer (FPI) (JRS Scientific Instruments) [111]. This complex tabletop instrument uses two FPIs in series and the frequency of the spin wave appears as the difference of frequencies between the incident and scattered beams. Two sets of measurements were performed, in the absence and in the presence of the magnetic field, respectively. In case of the former, the sample was measured in the demagnetized state in presence of the stripe domains for k parallel and perpendicular to the stripe axis. This was accomplished by rotating the sample around the film normal, *i.e.*, varying the angle φ between **k** and stripe axis. In the other measurement set, the sample was saturated by applying a strong in-plane magnetic field ($\mu_0 H = 0.35\,\text{T}$) and the wavevector dispersion was taken in the DE geometry, *i.e.*, $\mathbf{k} \perp \mathbf{H}$.

4.2.2 Simulation details

To validate the experimental results and to understand the physical characteristics of the resonant modes we exploited micromagnetic simulations using MuMax3 [86].

Micromagnetic simulations primarily were used to relax the stable stripe domain pattern of comparable lattice constant as one obtained experimentally, to compute dispersion relations for the uniformly in-plane magnetized structure and the stripe domain pattern, and finally, to visualize resonant mode profiles. In micromagnetic simulations, we assumed effective homogeneous material parameters across the whole thickness of the structure.

The saturation magnetization and anisotropy constants were set to the values obtained from the VSM measurements. To fit micromagnetic simulations results with the BLS data, we had to assume an anisotropic exchange constant [112], with $A_{ex,IP} = 1.45 \times 10^{-6}$ erg/cm and $A_{ex,OOP} = 1.05 \times 10^{-6}$ erg/cm along the in-plane and out-of-plane directions, respectively. Micromagnetic simulations were performed for the system of thickness $L_z = 42.5$ nm discretized with the unit cell of dimensions being much smaller than the exchange length or the width of domain walls, *i.e.*, $2 \times 1.56 \times L_z/16\,\text{nm}^3$. At the lateral edges of the simulated structure, which dimensions were $2.56\,\mu\text{m} \times 3.2\,\mu\text{m} \times L_z$ nm, the periodic boundary conditions were assumed.

To obtain a stable stripe domain pattern the following algorithm of relaxation was applied. In order to accelerate simulations and to obtain stripe domains along the x-axis, one unit cell with periodic boundary conditions along the x-axis and periodic boundary conditions along the y-axis were assumed. Along the y-axis, a sufficiently large number of unit cells should be assumed to enable relaxation of dozens of stripe domains. In the first stage of the simulations, the system was in-plane magnetized by a high magnetic field parallel to the x-axis. Then, the field strength was reduced (in simulations the field steps of value $\mu_0 \Delta H = 25$ mT has been used) and the magnetic configuration was slightly perturbed by a dynamic impulse of the orthogonal magnetic field (parallel to the z-axis) and again relaxed. This procedure was repeated until the external field strength was reduced to $H = 0$. In the end, a stripe domain pattern is obtained as shown in Fig. 12C. In the last stage, the system was stretched along the x-axis and cut to desired dimensions before starting simulations of spin wave dynamics.

The spin waves dispersion relation and modes profiles were calculated according to Ref. [113]. In the first stage a stripe domain pattern was perturbed by a small amplitude dynamic magnetic field of a broadband frequency and wavevector spectra, *i.e.*,

$$h_{mf} \propto \text{sinc}(2\pi f_{cut} t)\left[\text{sinc}(k_{cut} x) + \text{sinc}(k_{cut} y)\right],$$

where f_{cut} and k_{cut} are the cut-off frequency and wavenumbers, respectively, defining the maximal value of the frequency and wavevector being excited by the applied spin-wave source. The dispersion relations were then computed by transforming the time-dependent magnetization (all three components) into the frequency and wave-vector space using two dimensional fast Fourier transformation (FFT).

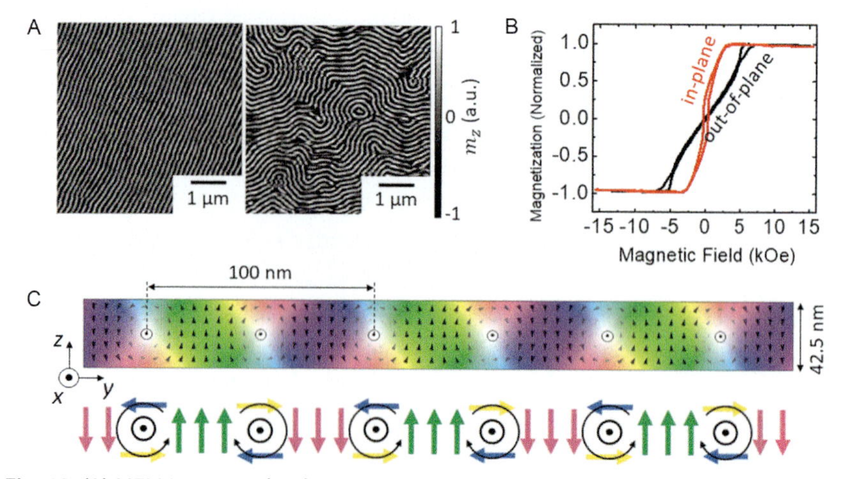

Fig. 12 (A) MFM image in the demagnetized state showing the parallel (left panel) and labyrinth (right panel) stripe domains of a Co/Pd multilayered system, obtained after IP and OOP AC demagnetization, respectively. (B) Magnetic hysteresis loops of the multilayers with IP and OOP applied magnetic fields. (C) The domain configuration (cross-sectional view) obtained from micromagnetic simulations. Arrows correspond to the magnetization vector. The bottom panel schematically visualizes static magnetization with clockwise and anticlockwise domain walls. The *x*-axis is assumed to be parallel to the domain wall whereas the *z*-axis is perpendicular Co/Pd stack.

4.2.3 Static magnetic configuration

The stripe domain patterns as measured by MFM in the demagnetized states are shown in Fig. 12A. After in-plane AC demagnetization, we observe well-defined parallel stripe domains aligned along the in-plane demagnetization direction. The half period of the pattern is $a/2 \approx 60\,\mathrm{nm}$ (a being the lattice constant), which is comparable to the thickness of the entire stack (42.5 nm). The competition between the perpendicular magnetocrystalline anisotropy and the thin film shape anisotropy (the presence of the magnetic charges on the surface of the film) leads to the formation of domains with up and down magnetization perpendicular to the film plane separated by "corkscrew"-like domain wall [114].

The magnetization reversal curve for in-plane magnetic field indicates two distinct phases of magnetization (Fig. 12B). The first phase is visible at field values near the coercive field. The in-plane component of magnetization quickly reverses at a field close to the coercive field, where the remanent magnetization is around 30%. It indicates, that the averaged magnetization is tilted with respect to the out-of-plane direction (has non-zero averaged in-plane component of magnetization). The second phase is visible as a linear approach to saturation. It is related to the rotation of the perpendicularly magnetized stripe domains under the application of the in-plane magnetic field.

The hysteresis loop for the out-of-plane magnetic field is characterized by domain nucleation, propagation, and annihilation characteristic of labyrinth domain structure [115]. The MFM image (Fig. 12A, right panel) taken after demagnetizing the film with perpendicular applied field confirms the labyrinth domain formation. Stripe domains are robust and stable under out-of-plane field exposures up to values about ± 0.3–$0.4\,\mathrm{T}$, that is in the region with the linear slope in the magnetization reversal curve for the out-of-plane magnetic field. The application of out-of-plane fields from this region alters only the relative widths of up and down domains preserving the overall domain topology. The uniaxial anisotropy constant associated with the perpendicular magnetocrystalline anisotropy and saturation magnetization were both, estimated from the hysteresis curves obtained by VSM as $K^V = 318.5\,\mathrm{kJ/m^3}$ and $M_S = 910\,\mathrm{kA/m}$ [116]. The quality factor obtained for these values is $Q = 2K^V/\mu_0 M_S^2 = 0.6$.

Fig. 12C presents the cross-sectional view of the simulated stripe domain structure where the magnetic state is shown with the arrows. The system consists of up and down domains separated by complex twisted domain walls occupying a significant region in each spatial period and the magnetization profile in each domain deviates from its rectangular shape. The spatial periodicity is estimated as $a = 100\,\mathrm{nm}$ which is in a reasonable agreement with the experimental MFM images (120 nm).

The magnetization profile in the domain wall at the film surfaces depicts a Néel wall like character, whereas the resultant magnetization at the center of the multilayer film is directed IP along the domain wall axis (Bloch wall like) [see the schematic representation of domain structure in Fig. 12C (lower panel)]. Therefore, we can distinguish a twisted pattern around the domain wall's axis (corkscrew type) [116], which is clockwise or anticlockwise for alternate domain walls and cores of off all domain walls are directed in the $+x$ direction (see Fig. 12C). These two types of domain walls with the same polarity (along the $+x$-axis) and two types of chirality (clockwise and counterclockwise) together with up and down domain walls create the spatial periodicity of the magnetization pattern. The unit cell, with the lattice constant of $a = 100\,\mathrm{nm}$, consists of two domain walls, a pair of "corkscrew"-like domain walls with opposite chirality.

4.2.4 BLS measurements: Experimentally measured spin wave dispersions

To understand how the presence of a periodic magnetization domain pattern modifies spin wave dynamics in the studied multilayers, first, we have measured spin wave dispersion in the uniformly magnetized system by applying an in-plane magnetic field of value $\mu_0 H = 0.35\,\text{T}$. This dispersion has been derived for DE geometry, *i.e.*, for the case of wavevectors perpendicular to the in-plane aligned magnetization (see Fig. 13A). The result of the BLS measurements revealing the dispersion of thermal spin waves is shown in Fig. 13D. In Fig. 13G, it is shown a typical spectrum taken at $k = 12\,\text{rad}/\mu\text{m}$ and in Fig. 14A results of BLS measurements (frequency peaks of thermal spin waves) are marked by the solid symbols labeled with capital letter "M" as a function for the wavevector. Two dispersive modes are visible. Mode M1 shows a maximum at $k = 10\,\text{rad}/\mu\text{m}$ whereas, the

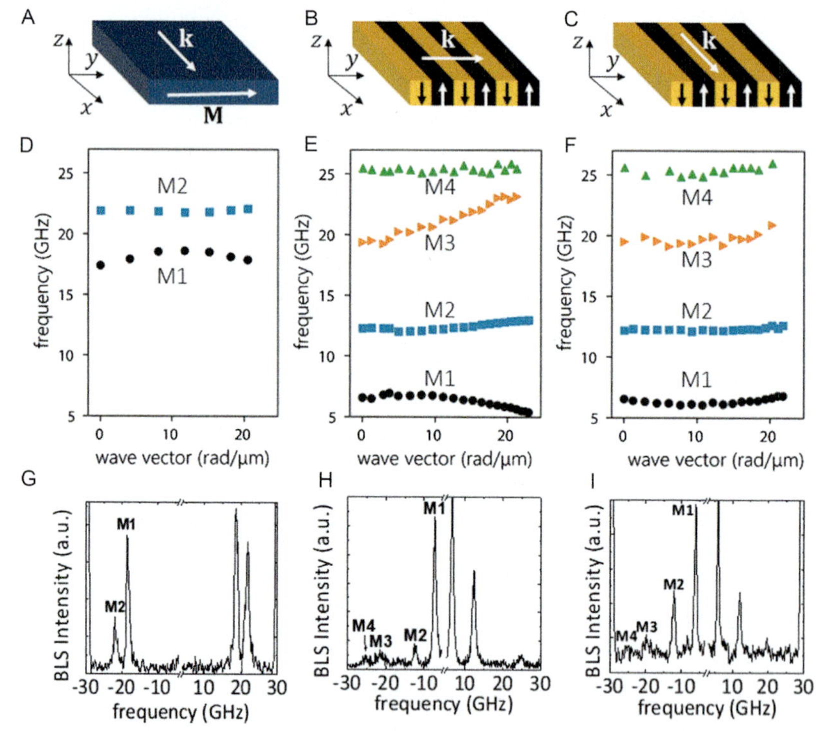

Fig. 13 (A–C) The measurement geometries for the saturated state in (A), $\varphi = 90°$ in (B), and $\varphi = 0°$ in (C), where φ as the angle between domain wall and wavevector. Measured dispersion relations for (D) the saturated states, (E) $\varphi = 90°$, and (F) $\varphi = 0°$. (G–I) Corresponding typical BLS spectra at $k = 1.2 \times 10^7$ rad/m.

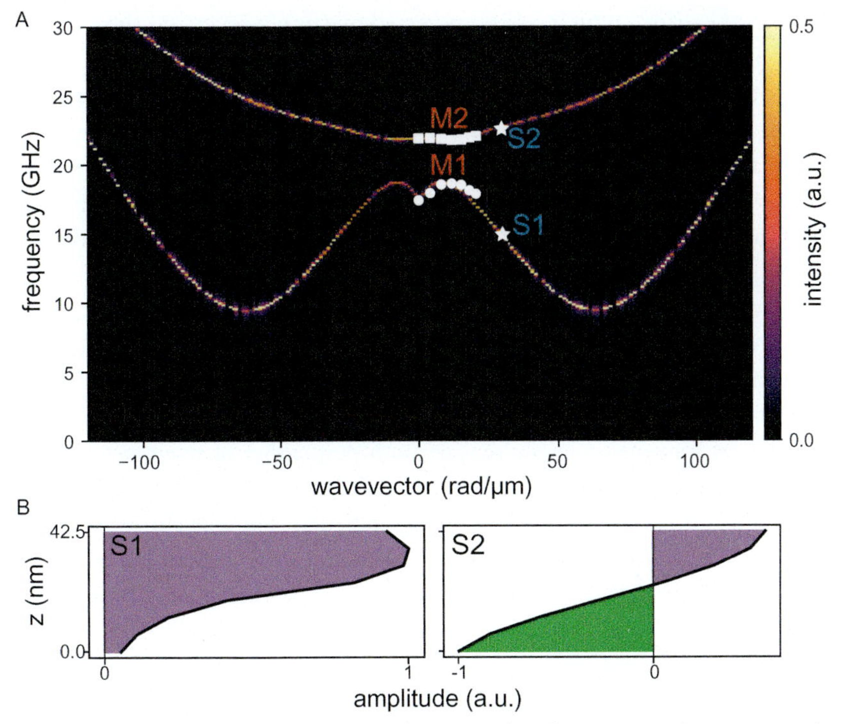

Fig. 14 (A) Simulated dispersion relation for the uniformly magnetized structure in the DE geometry. Measured in BLS dispersions and modes are labeled with "M" and marked with the color points. Simulation results are shown by the colormap, the stars labeled with "S" represent modes visualized in (B) for the wavevector $k = 30\,\text{rad}/\mu\text{m}$. The localization of the first two modes at the opposite surfaces of the sample is visible.

mode M2 shows there a minimum. It results in a 3-GHz-wide frequency gap at $k = 10\,\text{rad}/\mu\text{m}$.

To validate the experimental results and to understand the physical characteristics of the resonant modes we exploited micromagnetic simulations. The simulated mode frequency in dependence on the wavevector is presented by the grayscale map in Fig. 14A (marked with label S), which agree well with the experimental results. Further, we investigated the physical nature of the modes. The mode profiles (their real part) calculated for $k = 30\,\text{rad}/\mu\text{m}$ are displayed in Fig. 14B. It is clearly visible that the mode M1 (S1) is the surface spin wave mode, *i.e.*, its amplitude extends across the volume of the sample with the amplitude increased at one of the surfaces. This localization increases with increasing wavenumber and it can be accounted by the relative intensities of the Stokes/anti-Stokes peaks in the BLS spectrum, which can be reversed by simply reversing the direction of the magnetic field.

Noteworthy, despite the typical DE mode dispersion, in the case of the system with out-of-plane anisotropy, the dispersion is non-monotonic. Interestingly, for relatively small frequencies (just a few GHz above ferromagnetic resonance frequency) spin waves are characterized by very short wavelengths, *e.g.*, for $f = 21$ GHz corresponding wavelength is *ca.* 50 nm. It means that application of uniformly in-plane magnetized films with perpendicular magnetocrystalline anisotropy due to the presence of local dispersion minima could enable to excite very short spin waves at experimentally available frequencies. The problem in such kind of structures for magnonic applications, however, is related to typically high damping of spin waves in multilayers with perpendicular magnetocrystalline anisotropy.

The second mode (M2/S2) is a perpendicular standing spin wave mode. This is an excitation of the exchange character with one nodal plane, see Fig. 14B. Note, that the amplitude of spin waves for higher wavevectors is slightly higher on the opposite surface than in the case of M1 (S1) (compare modes S1 and S2 in see Fig. 14B).

In the next step, we investigated how the spin wave dynamics are modified in the presence of a stripe domain structure. Let us first compare the typical BLS spectra at $H = 0$ for the wavevector perpendicular ($\varphi = 90°$) and parallel ($\varphi = 0°$) to the stripe domain axis, respectively (Fig. 13E and F), with BLS spectra for the uniformly magnetized film (Fig. 13D). It is visible, that for the aligned stripe domain magnetization pattern the typical BLS spectra are more complex and characterized by the presence of four peaks. The dispersion relations for these two orthogonal configurations ($\varphi = 90°$ and $\varphi = 0°$) differs, nevertheless, dispersions merge to the same frequencies as the wavevector approaches 0.

As it will be described in the subsequent subsections utilizing micromagnetic simulations results, the increase in the number of peaks as compared to the saturated state signals results from the periodical magnetization texture. In the case of spin waves propagation across domain walls, those magnetization non-uniformities are treated as regular scattering centers, which are provided by the domain boundaries, similar to the geometric magnetic boundaries of a lithographically patterned magnonic crystal. In the case of spin waves propagating parallel to the domain walls, those inhomogeneities are treated as different channels defined by different magnetization orientation. In both cases, periodicity is crucial to collect strong enough BLS signal from

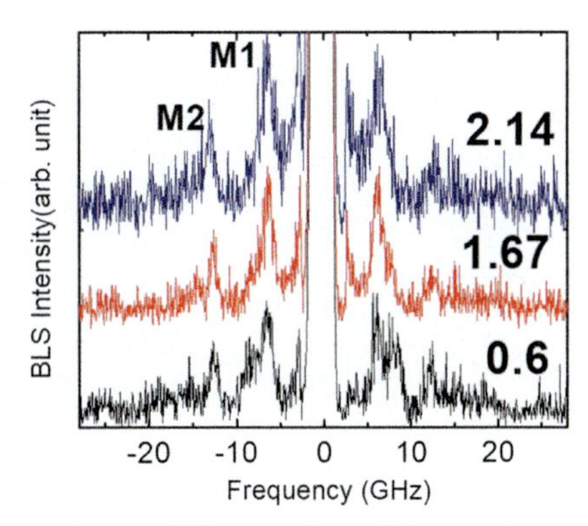

Fig. 15 The BLS spectra taken at different values of the in-plane transferred wavevector *k* (denoted in units of 10 rad/μm) for the sample with labyrinth stripe domains. The mode numbers are assigned in accordance with the aligned stripe domain sample.

the spin wave excitations and their influence on spin wave dynamics is described in subsequent paragraphs.

However, before that, we would like to briefly discuss BLS measurements of a labyrinth-like domain structure. BLS spectra measured for a few wavevectors (see Fig. 15) is very noisy and almost wavevector independent. The mode M1 splits for lower wavevectors and expect for that shoulder mode of M1, no other modes depends on the wavevector value. The asymmetry of the BLS intensities in the Stokes and anti-Stokes sides is not significantly affected by changing the wavevector. Moreover, modes M3 and M4 are not observed in the spectra. In consequence, spin-wave dispersion obtained for the periodically aligned stripe domain pattern is significantly different than recorded BLS spectra obtained for the labyrinth-like domain pattern, *i.e.*, it consists of more dispersive modes and the measured data is less noisy.

4.2.5 Spin waves propagation across stripe domain pattern: Micromagnetic study

To validate the experimental results and to understand the physical characteristics of the resonant modes we exploited micromagnetic simulations. Let us now consider the propagation of spin waves across domain walls, *i.e.*, for $\varphi = 90°$. The corresponding spin wave frequencies exhibit pronounced dispersion with wavevector, together with characteristic bandgaps (Fig. 16). The dispersion is periodic with the periodicity resulting from the magnetization pattern, *i.e.*, $G = 2\pi/a$ (in simulations the lattice constant was *ca.* 100 nm), see Fig. 16. The two, well pronounced, few-GHz-wide bandgaps are visible.

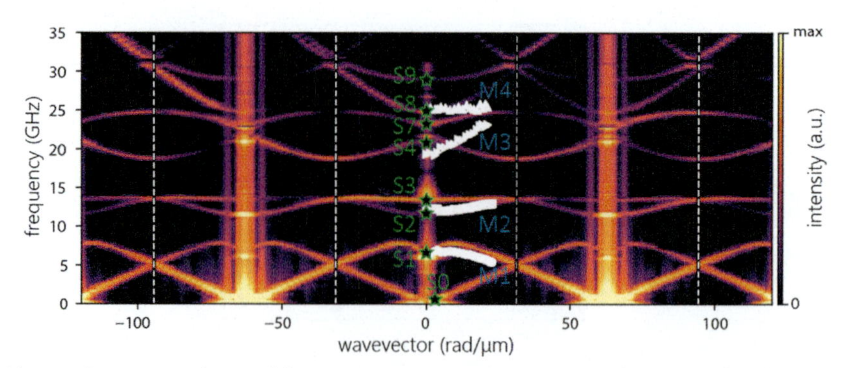

Fig. 16 Dispersion obtained for spin waves propagating across domain walls (along the *y*-axis). Colormap corresponds to the power of spin waves obtained from micromagnetic simulations. The white symbols represent results of BLS measurements and the black stars with captions mark spin wave modes plotted in Fig. 13E.

The nature of the above spin wave spectra can be better understood by carefully looking at the relevant cross-sectional spin precession profiles that are depicted in Fig. 17. The solid black lines represent static M_z component at the top surface of the film with the domain structure indicating the up and down domains for reference. The amplitude of the spin wave modes are localized mainly in the domain walls, the first two bands (S0 and S1) do not possess any nodal lines across the film's thickness, S2 and S3 have one nodal line, S8 and S9 have two nodal lines, and modes S4–S7 have more complex nature, probably resulting from their hybridization.

The dispersion line at low frequency for S0 mode is characterized by a linear slope approaching close to zero-frequency (its precise value is below frequency resolution) when the wavevector approaches $0 \pm \pi/a$. This points at a Goldstone mode connected with the domain wall oscillations [117] which is further confirmed by the dynamic magnetization profiles. The higher-order domain wall modes (S2, S3, S8, and S9) [118] or bulk spin waves modes (S4–S7) over periodic domain wall background are also present in the spectrum. Since the unit cell of the periodic structure is a host of two domain walls, the Goldstone mode is composed of two branches.

The zero modes of the domain wall oscillation (S0 and S1) can be understood considering the M_z component (Fig. 17). The S0 mode has dynamic magnetization z-component localized at the domain walls with opposite sign in neighboring walls. Since the walls limit domains with up or down orientation, the opposite sign of dynamic z-component indicates a uniform shift of the domains. The uniform shift of the domain does not change the total energy of the system.

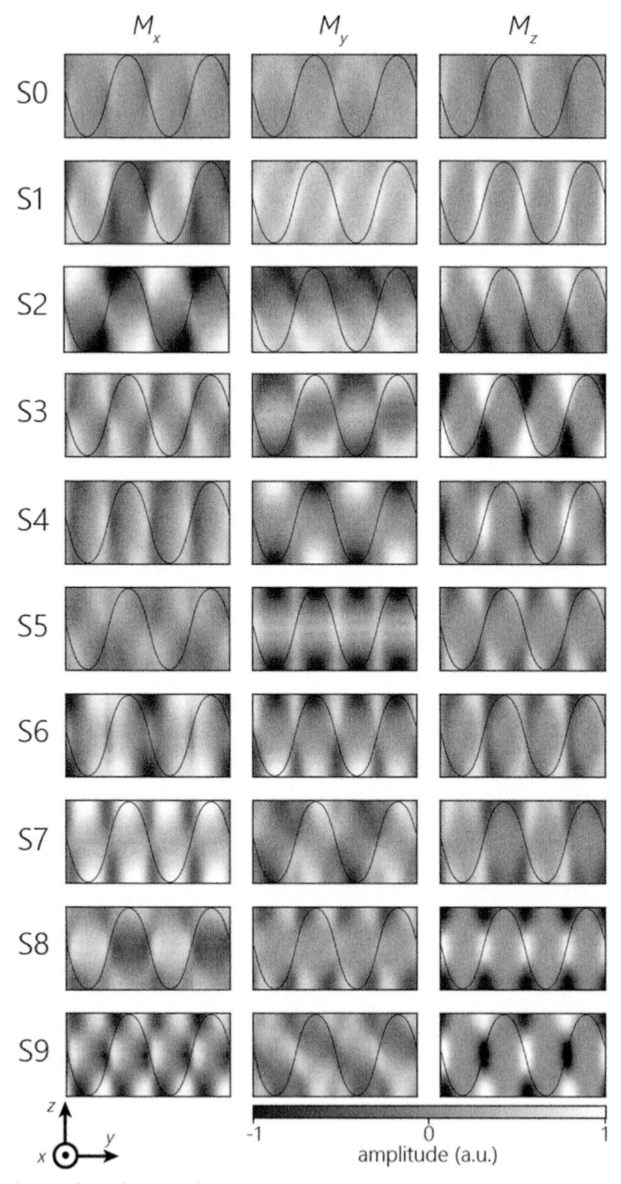

Fig. 17 Mode profiles for the first 9 modes calculated by micromagnetic simulations and plotted at $k=0$ (see black stars in Fig. 16) separately for all three components of the magnetization. Note, that due to numerical reasons, the S0 mode has been obtained for the first resolved wavevector greater than 0. The auxiliary sinusoidal black line corresponds to the out-of-plane component of the magnetization.

The S1 mode at $k=0$ is characterized by the in-phase oscillation of the magnetization z-component. Therefore, it is an oscillating shrinking and expanding of neighboring domains (breathing mode). The dispersion of S1 is non-monotonic, similarly to single domain state (Fig. 14). It can indicate a strong role of dipolar interactions and will be described in the next part.

Good agreement between BLS (M1, M2, and M4) and simulations (S1, S2, and S4) are present. The third measured band (M3), however, doesn't match so well with the simulated dispersion relation (S3), it is shifted down with respect to the nearest band with a similar slope. Simulations show also, that the horizontal dispersion line M4, being related to the perpendicular standing spin wave mode with two nodal lines across the thickness S8 is hybridized with the other dispersive modes S7. In BLS measurements, however, there is no indication for the hybridization. This discrepancy arises from a different lattice constant observed experimentally (*ca.* 120 nm) and relaxed in micromagnetic simulations (*ca.* 100 nm). It means that in the simulated structure (with the smaller lattice constant) the Brillouin zone is larger and, therefore, the back-folding takes place for the larger wavevectors. Consequently, the folded bands are shifted up with respect to the measured dispersion (for the system with smaller Brillouin zone). This explanation is schematically shown in Fig. 18.

4.2.5.1 The influence of domain wall structure

To demonstrate that spin wave band structure is determined not only by the lattice constant but also by the domain wall structure we have performed micromagnetic simulations for very similar system as in the previous subsection, *i.e.*, we have stabilized a periodic stripe domain pattern with the same lattice constant (*ca.* 100 nm), but with opposite polarity (see the cross-sectional view of the magnetic configuration in Fig. 19A). The subsequent domain walls have alternating polarity (they are directed in the $+x$ or $-x$ directions). In order to obtain such a configuration, the static magnetic configuration used in all the previous simulations has been numerically modified by flipping toward the $-x$ direction each second domain wall. Then, the obtained spin texture has been relaxed and the resulting system (with the same lattice constant) has been used in dynamic simulations leading to the dispersion relation.

The resulting band structure for this system with the thickness 38 nm is presented in Fig. 20A. We consider here a thinner sample than in the previous section in order to avoid overlapping with higher-order modes that are

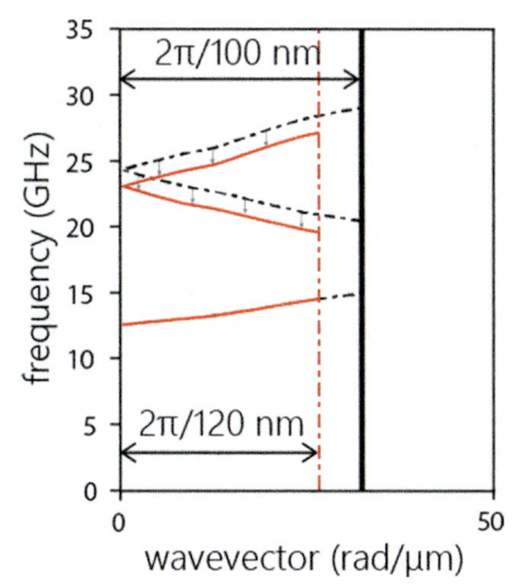

Fig. 18 Schematic representation of the down-shift of the folded bands resulting from different lattice constant in simulations and experimentally measured stripe domain pattern. The dash-dotted line represents dispersion relation of the band with one nodal line across the film's thickness for the simulated lattice constant (this line is taken directly from the simulations). The solid, red line represents the same dispersion but folded for shorter Brillouin zone (2π/120 nm instead of the simulated 2π/100 nm). This earlier back-folding causes down-shift of the frequencies of the folded band.

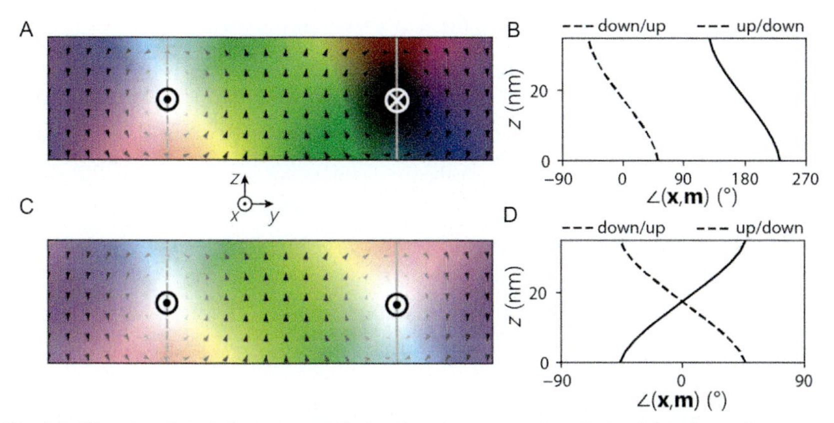

Fig. 19 The simulated domain configuration (cross-sectional view) for the subsequent domain walls with (A) the alternating and (B) the same polarities. (B) and (D) the rotation of the magnetic moments with respect to the *x*-axis plotted across the thickness (see the vertical lines in (A) and (C)) for "down/up" (dashed line) and "up/down" (solid lines) domains for the case from (A) and (C), respectively.

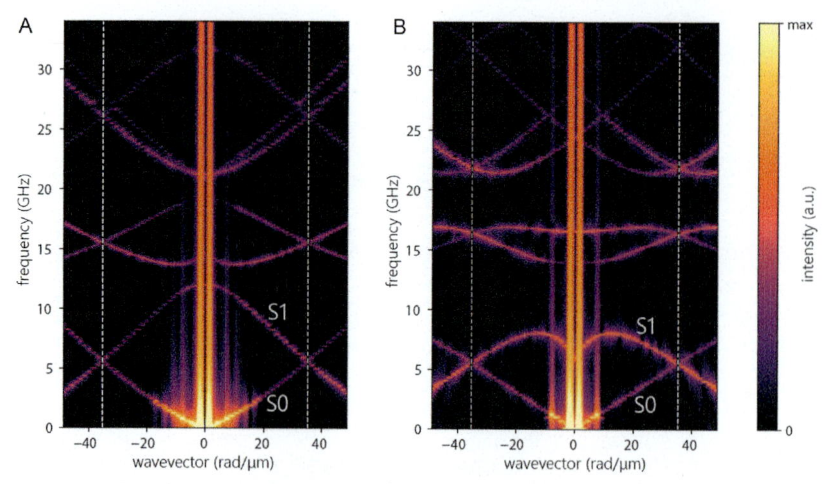

Fig. 20 Simulated dispersion relations obtained for stripe domain pattern with domain walls of (A) the alternating and (B) the same polarities with marked bands S0 and S1.

present for 42.5 nm sample, when the polarity of neighboring walls is set opposite. For the sake of comparison, in Fig. 20B is shown the dispersion obtained for stripe domain pattern of the same thickness (38 nm) but with the same polarity of all domain walls. It is visible, that for the case of the structure with the same polarity of the domain walls (Fig. 20B) the non-monotonic dispersion of the S1 mode is present, whereas for the case with the opposite polarity of the domain walls S1 band is monotonic (Fig. 20A). We also note that the mode S0 in the structure with the same polarity (Fig. 20B) has a finite frequency at $k=0$. This effect can be associated with the intermediate static configuration, *i.e.*, since the polarity of the neighboring walls is the same and domain walls are relatively close to each other, there is always a finite static x-component of the magnetization (*i.e.*, M_x never reaches zero). The static configuration is an intermediate state between the single domain state and the domain wall. Therefore, we expect that the frequency of the mode will not be determined by the zero-mode related to the translation of the domain walls, but it can exhibit some properties of the single-domain excitation.

We can infer that the dipole interactions play an important role in separation between the S0 and S1 modes at $k=0$. As discussed in [95], when the dipole interactions are not present, these two modes should be degenerated, since two modes are symmetric to each other. The degeneracy is lifted by introducing dipole interactions since frequency repulsion between these modes occurs. We observe well-separated frequencies for antiparallel configuration and smaller separation at $k = 0$ when polarizations are parallel. This suggests, that the dipole interactions are larger when domain walls are oppositely polarized. In Fig. 19B and D, the angle of the magnetization moments with respect to the x-axis at the center of the domain walls across the thickness are plotted for both structures, respectively. The important difference is the relative alignment of the spins. In the case of the structure with domain walls of the same polarity, the relative spin orientation in the neighboring with respect to each other domains walls is changing and the magnetic moments are parallel only in the domain wall center. On the contrary, in the case of the oppositely polarized domain walls, the magnetic moments across the domain wall's cross-section are antiparallel (shifted by 180 degree). As a result, the dipole interaction strength is higher for the parallel alignment of spins than any other alignment. Therefore, the dipole interactions between the domain walls are stronger when the neighboring wall polarity is opposite and it causes a larger frequency difference between S0 and S1 at $k = 0$.

4.2.6 Spin waves propagation along stripe domain pattern: Micromagnetic study

Let us now consider the propagation of spin waves along domain walls, *i.e.*, for $\varphi = 0°$, see Fig. 21A. Very good agreement between BLS (M1, M2) and simulation (S1 and S2) results for the first two bands is found, also the fourth band (M4 with S4) match well, apart from the small wavenumbers. Noteworthy, the band M3 doesn't match with simulations results, however, as it is described in the previous subsection, it is related to the discrepancy between the lattice constant of experimentally measured structure and the spin texture relaxed in simulations, *i.e.*, 120 nm and 100 nm, respectively.

The spatial distribution of the dynamic magnetization is illustrated in Fig. 21B. The modes are localized in the domain walls and in some cases also have amplitude at the top and bottom surfaces of the structure. The nature of these spin-wave excitations can be clarified by looking at the relevant cross-sectional spin precession profiles calculated at $k = 30\,\mathrm{rad}/\mu\mathrm{m}$, which are depicted in Fig. 21B. The solid black lines represent static M_z component at the top surface of the film with the domain structure, indicating the up and down domains for reference. The amplitude of the first two modes (S0 and S1), which are uniform across the thickness, is concentrated in alternate domain walls. Although the mode profiles of spin waves from bands S0 and S1 are similar, their dispersion, especially for small wavevectors, differs very much. The dispersion of S1 is very similar to

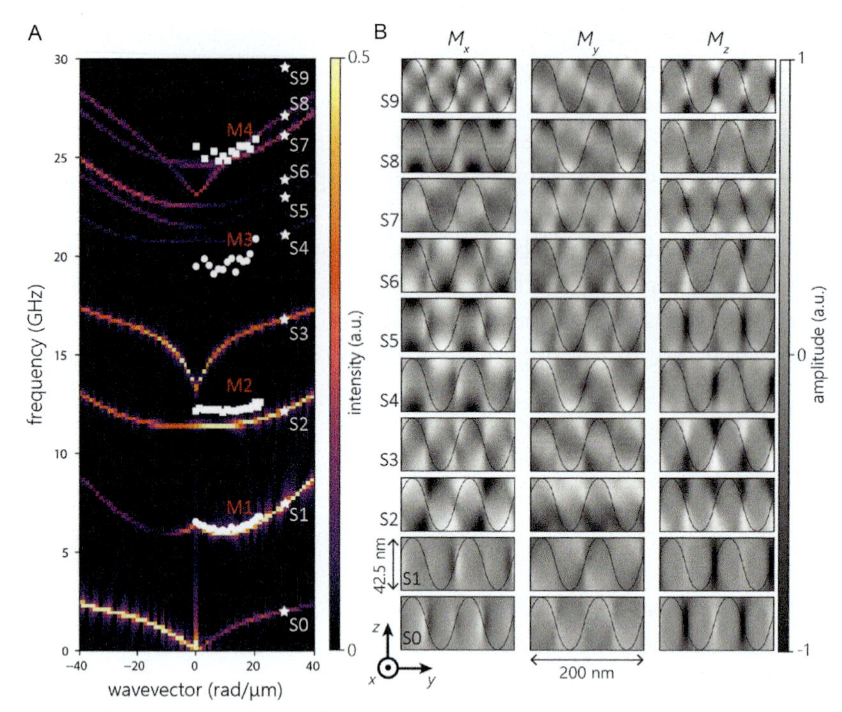

Fig. 21 (A) Dispersion relation for spin waves propagation along domain walls. The intensity of colors corresponds to the amplitude of spin waves. (B) The cross-sectional view of the first 10 mode profiles obtained for spin waves propagation parallel to the domain walls presented for all three components of the magnetization calculated at the wavevector equal $k = 30\,\mathrm{rad}/\mu\mathrm{m}$. The grayscale corresponds to the real part of the amplitude and the dark solid line (sin function) presents the z-component of the static magnetization at the top surface of the structure. It is visible that in the presented range of frequencies all modes are localized in the region of domain walls (some modes possess also amplitude at the top/bottom surfaces of the structure).

the magnetostatic backward volume spin waves with the local minima for finite wavevectors. S0 mode, from the other hand, for $k = 0$ approaches with the linear slope zero-frequency (see dispersion Fig. 22A). It is also supported by the asymmetry of the intensities of the first two branches in the dispersion relation (derived along one of the domain walls), S0 is more intense for the negative wavevectors, whereas S1 is brighter for positive. This localization in the alternating domain walls is also present for modes S4–S6. Modes S2, S3, S6, and S8 possess one nodal line, whereas mode S7 and S9 have two nodal lines. Modes S2 and S3 (with a single nodal plane) similarly like modes S7 and S9 (with two nodal planes) can be classified as symmetric and antisymmetric, respectively. Overall, the non-collinearities of the magnetization

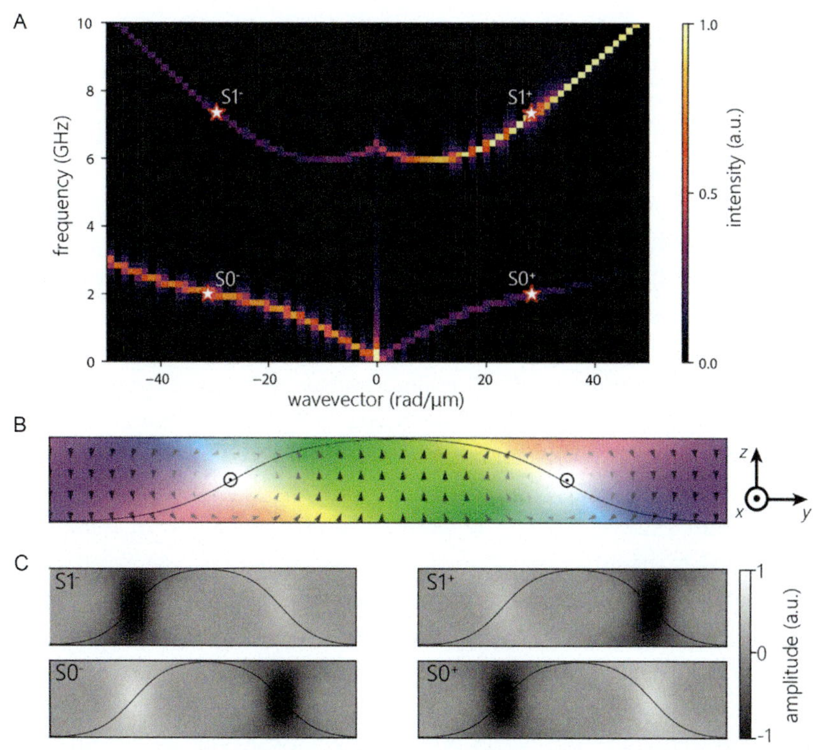

Fig. 22 (A) Dispersion relation obtained from magnetostatic simulations for the spin waves propagating along domains (along the *y*-axis). (B) Cross-sectional view across the film's thickness showing the static magnetic configuration. The left domain wall possesses clockwise chirality whereas the right one the counterclockwise chirality. (C) The mode profiles (real part) of spin waves from the first and second band (S0 and S1) calculated at $k = 30\,\mathrm{rad\,\mu m^{-1}}$. The black solid sinusoidal lines in (B) and (C) show M_z dependence on *x* plotted for the top surface of the system.

texture induced by domain walls serve as different channels defined by different magnetization orientation through which multimodal transmission is possible.

Let us now focus on the first two modes, *i.e.*, S0 and S1 that have similar amplitude profiles. Profiles of the M_z component of these modes are depicted in Fig. 22C for $k = -30\,\mathrm{rad\,\mu m^{-1}}$ and $k = 30\,\mathrm{rad\,\mu m^{-1}}$ and labeled as $\mathrm{S0}^-(\mathrm{S1}^-)$ and $\mathrm{S0}^+(\mathrm{S1}^+)$, respectively. It is visible, that these modes have the opposite signs of amplitude in the subsequent domain walls (intense black and slightly brighter regions in domain wall) and the amplitude is higher in a domain wall of one type of chirality, *i.e.*, spin waves from $\mathrm{S0}^-$ band prefers to propagate in clockwise type domain walls (in the $-y$

direction) whereas the spin waves from $S0^+$ band prefers to propagate in the counterclockwise type domain walls (in the $+y$ direction). In the case of $S1^-$ and $S1^+$ it is the opposite scenario, spin waves from $S1^-$ are confined in the counterclockwise type of the domain walls whereas from $S1^+$ in the clockwise type domain walls.

In order to demonstrate some functionalities related to the localization properties of the first two bands let us show the results of the micromagnetic simulation. In the simulations, the spin wave source working at frequency 3 and 7 GHz has been employed to excite waves propagating along the domain walls, see Fig. 23A and B, respectively. In these figures is displayed the dynamic x-component of the magnetization after 1 ns of continuous spin wave excitation, *i.e.*, $M_x(\mathbf{r};t=1\text{ ns})-M_x(\mathbf{r};t=0)$. The antenna used for the excitation is placed in the central part of the simulated region (indicated by the bright vertical line).

For spin waves excited at frequency 3 GHz unidirectional (diode-like) propagation in the subsequent domain walls is visible (Fig. 23A). Similar localization in every second domain wall has been also observed in Refs. [95,110]. For the higher frequency (7 GHz) there are two modes (S0 and S1) which, however, propagate in the opposite directions for domain walls of the same chirality. For instance, it is visible that for the domain wall located at the bottom of the figure spin waves propagating leftward (in the $-x$ direction)

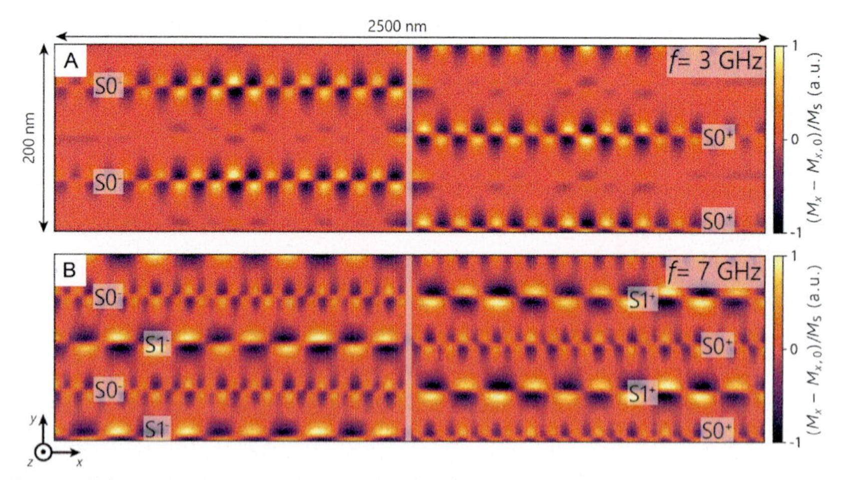

Fig. 23 Colormaps depicting the amplitude of spin waves at $z=21$ nm (middle cross-section) excited at the frequency (A) 3 GHz and (B) 7 GHz in the region marked as the bright vertical line. Unidirectional propagation is visible.

are longer than the waves propagating in the opposite direction. The situation is opposite for the nearest neighboring domain walls, the waves propagating rightward are longer. Schematically, it can be described that the wavenumbers of spin waves propagating in alternating domain walls obey $k_{CW}(f) = -$-$k_{CCW}(f)$, where k_{CW} and k_{CCW} are wavenumbers of dominating modes of spin waves in clockwise and counterclockwise domain walls, respectively. It is important to point out, that this feature doesn't arise from the dispersion nonreciprocity but from localization properties of the first two bands.

The spin wave confinement is explained in details in Ref. [95], where the nonreciprocity of spin waves propagating along a single domain wall confined in a narrow stripe[a] is theoretically found. According to Henry et al., the conditions similar to DE nonreciprocal wave can arise there. That is, spin waves propagate perpendicularly to the magnetization orientation in the domain wall and breaking of the symmetry occurs. The wave propagating in opposite directions along the domain wall generates dynamic magnetization components that are oriented differently with respect to the domain wall (see fig. 2B in Ref. [95]). In result, the amount of volume charges generated by the oppositely propagating spin waves along the Bloch domain wall is different (that breaks symmetry perpendicular to both magnetization and direction of propagation) what induces the nonreciprocal effect.

In the same paper [95], a pair of domain walls with an opposite chirality in the confined stripe is considered. In the limit of non-interacting domain walls, it leads to the presence of two branches of nonreciprocal dispersions that are mirror images at $k = 0$ (see fig. 3B in ref. [95]). However, when dipole interactions are included, the bandgap at the crossing of the branches at $k = 0$ occurs and it allows unidirectional propagation within some frequency range.

The main difference between our system and the structure studied by Henry et al. is that in our Co/Pd stack the domains are not confined in a finite strip. In result, the lowest mode starts at a frequency close to zero and its dispersion is monotonic. Therefore, the region of the unidirectional channel formation spans between zero and the lowest frequency of S1 mode.

[a] Henry et al. [95] consider a thinner material (15 nm) with higher quality factor ($Q = 1.6$). In consequence, the stabilized domain walls were more uniform across the thickness than in our case, reminding Bloch-type domain walls.

5. Summary

We summarized studies on? the magnetization dynamics in periodic magnetic stripe domain patterns in thin ferromagnetic films. First, we have described how magnetic interactions can lead to domain wall formation. The presence of this non-collinear spin texture leads to a decrease in the effective magnetic field. That in turn, acts on the spin wave as a potential well and results in spin wave localization within the domain wall. The magnetization dynamics in a single domain wall was described for various configurations of magnetic domains (in-plane and out-of-plane) and domain walls (Bloch and Néel-type domain walls). In some of the studied cases, strong nonreciprocal propagation is exhibited. Further, the spin wave propagation across the wall was discussed, in particular, reflectionless propagation across the Bloch wall.

The contributions of magnetic anisotropy and dipole interactions can also favor the presence of multiple domains, with characteristic width of the wall and width of the domain. We have shown that gradually reducing the in-plane magnetic field applied to a multilayer system composed of Pd/Co can lead to the formation of magnetic domains aligned along the same axis. The magnetization configuration takes the form of a periodically repeated array of domains, separated by domain walls. We have discussed the static and dynamic properties of these magnetic states. It has been shown that the magnetization configuration is non-collinear across the domain walls and across the thickness of the film. It has the form of a "corkscrew"-like structure, with Néel-type wall near the film surfaces and Bloch-type wall at the center. The history of the magnetic field influences the polarity of the Bloch-type walls, which is the same for all walls. The BLS measurements were performed to study magnetization dynamics experimentally and the results were interpreted with the use of micromagnetic simulations (Mumax3). The periodic arrangement of the magnetization increases the number of spin wave bands in the system. In the case of spin waves propagation across domain walls, those magnetization non-uniformities are treated as regular scattering centers, which are provided by the domain boundaries, similar to a one-dimensional magnonic crystal. In the case of spin waves propagating parallel to the domain walls, those inhomogeneities are treated as different channels defined by different magnetization orientation. We have shown the properties of the dynamical excitation, related to translational motion of the domain wall (zero-frequency Goldstone modes). Further, we looked at the dynamics of the magnetization configurations where the polarity of the domain wall is alternating in each neighboring wall. The simulations show that the dynamical coupling between the walls is enhanced as compared to domain walls with the same polarity. We also show the domain wall dynamics along its direction. Here, the interaction between the walls and nonreciprocal properties (described earlier for the single wall) result in the formation of unidirectional channels, where, waves travel in every second wall in the opposite direction.

The non-collinear periodic structures are of great interest for the magnonics community. They allow forming reconfigurable, nanosize magnonic crystals in inhomogeneous ferromagnetic films. Since the periodic structure is generated spontaneously (or by external field manipulation), the magnonic crystal is quasi self-organized without the use of modern patterning techniques. This allows generating structures with lattice constant below the lithography limit, without defects. The spin waves localized in domain walls can exhibit unidirectional propagation, which is an important step toward information processing based on a robust host architecture for spin waves.

Acknowledgments

We acknowledge funding from the EU Horizon 2020 project MagIC No. 644348 and National Science Centre of Poland Grant No. 2018/30/Q/ST3/00416. The simulations were partially performed at the Poznan Supercomputing and Networking Center (Grant No. 398). M.M. thanks to Alexander Sadovnikov for discussion and acknowledge the funding from the project: Era.Net RUS Plus (TSMFA).

References

[1] J. Bardeen, W.H. Brattain, The transistor, a semi-conductor triode, Phys. Rev. 75 (1948) 1208.
[2] Semiconductor, Industry, and Association, International Technology Roadmap for Semiconductors, http://www.itrs.net/Links/2011ITRS/2011Chapters/2011ERD.pdf, 2011.
[3] S.S. Oh, C.S. Kee, J.-E. Kim, et al., Duplexer using microwave photonic band gap structure, Appl. Phys. Lett. 76 (2000) 2301.
[4] A. D'Orazio, M.D. Sario, V. Petruzzelli, F. Prudenzano, Photonic band gap filter for wavelength division multiplexer, Opt. Express 11 (2003) 230.
[5] M. Loncar, T. Doll, J. Vuckovic, A. Scherer, Design and fabrication of silicon photonic crystal optical waveguides, J. Lightwave Technol. 18 (10) (2000).
[6] J. Li, K.S. Chiang, Guided modes of one-dimensional photonic bandgap waveguides, J. Opt. Soc. Am. B 24 (2007) 1942.
[7] J. Li, K.S. Chiang, Light guidance in a photonic bandgap slab waveguide consisting of two different Bragg reflectors, Opt. Commun. 281 (2008) 5797.
[8] J. Li, T.P. White, L. O'Faolain, et al., Systematic design of flat band slow light in photonic crystal waveguides, Opt. Express 16 (2008) 6227.
[9] J.C. Knight, Photonic crystal fibres, Nature 424 (2003) 847.

[10] A. Petrov, M. Eich, Dispersion compensation with photonic crystal line-defect waveguides, IEEE J. Sel. Areas Commun. 23 (2005) 1396.
[11] S.J. McNab, N. Moll, Y.A. Vlasov, Ultra-low loss photonic integrated circuit with membrane-type photonic crystal waveguides, Opt. Express 11 (2003) 2927.
[12] F.V. Laere, T. Stomeo, C. Cambournac, et al., Nanophotonic polarization diversity demultiplexer chip, J. Lightwave Technol. 27 (2009) 17.
[13] F.V. Laere, G. Roelkens, M. Ayre, et al., Compact and highly efficient grating couplers between optical fiber and nanophotonic waveguides, J. Lightwave Technol. 25 (2007) 151.
[14] A. Barman, J. Sinha, Spin Dynamics and Damping in Ferromagnetic Thin Films and nanostructures, Springer, Switzerland, 2018.
[15] S.A. Wolf, D.D. Awschalom, R.A. Buhrman, et al., Spintronics: a spin-based electronics vision for the future, Science 294 (2001) 1488.
[16] S. Tehrani, E. Chen, M. Durlam, et al., High density submicron magnetoresistive random access memory, J. Appl. Phys. 85 (1999) 5822.
[17] A. Imre, G. Csaba, L. Ji, et al., Majority logic gate for magnetic quantum-dot cellular automata, Science 311 (2006) 205.
[18] D.A. Allwood, G. Xiong, C.C. Faulkner, et al., Magnetic domain-wall logic, Science 309 (2005) 1688.
[19] K.-S. Lee, D.-S. Han, S.-K. Kim, Physical origin and generic control of magnonic band gaps of dipole-exchange spinwaves in width-modulated nanostrip waveguides, Phys. Rev. Lett. 102 (2009) 127202.
[20] Y.L. Etko, A.B. Ustinov, Broadband spin-wave delay lines with slot antennas, Tech. Phys. Lett. 37 (2011) 1015.
[21] D. Kumar, S. Barman, A. Barman, Magnetic vortex based transistor operations, Sci. Rep. 4 (2014) 4108.
[22] S.S.P. Parkin, M. Hayashi, L. Thomas, Magnetic domain-wall racetrack memory, Science 320 (2008) 190.
[23] T. Thomson, G. Hu, B.D. Terris, Intrinsic distribution of magnetic anisotropy in thin films probed by patterned nanostructures, Phys. Rev. Lett. 96 (2006) 257204.
[24] M.N. Baibich, J.M. Broto, A. Fert, et al., Giant magnetoresistance of (001) Fe/(001) Cr magnetic superlattices, Phys. Rev. Lett. 61 (1988) 2472.
[25] G. Binasch, P. Grünberg, F. Saurenbach, W. Zinn, Enhanced magnetoresistance in layered magnetic structures with antiferromagnetic interlayer exchange, Phys. Rev. B 39 (1989) 4828.
[26] L. Berger, Emission of spin waves by a magnetic multilayer traversed by a current, Phys. Rev. B 54 (1996) 9353.
[27] J. Slonczewski, Current-driven excitation of magnetic multilayers, J. Magn. Magn. Mater. 159 (1996) L1.
[28] J.E. Hirsch, Spin Hall effect, Phys. Rev. Lett. 83 (1999) 1834.
[29] Y.K. Kato, R.C. Myers, A.C. Gossard, D.D. Awschalom, Observation of the Spin Hall effect in semiconductors, Science 306 (2004) 1910.
[30] A. Ganguly, R.M. Rowan-Robinson, A. Haldar, S. Jaiswal, J. Sinha, A.T. Hindmarch, D.A. Atkinson, A. Barman, Time-domain detection of current controlled magnetization damping in Pt/Ni81Fe19 bilayer and determination of Pt spin hall angle, Appl. Phys. Lett. 105 (2014) 112409.
[31] A. Brataas, A.D. Kent, H. Ohno, Current-induced torques in magnetic materials, Nat. Mater. 11 (2012) 372.
[32] S.N. Panda, S. Mondal, J. Sinha, S. Choudhury, A. Barman, All-optical detection of interfacial spin transparency from spin pumping in β-Ta/CoFeB thin films, Sci. Adv. 5 (2019) eaav7200.
[33] K. Uchida, S. Takahashi, K. Harii, et al., Observation of the spin Seebeck effect, Nature 455 (2008) 778.

[34] I.E. Dzyaloshinskii, Thermodynamic theory of weak ferromagnetism in antiferromagnetic substances, Sov. Phys. - JETP 5 (1957) 1259.
[35] T. Moriya, Anisotropic superexchange interaction and weak ferromagnetism, Phys. Rev. 120 (1960) 91.
[36] S. Emori, U. Bauer, S.-M. Ahn, et al., Current-driven dynamics of chiral ferromagnetic domain walls, Nat. Mater. 12 (2013) 611.
[37] A.I. Akhiezer, V.G. Bar'yakhtar, S.V. Peletminskii, Spin Waves, North-Holland, Amsterdam, 1968.
[38] A.G. Gurevich, G.A. Melkov, Magnetization Oscillations and Waves, Chemical Rubber Corp, New York, 1996.
[39] F. Bloch, Zur theorie des ferromagnetismus, Z. Phys. 61 (1930) 206.
[40] S.A. Nikitova, P. Tailhadesa, C.S. Tsai, Spin waves in periodic magnetic structures-magnonic crystals, J. Magn. Magn. Mater. 236 (2001) 320.
[41] J. Jorzick, S.O. Demokritov, B. Hillebrands, et al., spin wave wells in nonellipsoidal micrometer size magnetic elements, Phys. Rev. Lett. 88 (4) (2002).
[42] G. Gubbiotti, L. Albini, G. Carlotti, et al., Finite size effects in patterned magnetic permalloy films, J. Appl. Phys. 87 (2000) 5633.
[43] B. Hillebrands, C. Mathieu, C. Hartmann, et al., Static and dynamic properties of patterned magnetic permalloy films, J. Magn. Magn. Mater. 175 (1997) 10.
[44] M. Grimsditch, Y. Jaccard, I.K. Schuller, Magnetic anisotropies in dot arrays: shape anisotropy versus coupling, Phys. Rev. B 58 (1998) 11539.
[45] S. Saha, R. Mandal, S. Barman, D. Kumar, B. Rana, Y. Fukuma, S. Sugimoto, Y. Otani, A. Barman, Tunable magnonic spectra in two-dimensional magnonic crystals with variable lattice symmetry, Adv. Funct. Mater. 23 (2013) 2378–2386.
[46] S. Choudhury, S. Barman, Y. Otani, A. Barman, Efficient modulation of spin waves in two-dimensional octagonal magnonic crystal, ACS Nano 11 (2017) 8814–8821.
[47] S. Mühlbauer, B. Bin, F. Jonietz, et al., Skyrmion lattice in a chiral magnet, Science 323 (2009) 915.
[48] A. Fert, V. Cros, J. Sampaio, Skyrmions on the track, Nat. Nanotechnol. 8 (2013) 152.
[49] R.F. Wang, C. Nisoli, R.S. Freitas, et al., Artificial 'spin ice' in a geometrically frustrated lattice of nanoscale ferromagnetic islands, Nature 439 (2006) 303.
[50] G. Moller, R. Moessner, Artificial square ice and related dipolar nanoarrays, Phys. Rev. Lett. 96 (2006) 237202.
[51] A.V. Chumak, A.A. Serga, B. Hillebrands, Magnon transistor for all-magnon data processing, Nat. Commun. 5 (2014) 4700.
[52] A.A. Nikitin, A.B. Ustinov, A.A. Semenov, et al., A spin-wave logic gate based on a width-modulated dynamic magnonic crystal, Appl. Phys. Lett. 106 (2015) 102405.
[53] K. Wagner, A. Kákay, K. Schultheiss, et al., Magnetic domain walls as reconfigurable spin-wave nanochannels, Nat. Nanotechnol. 11 (2016) 432.
[54] R. Hertel, W. Wulfhekel, J. Kirschner, Domain-wall induced phase shifts in spin waves, Phys. Rev. Lett. 93 (2004) 257202.
[55] C. Bayer, H. Schultheiss, B. Hillebrands, R. Stamps, Phase shift of spin waves traveling through a 180° Bloch-domain wall, IEEE Trans. Magn. 41 (2005) 3094.
[56] P. Pirro, T. Koyama, T. Brächer, et al., Experimental observation of the interaction of propagating spin waves with Néel domain walls in a Landau domain structure, Appl. Phys. Lett. 106 (2015) 232405.
[57] K. Di, V.L. Zhang, H.S. Lim, et al., Asymmetric spin-wave dispersion due to Dzyaloshinskii-Moriya interaction in an ultrathin Pt/CoFeB film, Appl. Phys. Lett. 106 (2015) 052403.
[58] A.K. Chaurasiya, C. Banerjee, S. Pan, et al., Direct observation of interfacial Dzyaloshinskii-Moriya interaction from asymmetric spin-wave propagation in

W/CoFeB/SiO_2 heterostructures down to sub-nanometer CoFeB thickness, Sci. Rep. 6 (2016) 32592.
[59] P. Weiss, L'hypothèse du Champ Moléculaire et la Propriété Ferromagnétique, J. Phys. Theor. Appl. 6 (1907) 661.
[60] C. Kittel, On the theory of ferromagnetic resonance absorption, Phys. Rev. 73 (1948) 155.
[61] A.H. Morrish, The Physical Principles of Magnetism, IEEE Press, New York, 2001.
[62] T. Moriya, Anisotropic superexchange interaction and weak ferromagnetism, Phys. Rev. 120 (1960) 91.
[63] P. Bruno, Physical origins and theoretical models of magnetic anisotropy, Magnetismus von Festkörpern und grenzflächen 24 (1993) 1–28.
[64] N. Nakajima, T. Koide, T. Shidara, et al., Perpendicular magnetic anisotropy caused by interfacial hybridization via enhanced orbital moment in Co/Pt multilayers: magnetic circular X-ray dichroism study, Phys. Rev. Lett. 81 (1998) 5229.
[65] S. Pal, B. Rana, O. Hellwig, et al., Tunable magnonic frequency and damping in $[Co/Pd]_8$ multilayers with variable Co layer thickness, Appl. Phys. Lett. 98 (2011) 082501.
[66] A. Hubert, Domain wall structures in thin magnetic films, IEEE Trans. Magn. 11 (1975) 1285.
[67] D. Navas, C. Redondo, G.A.B. Confalonieri, et al., Domain-wall structure in thin films with perpendicular anisotropy: magnetic force microscopy and polarized neutron reflectometry study, Phys. Rev. B 90 (2014) 054425.
[68] C. Banerjee, P. Gruszecki, J.W. Klos, et al., Magnonic band structure in a Co/Pd stripe domain system investigated by Brillouin light scattering and micromagnetic simulations, Phys. Rev. B 96 (2017) 024421.
[69] A. Thiaville, S. Rohart, É. Jué, et al., Dynamics of Dzyaloshinskii domain walls in ultrathin magnetic films, Europhys. Lett. 100 (2012) 57002.
[70] L. Landau, E. Lifshits, On the theory of the dispersion of magnetic permeability in ferromagnetic bodies, Phys. Z. Sowjet 8 (1935) 153.
[71] T. Gilbert, A phenomenological theory of damping in ferromagnetic materials, IEEE Trans. Magn. 40 (2004) 3443.
[72] H. Suhl, Theory of the magnetic damping constant, IEEE Trans. Magn. 34 (1998) 1834.
[73] V. Kambersky, C. Patton, Spin-wave relaxation and phenomenological damping in ferromagnetic resonance, Phys. Rev. B 11 (1975) 2668.
[74] B. Heinrich, R. Urban, G. Woltersdorf, Relaxation in metallic films: single and multilayer structures, J. Appl. Phys. 91 (2002) 7523.
[75] J.Z. Sun, B. Ozyilmaz, W. Chen, et al., Spin-transfer-induced magnetic excitation: the role of spin-pumping induced damping, J. Appl. Phys. 97 (2005) 10C714.
[76] R. Arias, D.L. Mills, Extrinsic contributions to the ferromagnetic resonance response of ultrathin films, Phys. Rev. B 60 (1999) 7395.
[77] H. Schultheiss, K. Vogt, B. Hillebrands, Direct observation of nonlinear four-magnon scattering in spin-wave microconduits, Phys. Rev. B 86 (2012) 054414.
[78] R.W. Damon, J.R. Eshbach, Magnetostatic modes of a ferromagnetic slab, J. Phys. Chem. Solids 19 (1961) 308.
[79] S.O. Demokritov, B. Hillebrands, A.N. Slavin, Brilluoin light scattering studies of confined spin waves: linear and non linear confinement, Phys. Rep. 348 (2001) 441.
[80] B.A. Kalinikos, A.N. Slavin, Theory of dipole-exchange spin wave spectrum for ferromagnetic films with mixed exchange boundary conditions, J. Phys. C 19 (1986) 7013.
[81] M. Cottam, D. Lockwood, Light Scattering in Magnetic Solids, Wiley, New York, 1986.
[82] K.Y. Guslienko, A.N. Slavin, Boundary conditions for magnetization in magnetic nanoelements, Phys. Rev. B 72 (2005) 014463.

[83] R. Mandal, S. Saha, D. Kumar, et al., Optically induced tunable magnetization dynamics in nanoscale co antidot lattices, ACS Nano 6 (2012) 3397.
[84] C. Yu, M.J. Pechan, W.A. Burgei, G.J. Mankey, Lateral standing spin waves in permalloy antidot arrays, J. Appl. Phys. 95 (2004) 6648.
[85] Z.K. Wang, V.L. Zhang, H.S. Lim, et al., Nanostructured magnonic crystals with size-tunable bandgaps, ACS Nano 4 (2010) 643.
[86] A. Vansteenkiste, J. Leliaert, M. Dvornik, M. Helsen, F. Garcia-Sanchez, B. Van Waeyenberge, The design and verification of MuMax3, AIP Adv. 4 (2014) 107133.
[87] S. Saha, S. Barman, Y. Otani, A. Barman, All-optical investigation of tunable picosecond magnetization dynamics in ferromagnetic nanostripes with a width down to 50 nm, Nanoscale 7 (2015) 18312–18319.
[88] L. Körber, K. Wagner, A. Kákay, H. Schultheiss, Spin-wave reciprocity in the presence of Néel walls, IEEE Magn. Lett. 8 (2017) 1–4.
[89] K. Wagner, Spin-wave Generation and Transport in Magnetic Microstructures, PhD diss, Technische Universität Dresden, 2018.
[90] R. Sbiaa, H. Meng, S.N. Piramanayagam, Materials with perpendicular magnetic anisotropy for magnetic random access memory, Phys. Status Solidi 5 (2011) 413–419.
[91] A. Hubert, R. Schäfer, Magnetic Domains: The Analysis of Magnetic Microstructures, Springer Science & Business Media, Berlin, 2008.
[92] F. Garcia-Sanchez, P. Borys, R. Soucaille, J.-P. Adam, R.L. Stamps, J.-V. Kim, Narrow magnonic waveguides based on domain walls, Phys. Rev. Lett. 114 (2015) 247206.
[93] J.M. Winter, Bloch wall excitation. Application to nuclear resonance in a Bloch wall, Phys. Rev. 124 (1961) 452.
[94] H. De Leeuw, R. van den Doel, U. Enz, Dynamic properties of magnetic domain walls and magnetic bubbles, Rep. Prog. Phys. 43 (1980) 689.
[95] Y. Henry, D. Stoeffler, J.-V. Kim, M. Bailleul, Unidirectional spin-wave channeling along magnetic domain walls of Bloch type, Phys. Rev. B 100 (2019) 024416.
[96] R. Hertel, W. Wulfhekel, J. Kirschner, Domain-wall induced phase shifts in spin waves, Phys. Rev. Lett. 93 (2004) 257202.
[97] C. Bayer, H. Schultheiss, B. Hillebrands, R.L. Stamps, Phase shift of spin waves traveling through a 180° Bloch-domain wall, IEEE Trans. Magn. 41 (2005) 3094–3096.
[98] P. Borys, F. Garcia-Sanchez, J.-V. Kim, R.L. Stamps, Spin-wave eigenmodes of Dzyaloshinskii domain walls, Adv. Electron. Mater. 2 (2016) 1500202.
[99] S. Macke, D. Goll, Transmission and reflection of spin waves in the presence of Néel walls, J. Phys. Conf. Ser. 200 (2010) 042015.
[100] N.J. Whitehead, S.A.R. Horsley, T.G. Philbin, A.N. Kuchko, V.V. Kruglyak, Theory of linear spin wave emission from a Bloch domain wall, Phys. Rev. B 96 (2017) 064415.
[101] X.S. Wang, X.R. Wang, Spin wave emission in field-driven domain wall motion, Phys. Rev. B 90 (18) (2014) 184415.
[102] B. Van de Wiele, S.J. Hämäläinen, P. Baláž, F. Montoncello, S. Van Dijken, Tunable short-wavelength spin wave excitation from pinned magnetic domain walls, Sci. Rep. 6 (2016) 21330.
[103] N.N. Dadoenkova, Y.S. Dadoenkova, I.L. Lyubchanskii, M. Krawczyk, K.Y. Guslienko, Inelastic spin-wave scattering by Bloch domain wall flexure oscillations, Phys. Status Solidi RRL 2019 (2019) 1800589.
[104] B. Zhang, Z. Wang, Y. Cao, P. Yan, X.R. Wang, Eavesdropping on spin waves inside the domain-wall nanochannel via three-magnon processes, Phys. Rev. B 97 (2018) 094421.
[105] W. Yu, J. Lan, R. Wu, J. Xiao, Magnetic Snell's law and spin-wave fiber with Dzyaloshinskii-Moriya interaction, Phys. Rev. B 94 (2016) 140410.
[106] M. Garst, J. Waizner, D. Grundler, Collective spin excitations of helices and magnetic skyrmions: review and perspectives of magnonics in non-centrosymmetric magnets, J. Phys. D 50 (2017) 293002.

[107] A. Belabbes, et al., Oxygen-enabled control of Dzyaloshinskii-Moriya interaction in ultra-thin magnetic films, Sci. Rep. 6 (2016) 24634.
[108] S. Jekal, A. Danilo, D. Phuong, X. Zheng, First-principles prediction of skyrmionic phase behavior in GdFe2 films capped by 4d and 5d transition metals, Appl. Sci. 9 (2019) 630.
[109] T.H. O'Dell, Ferromagnetodynamics, in: The Dynamics of Magnetic Bubbles, Domains, and Domain Walls, Wiley, New York, 1981.
[110] C. Liu, S. Wu, J. Zhang, J. Chen, J. Ding, J. Ma, Y. Zhang, et al., Current-controlled propagation of spin waves in antiparallel, coupled domains, Nat. Nanotechnol. 2019 (2019) 1.
[111] J.R. Sandercock, M. Cardona, G. Guntherodt (Eds.), Light Scattering in Solids III, Springer, Berlin, 1982.
[112] D.M. Schaadt, R. Engel-Herbert, T. Hesjedal, Effects of anisotropic exchange on the micromagnetic domain structures, Phys. Status Solidi B 244 (2007) 1271.
[113] J. Rychly, P. Gruszecki, M. Mruczkiewicz, J.W. Klos, S. Mamica, M. Krawczyk, Magnonic crystals—prospective structures for shaping spin waves in nanoscale, Low Temp. Phys. 41 (2015) 745.
[114] O. Hellwig, J.B. Kortright, A. Berger, E.E. Fullerton, Domain structure and magnetization reversal of antiferromagnetically coupled perpendicular anisotropy films, J. Magn. Magn. Mater. 319 (2007) 13.
[115] O. Hellwig, G.P. Denbeaux, J.B. Kortright, E.E. Fullerton, X-ray studies of aligned magnetic stripe domains in perpendicular multilayers, Physica B 336 (2003) 136.
[116] M.T. Johnson, P.J.H. Bloemen, F.J.A.d. Broeder, J.J.d. Vries, Magnetic anisotropy in metallic multilayers, Rep. Prog. Phys. 59 (1996) 1409.
[117] B.A. Kalinikos, M.P. Kostylev, N.V. Kozhus, A.N. Slavin, The dipole-exchange spin wave spectrum for anisotropic ferromagnetic films with mixed exchange boundary conditions, J. Phys. Condens. Matter 2 (1990) 9861.
[118] J. Ding, et al., Higher order vortex gyrotropic modes in circular ferromagnetic nanodots, Sci. Rep. 4 (2014) 4796.

Further reading

[119] A.P. Malozemoff, J.C. Slonczewski, Magnetic Domain Walls in Bubble Materials, Academic Press, New York, 1979.

CHAPTER THREE

Functional domain walls: Concepts and perspectives

Jan Seidel*
School of Materials Science and Engineering, UNSW Sydney, Sydney, NSW, Australia
ARC Centre of Excellence in Future Low-Energy Electronics Technologies, UNSW Sydney, Sydney, NSW, Australia
*Corresponding author: e-mail address: jan.seidel@unsw.edu.au

Domain walls in ferroic materials have been recognized and investigated as functional nanoscale entities over the last few years. Their intrinsic size of the order of 1 nm in ferroelectric and ferroelastic materials make them an attractive feature that can be utilized in nanoelectronics and spintronics concepts [1]. While some of the fundamental properties of these topological defects have been explored, there is still a large amount of questions that remains unanswered, for example, the precise nature of nanoscale spin textures in multiferroic domain walls or coupling of order parameters at walls in experimental investigations. Progress has been achieved, however, with respect to prototype nanoelectronics devices, where the domain wall is the active part. These include simple switches, memristive components, memory cells, diodes, tunnel junctions and others [2]. The ability to store or transport information in ~1 nm wide domain walls in ferroic systems is highly attractive for future nanoelectronics [3].

Magnetic domain wall memory and logic has been investigated following initial work on race track geometries by Parkin et al. [4]. More recently other types of topological defects [5], such as skyrmions [6] (and derived structures such as hopfions, merons, etc.) have been heavily investigated with respect to their nanoelectronic functionality. Similar structures have now also been found experimentally in ferroelectric systems [7]. The specific type of topological defect, i.e., domain wall or skyrmion, for example, is formed in the elastic distortion (ferroelastics), electric dipoles (ferroelectrics), or spin arrangement (magnetic materials) in materials with such types of order, i.e., ferroic systems, upon symmetry lowering phase transitions below a critical temperature. Various types of complex patterns that form this way have been discussed previously by Kittel [8,9], Mermin [10] and in other works, and depend on several energetic factors, for example, the specific type of exchange in the material, crystal anisotropies, various types of surface

Solid State Physics, Volume 70
ISSN 0081-1947
https://doi.org/10.1016/bs.ssp.2019.09.004

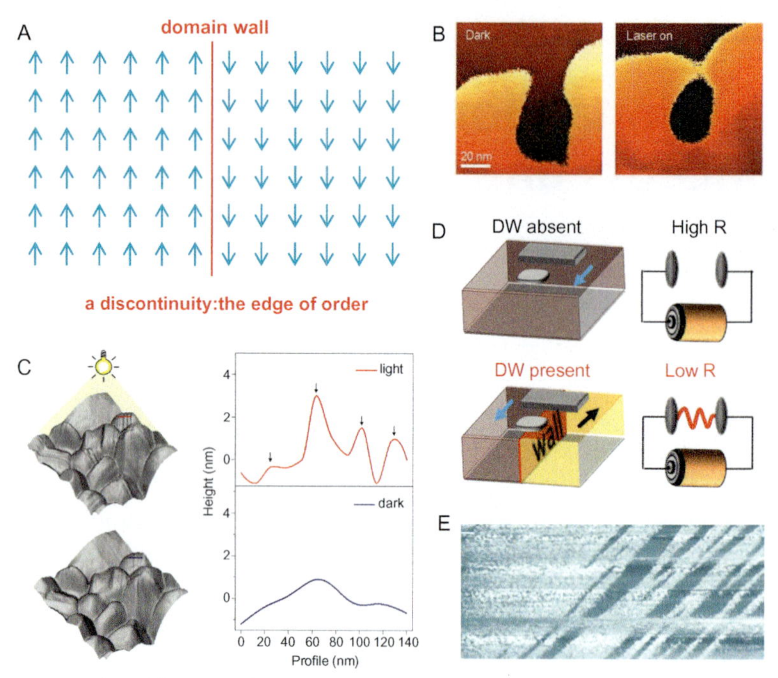

Fig. 1 Examples of functional domain walls. (A) schematic of the topological defect that is a domain wall and associated discontinuity of order, (B) example of optical control of domain walls (pinched wall segment) in ferroelectric KNN-BNNO [15] by light illumination, (C) surface corrugation enhancement (height change) at ferroelastic domain walls (stripes inside grains of the polycrystalline material) in halide perovskites due to structural disorder that is induced optically or electrically [16], (D) concept of ferroelectric domain wall memory, i.e., high and low resistance states mediated by the absence or presence of a wall, (E) in-plane ferroelectric domains observed in layered van der Waals In_2Se_3 by piezoresponse force microscopy (PFM).

or interfaces, and local defects (such as dopants [11] or oxygen vacancies in oxides [12–14]), among others. A few examples of different types of functional domain walls are shown in Fig. 1.

Fundamental investigations on domain walls start with their atomic structure, which can be visualized, for example, by electron microscopy, i.e., high resolution HAADF-STEM measurements [17,18] The ability to image atomic-scale structure with a resolution better than 0.5 Å by aberration-corrected imaging has provided significant new insight over the last few years. In addition to structural changes, the reorientation of polarization and oxygen octahedral tilts can nowadays be ascertained across domain walls [19], even for minute displacements of ions in oxide perovskite ferroelectrics [20]. Additional capabilities include elemental and electronic structure analysis by electron energy-loss spectroscopy (EELS) that have also been applied to the study of domain walls. One example is the study of chemically distinct domain walls in terbium manganite [21]. HAADF-STEM was also used to compare domain wall structure details with previous ab initio predictions used to explain increased domain wall conductivity. Lubk et al. showed that chemical pressure introduced by La-doping in bismuth ferrite can modify the domain wall angle in accordance with the rhombohedral lattice angle and discusses diffuse domain walls proposed to explain experimentally found domain wall mobilities. Furthermore, unit-cell-wise resolved strain and polarization profiles around edge dislocations reveal their interaction with domain walls in ferroelectric thin films. There, a wealth of material states including polymorph nanodomains and multiple domain walls characteristically pinned to the dislocation are found. Piezoelectric coupling was identified as the driving force for the observed phenomena, explaining, for example, the specific orientation of domain walls with respect to the edge dislocation in agreement with atomic models for the dislocation core.

The interaction of topological structures with external stimuli is at the foundation of nanoelectronics concepts based on these features [22,23]. This includes electronic conductivity [24], which can be influenced by electric fields as the most common example. Quantitative measurements of specific wall conductivity have remained a significant experimental challenge though. The reason for this is that the nanoscale contacts used in such measurements through lithographic processes or scanning probes typically do not provide Ohmic properties (linear I–V in small bias range), but rather form Schottky barriers between metallic electrodes and semiconducting wall or bulk materials. Another severely limiting factor is the usually very small current levels associated with such measurements which rule out standard procedures for quantifying specific conductivities such as Hall measurements, or even Seebeck measurements for that matter to discern carrier types in domain walls, although currents on the order of hundreds of nA are feasible [25] and some progress has been made recently with respect to Hall measurements at domain walls [26].

Another type of stimulus that can interact with domain walls can be found in magnetic fields for magnetic or multiferroic materials systems (coupling of spin charge and lattice properties of the material [27,28]). Examples include hexagonal rare earth manganites ($RMnO_3$) in which ferroelectric domain walls are observed alongside antiferromagnetic domain walls [29,30], and also bismuth ferrite, in which ferroelectric and antiferromagnetic walls have been shown to coincide [31]. Control over domain wall and associated charge states has been demonstrated for $TbMnO_3$ [32] or $MnWO_4$ [33]. Magnetic fields applied to these materials can induce a spin flop transition that also rotates the polarization by 90° (from neutral to charged). One of the few experimental examples of a finite magnetic component at a domain walls was reported by Geng et al. using magnetic force microscopy (MFM) for $ErMnO_3$. More indirect measurements of magnetic structure and associated properties at multiferroic domain walls in bismuth ferrite thin films have also been reported [34–36]. These studies mainly focus on electrical transport measurements, i.e., magnetoresistance, that can provide some insight into the magnetic state of the investigated material, but do not directly resolve the complex magnetic spin structure at the domain wall. Some results related to magnetism at domain walls have also been reported for $TbMnO_3$ thin films [21].

Of considerable interest is also the interaction of domain walls with light illumination or other electromagnetic radiation in optically active devices. This includes photo-induced ferroelectric domain switching in various materials, e.g., $Pb(Zr_{0.2}Ti_{0.8})O_3/La_{0.7}Sr_{0.3}MnO_3$ (PZT/LSMO) bilayer thin films with varying surface morphologies, which have been studied by piezoresponse force microscopy under light illumination [37]. Wang et al. demonstrate that reverse poled ferroelectric regions can be almost fully recovered under laser irradiation of the PZT layer and that the recovery process is dependent on the surface morphology on the nanometer scale. The optical recovery process is well described by the standard Kolmogorov-Avrami-Ishibashi model, and the evolution speed is controlled by light intensity, sample thickness, and initial write voltage for ferroelectric domain switching. Their findings shed light on optical control of the domain structure in ferroelectric thin films with different surface morphologies. The role of domain walls, i.e., their optical [15,38,39] and photovoltaic properties [40] are still being investigated and debated widely. For example, the interplay of bulk photovoltaic effects and domain wall related properties is still unresolved. One recent example of the influence of domain wall properties on solar cell performance is found in organic-inorganic metal halide perovskites, that have gained considerable attention for next-generation photovoltaic cells due to rapid improvement in power conversion efficiencies [41–46]. Fundamental understanding of underlying mechanisms related to light- and bias-induced effects at the nanoscale in these materials is still heavily investigated. Kim et al. [16] showed structural variations of the perovskites that are induced by light and bias through systematic investigations using scanning probe microscopy techniques. Periodically striped ferroelastic domains, spacing between 40 and 350 nm, that exist within grains could be modulated significantly under illumination as well as by electric bias (Fig. 1C). Williamson-Hall analysis of X-ray diffraction results shows that strain disorder is induced by these applied external stimuli, which can be observed as surface corrugations at the ferroelastic domain walls in the material. These findings point to potential origins of I—V hysteresis in halide perovskite solar cells. Furthermore, Vats et al. recently showed reversible control over domains and domain walls in $((K_{0.5}Na_{0.5})NbO_3$-2 mol% $Ba(Ni_{0.5}Nb_{0.5})O_{3-\delta}$ (KNBNNO, or KNN-BNNO)) [15]. A clearly reversible and collapse and splitting of two domain walls was observed on exposure to light (Fig. 1B). Further improvement could help to achieve precise optical control over domain walls that can be used for an optically controlled ferroelectric counterpart to magnetic domain wall race-track memory and other advanced optoelectronic devices based on domain walls.

Domain wall logic and memory based on the electronic conduction properties in ferroelectrics and multiferroics is one of the key concepts that has been investigated over the last 10 years. The absence or presence of walls in this case defines the resistance state (high or low R) of the specific device element (Fig. 1D). Manipulation of domain wall properties by electric fields is especially interesting in this respect, as it typically requires less energy than current-driven circuits, which is a key aspect for the development of ultra-low power electronics technology based on novel materials. Some heterogeneity of electronic transport has been reported in domain walls reflecting the importance of defect chemistry, charged defects, and growth conditions of the materials involved. Here, oxygen vacancies acting as electron donors are one of the most commonly occurring defects in thin film synthesis of ferroelectric oxides and play a critical role in domain wall conduction. Nevertheless, domain walls can be reconfigured using external electric fields, which is one of their key advantages over fixed CMOS architectures, for example. The recent demonstration of a working prototype of a nonvolatile domain wall memory device is based on precise injection of a domain wall pair between two metal electrodes [47]. The device exhibited relatively high on-off ratios ($\sim 10^3$) with excellent endurance and retention characteristics. The encoding and non-destructive retrieval of information was done at moderate biases, thus enabling low-energy operation of the memory cell. This study provided direct experimental proof of a working non-volatile solid-state domain wall device, wherein information is processed directly by ferroelectric domain walls.

Precise control over the conductivity state at a domain wall in principle also allows for multi-level states that can also be realized in such devices. Recent developments toward such devices have been demonstrated by tuning the domain wall shape with respect to the polarization direction in the material and its associated charge state [48]. The resistance of ferroelectric domain walls is directly proportional to the wall length, i.e., walls of varying lengths can be thought of as the nanoscale analog of resistors of different resistances. This simple concept enables multi-level states in a two-terminal device. The wall length can be altered, for example, by electrical bias pulses of varying amplitude and duration. Another method by which multi-level states can be designed is by control of the charge state of the domain wall. Using electric fields, domain walls can be 'folded' into various geometric shapes or states resembling stable charged, neutral or no domain wall states in a memory cell. These states can be selectively and alternatively injected or erased between two metallic electrodes, giving a reversible precise stable conformational control over the charge state of ferroelectric domain walls. This concept is analogous to conformational switches in molecular electronics. Conformational control of domain walls in this scenario depends on the strength and/or temporal-profile of the applied inhomogeneous electric field distribution between metal contacts and majority carrier type to stabilize charged domain walls. Such a device transitions from an OFF to a low resistance ON state upon the injection of a neutral domain wall, then to a charged domain wall based even lower resistance ON state, in which the resistance is lowered by another order of magnitude, presenting sufficient fidelity and dynamic range of the memory system.

Another interesting aspect are applied force or strain in ferroelastic or multiferroic systems, especially in conjunction with ferroelectricity, but also electro- and magnetostrictive systems, with some work reported on conventional and morphotropic [49–57] ferroelectric systems. The widely investigated mixed-phase structure in bismuth ferrite thin films presents a new concept, namely, that of hybrid domain walls, i.e., a transition of the order parameter coupled to a structural transition of the lattice of the material [58–60]. After an initial report of the stabilization of such morphotropic phase boundaries in bismuth ferrite through epitaxial thin film growth on highly compressive substrates [61], these features have seen a great deal of attention. The original work on tetragonal-like and rhombohedral-like phases in that system has been extended to also include orthorhombic phases [62] with additionally altered magnetic spin structure [42]. These phases have been shown to reversibly convert into each other by electric fields and pressure, accompanied by measurable displacements of the surface in AFM measurements. Very high reported piezoelectric responses of these systems because of phase competition and movement of domain walls make them a new lead-free ferroelectric system of interest for actuator applications. The study of mechanical properties of nanoscale ferroelectrics using the above methods provides a unique tool to access and study complementary information to the otherwise already accessible electronic and structural properties of these materials with nanometer spatial resolution. The sensitivity and selectivity of the mechanical interaction opens new pathways for real space probing of multiple order parameters and phases of complex ferroic materials.

Apart from quasi-static electronic properties, domain walls have also been investigated with respect to their response to varying, i.e., AC electromagnetic fields, especially in the device relevant GHz frequency range. Here, recent work shows that domain walls can be engineered to achieve tailored dynamic responses [63]. Gu et al. investigate the dielectric response of strained $BaxSr_{1-x}TiO_3$ thin films with specific compositions to study the influence of high densities of domain walls in their samples. The experimental results showed wide microwave tunability (1–10 GHz) and low dielectric loss associated with their materials, i.e., properties that are indeed superior by a few orders of magnitude to state-of-art thin film devices reported so far. Resonant domain wall movement, i.e., oscillations in a domain wall rich sample were determined as the main factor for this behavior. Another example is the observation of frequency-dependent decoupling of domain wall motion and lattice strain investigated recently in bismuth ferrite [64]. Liu et al. show that the domain wall-controlled piezoelectric behavior in multiferroic $BiFeO_3$ is distinct from that reported in other classical ferroelectrics. Using in-situ X-ray diffraction they were able to show the separate nature of the electric-field-induced lattice strain and strain due to displacements of non-180° domain walls in polycrystalline samples over a wide frequency range. Interestingly, these piezoelectric strain mechanisms have opposite trends as a function of frequency, lattice strain increases with increasing frequency while domain wall motion exhibits the opposite behavior, i. e., decreases with increasing frequency. The cause for this unusual behavior was found in charge redistribution at conducting domain walls.

Another highly interesting field of domain wall related research is the recently brought forward concept of negative capacitance [65,66]. This aspect has high relevance for future developments in nanoelectronics. Several approaches have been discussed, one of them involving domain wall creation and annihilation, resulting in a temporary negative capacitance of the associated device structure, which in the future could lead to more efficient transistor technology. Other concepts for negative capacitance rely on highly engineered oxide heterostructures [67,68] that has enabled strain control of topological structures, as seen, for example, in highly ordered ferroelectric vortex arrays in $PbTiO_3/SrTiO_3$ systems [69]. Such precisely designed model material systems suggest themselves also for vortex based memory applications. The controlled utilization of domain walls around vortex-antivortex pair structures has also been discussed in $BiFeO_3$ [70], where by electric field writing in various geometries combined with electric readout through conductive domain walls or vortex cores enables new functionality.

An emerging research field is the investigation of domain walls in 2D and layered materials, that are found, for example, in transition metal chalcogenides and other currently investigated compounds [71,72] (Fig. 1E). Only very recently it has been shown that 2D materials with intrinsic magnetism exist, such as transition metal halides [73] and that their magnetic properties can be influenced electrically. In addition, interesting recent developments in this regard are the prediction of multiferoic 2D or layered materials [74]. A variety of unusual features found in such atomically thin materials could potentially be exploited for atomic scale magnetoelectronics and engineered interface phenomena in multilayer devices.

2D and layered materials also provide other interesting aspects, for example, that of significantly anisotropic electronic properties with respect to crystallographic directions in the materials. One specific example is the highly interesting concept of polar and ferroelectric metals recently discussed by Sharma et al. [75] Conventionally, ferroelectricity has been observed in materials that are insulating or semiconducting rather than metallic, because conduction electrons in metals screen-out the static internal fields arising from the dipole moment, although the coexistence of reversible polar distortions and metallicity leading to a ferroelectric metal was postulated by Anderson and Blount [76]. Sharma et al. investigated bulk-crystalline tungsten ditelluride (WTe_2), which belongs to a class of materials known as transition metal dichalcogenides (TMDCs). Spectroscopic electrical transport measurements and conductive-atomic force microscopy (c-AFM) confirmed its metallic behavior, together with piezoresponse force microscopy (PFM) to map the polarization and detect lattice deformation due to an applied electric field. The investigation of domain walls in such a ferroelectric metallic material system is highly interesting and might lead to new effects that can be observed experimentally.

We have seen a significant amount of emerging aspects of domain wall functionality over the last years. Key factors here include conduction properties, charge, and electronic structure. Some progress has also been made with respect to domain wall interaction with defects, which, for example, can significantly alter domain wall mobilities and polarization retention properties [77] and enable new tailored functionality in ferroelectrics. The investigation of magnetism and magnetoelectric properties of multiferroic domain walls remains highly interesting and challenging and together with optical property investigation and manipulation of domain walls will drive future research in this direction, not only for fundamental property investigations but also for prototype domain wall device concepts.

References

[1] G. Catalan, J. Seidel, R. Ramesh, J.F. Scott, Rev. Mod. Phys. 84 (2012) 119.
[2] P. Maksymovych, J. Seidel, Y.H. Chu, P. Wu, A.P. Baddorf, L.Q. Chen, S.V. Kalinin, R. Ramesh, Nano Lett. 11 (2011) 1906–1912.
[3] J. Seidel, Nat. Mater. 18 (2019) 188.
[4] S.S.P. Parkin, M. Hayashi, L. Thomas, Science 320 (2008) 190.
[5] J. Seidel, et al., Adv. Electron. Mater. 2 (2016) 1500292.
[6] P. Milde, et al., Science 340 (2013) 1076.
[7] S. Das, et al., Nature 568 (2019) 368.
[8] C. Kittel, Phys. Rev. 70 (1946) 965.
[9] C. Kittel, Rev. Mod. Phys. 21 (1949) 541.
[10] N.D. Mermin, Rev. Mod. Phys. 51 (1979) 591.
[11] C.-H. Yang, et al., Phys. Chem. Chem. Phys. 14 (2012) 15953.
[12] M.L. Scullin, et al., Acta Mater. 58 (2010) 457.
[13] M.L. Scullin, et al., Appl. Phys. Lett. 92 (2008) 202113.
[14] S. Kaya, et al., J. Alloys Compd. 583 (2014) 476.
[15] G. Vats, et al., Adv. Opt. Mater 7 (2019) 1800858.
[16] D. Kim, Nat. Commun. 10 (2019) 444.
[17] A. Lubk, et al., Phys. Rev. Lett. 109 (2012) 047601.
[18] A. Lubk, et al., Nano Lett. 13 (2013) 1410.
[19] C.-L. Jia, S.-B. Mi, K. Urban, I. Vrejoiu, M. Alexe, D. Hesse, Nat. Mater. 7 (2008) 57.
[20] J. Seidel, et al., Nat. Commun. 3 (2012) 799.
[21] S. Farokhipoor, et al., Nature 515 (2014) 379–383.
[22] J. Seidel, J. Phys. Chem. Lett. 3 (2012) 2905.
[23] J. Seidel, G. Singh-Bhalla, Q. He, S.-Y. Yang, Y.-H. Chu, R. Ramesh, Phase Transit. 86 (2013) 53.
[24] J. Seidel, et al., Nat. Mater. 8 (2009) 229–234.
[25] Z.L. Bai, et al., Adv. Funct. Mater. 28 (2018) 1801725.
[26] M.P. Campbell, et al., Nat. Commun. 7 (2016) 13764.
[27] M. Ramirez, et al., Appl. Phys. Lett. 94 (2009) 161905.
[28] B.-K. Jang, et al., Nat. Phys. 13 (2017) 189.
[29] Y. Geng, et al., Nat. Mater. 13 (2014) 163–167.
[30] M. Fiebig, T. Lottermoser, D. Fröhlich, A.V. Goltsev, R.V. Pisarev, Nature 419 (2002) 818.
[31] Z. Gareeva, et al., Phys. Rev. B 91 (2015) 4.
[32] M. Matsubara, et al., Science 348 (2015) 1112–1115.
[33] N. Leo, et al., Nat. Commun. 6 (2015) 4.
[34] N. Domingo, et al., J. Phys. Condens. Matter 29 (2017) 334003.

[35] Q. He, et al., Phys. Rev. Lett. 108 (2012) 067203.
[36] J.H. Lee, et al., Adv. Mater. 26 (2014) 7078–7082.
[37] J. Wang, et al., Appl. Phys. Lett. 111 (2017) 092902.
[38] Y. Bai, et al., Adv. Mater. 30 (2018) 1803821.
[39] J. Seidel, L.M. Eng, Curr. Appl. Phys. 14 (2014) 1083.
[40] J. Zhang, et al., Nat. Nanotechnol. 6 (2011) 98.
[41] R. Pandey, Adv. Mater. (2019). https://doi.org/10.1002/adma.201807376.
[42] Y. Cho, Adv. Energy Mater. 8 (2018) 1703392.
[43] X. Liu, Adv. Energy Mater. 8 (2018) 1800138.
[44] D.S. Lee, ACS Energy Lett. 3 (2018) 647.
[45] N. Faraji, J. Phys. Chem. C 122 (2018) 4817.
[46] J.S. Yun, Adv. Funct. Mater. 28 (2018) 1705363.
[47] P. Sharma, et al., Sci. Adv. 3 (2017) e1700512.
[48] P. Sharma, et al., Adv. Funct. Mater. 29 (2019) 1807523.
[49] Y. Heo, et al., Adv. Mater. 26 (2014) 7568.
[50] P. Sharma, et al., Adv. Mater. Interfaces 3 (2016) 1600033.
[51] P. Sharma, et al., Adv. Electron. Mater. 2 (2016) 1600283.
[52] S. Hu, et al., Phys. Status Solidi A 214 (2017) 1600356.
[53] A. Alsubaie, et al., Nanotechnology 28 (2017) 075709.
[54] Y. Heo, et al., ACS Nano 11 (2017) 2805.
[55] P. Sharma, et al., Nanotechnology 29 (2018) 205703.
[56] A. Alsubaie, et al., ACS Appl. Mater. Interfaces 10 (2018) 11768.
[57] J. Zhou, et al., J. Appl. Phys. 112 (2012) 064102.
[58] J.H. Lee, et al., Phys. Rev. B 96 (2017) 064402.
[59] K.-E. Kim, et al., NPG Asia Mater. 6 (2014) e81.
[60] J. Seidel, et al., Adv. Mater. 26 (2014) 4376.
[61] R.J. Zeches, et al., Science 326 (2009) 977.
[62] Y. Heo, et al., NPG Asia Mater. 8 (2016) e297.
[63] Z. Gu, et al., Nature 560 (2018) 622.
[64] L. Liu, et al., Nat. Commun. 9 (2018) 4928.
[65] S. Salahuddin, S. Datta, Nano Lett. 8 (2008) 405.
[66] T. Sluka, P. Mokry, N. Setter, Appl. Phys. Lett. 111 (2017) 152902.
[67] M. Lorentz, et al., J. Phys. D Appl. Phys. 49 (2016) 433001.
[68] P. Zubko, J.C. Wojdeł, M. Hadjimichael, S. Fernandez-Pena, A. Sené, I. Luk'yanchuk, J.-M. Triscone, J. Íñiguez, Nature 534 (2016) 524.
[69] A.K. Yadav, et al., Nature 530 (2016) 198–201.
[70] K.-E. Kim, S. Jeong, K. Chu, J.H. Lee, G.-Y. Kim, F. Xue, T.Y. Koo, L.-Q. Chen, S.-Y. Choi, R. Ramesh, C.-H. Yang, Nat. Commun. 9 (2018) 403.
[71] K. Yasuda, M. Mogi, R. Yoshimi, A. Tsukazaki, K.S. Takahashi, M. Kawasaki, F. Kagawa, Y. Tokura, Science 358 (2017) 1311–1314.
[72] C. Zheng, et al., Sci. Adv. 4 (2018) eaar7720.
[73] B. Huang, G. Clark, E. Navarro-Moratalla, D.R. Klein, R. Cheng, K.L. Seyler, D. Zhong, E. Schmidgall, M.A. McGuire, D.H. Cobden, W. Yao, D. Xiao, P. Jarillo-Herrero, X. Xu, Nature 546 (2017) 270–273.
[74] C. Huang, Y. Du, H. Wu, H. Xiang, K. Deng, E. Kan, Phys. Rev. Lett. 120 (2018) 147601.
[75] P. Sharma, Sci. Adv. 5 (2019) eaax5080.
[76] P.W. Anderson, E. Blount, et al., Phys. Rev. Lett. 14 (1965) 217.
[77] D. Zhang et al., Nat. Commun. n.d (submitted for publication).

CHAPTER FOUR

Skyrmions in ferroelectric materials

Jiri Hlinka*, Petr Ondrejkovic
Institute of Physics of the Czech Academy of Sciences, Prague, Czech Republic
*Corresponding author: e-mail address: hlinka@fzu.cz

Contents

There is no doubt that there are many analogies between classical electricity and magnetism. Although these analogies are not always perfect, they have been often useful in driving the intuition of physicists. Perhaps the most concise expression of such an analogy is encapsulated in Maxwell equations where the absence of magnetic monopole density in vacuum constitutes the essential ingredient of its beauty and imperfection at the same time. Also the early phenomenological theory of ferroelectric materials has been developed on the footsteps of the theory of ferromagnetism, and many technical terms have been adapted or adopted from there. At the phenomenological level at least, both have vector order parameters and allow some level of description in terms of the continuous Ginzburg-Landau theory. It is therefore quite natural that very soon after the discovery of ferromagnetic skyrmions in a chiral magnet MnSi in 2009 [1], similar topological defects were also sought in ferroelectric materials. This chapter aims to review the latest developments and observations of topological defects of electric

Solid State Physics, Volume 70
ISSN 0081-1947
https://doi.org/10.1016/bs.ssp.2019.09.005

polarization fields that bear some analogy with magnetic skyrmions or skyrmion bubbles. This discipline is rather quickly developing and results are often presented from different standpoints and in different terms. In order to facilitate the summary of various recent observations and predictions, we also present some essential concepts and aspects of ferroelectric materials that are related to a formation of topological defects in ferroelectrics in general. At the end we will try to propose what are current perspectives and essential research targets to be challenged in the future.

1. Ferroelectric domains and domain boundaries

The concept of ferroelectric substances in one way or another refers to ferroelectric domains and thus to the ferroelectric phase transition. At a ferroelectric phase transition, either really accessible one or a fictive one, some symmetry operation of the paraelectric crystalline phase are broken by the appearance of the spontaneous polarization in the ferroelectric phase. Because spontaneous polarization is a macroscopic quantity, there is a decrease in the macroscopic symmetry of the material at this transition. The lost symmetry operations imply that there are at least two energetically equivalent domain states, which are structurally identical up to an isometry operation like rotation or inversion, but which differ by the direction of their spontaneous polarization. In brief, in terms of the Landau theory of phase transitions, ferroelectric materials can be defined as crystalline materials with a broken macroscopic symmetry, where electric polarization or at least one of its Cartesian component acts as an order parameter.

The order parameter concept also allows to distinguish proper and improper ferroelectric phase transitions, based on the polarization being a primary order parameter or a secondary one. Here, the primary order parameter is usually defined solely by its symmetry aspect, requiring that it transforms as a physically irreducible representation of the high-symmetry group, which fully determines possible symmetries of the low-symmetry group. However, in many practical situations, one has to invoke also its physical aspect, assuming divergence of the associated dielectric susceptibility, at some finite temperature, due to the quasiharmonic instability of the paraelectric phase along the critical configuration coordinate [2]. In this sense it is also convenient to employ the full/partial classification. A ferroelectric phase transition is denoted as a full or partial one, depending whether any pair of its macroscopic domain states differ by the polarization vector or not. This classification relies only on the type of macroscopic symmetry breaking, and can be therefore uniquely determined on the basis of the macroscopic symmetry alone, relying on the concept of symmetry breaking species [3]. Although we largely disregard the subtleties of improper ferroelectrics and partial ferroelectrics in this chapter, it is worth to stress that in the case of improper ferroelectrics a full enumeration of domain states may be required to consider microscopic (space group) symmetry of the crystal and in the case of partial ferroelectrics, it is convenient to enhance the description of available domain states with one or more additional macroscopic order parameters, in particular the primary one.

Ferroelectric ordering is a rather robust phenomenon against thermal fluctuations in the sense that in many substances the ordering persists well above the room temperature. A considerable crystalline anisotropy and electrostriction interactions in typical ferroelectric substances imply that most of the sample is locally structurally identical to one of the allowed domain states. Inhomogeneity of the order parameter is limited to rather narrow domain walls with a typical thickness of the order of 1 nm.

Orientation of ferroelectric domain boundaries in a usual dielectric ferroelectric is considerably limited by electromechanical compatibility of the domain states [4–6]. In the absence of charge-compensating mechanisms, the interface between ferroelectric domains always tends to orient in a direction that ensures that the normal components of the polarization in adjacent domains are equal. In proper ferroelectrics, where polarization is the primary order parameter, domain boundaries oriented only a few degrees away from the ideal neutrality have usually quite different properties already and which implies some local compensating charge transport to be involved. Observation of fully head-to-head or tail-to-tail domain walls normal or close to normal to the polarization in insulator ferroelectrics is rare and formation of such so-called charged walls in good insulators usually requires sophisticated engineering approaches.

Very often domain states also differ in the spontaneous strain tensor orientation. In this case, domain boundaries are denoted as ferroelastic ones and there are additional mechanical compatibility conditions to be met. The mechanical compatibility conditions ensure that the atomic lattices of the adjacent domain states match coherently, or in other words, without necessity to form dislocations or other disruptions of the translational lattice symmetry in directions within the domain boundary interface. Mechanical compatibility conditions impose even more severe restrictions on the orientation of the ferroelastic-ferroelectric domain boundaries. In general, permissible macroscopic domain boundaries are therefore locally of either planar shape (parametrically dependent S-walls and crystallographically prominent W_f walls) or of cylindrical shape (W_∞ walls). Permissible crystallographic orientation of each domain state pair can be derived using well established procedures or it can be found for example in the International Tables of Crystallography, where also the symmetry of permissible planar domain boundaries has been derived and tabulated [7].

In the continuous approach, infinitely sharp domain boundaries with fixed permissible domain wall normals are objects completely specified by the pair of adjacent domain states and the oriented domain wall normal. Microscopic description of infinitely sharp domain boundaries could be possibly complemented by specifying the exact termination of each domain state at the domain boundary. In both cases, crystallographic symmetry of such walls can be described in terms of crystallographic layer groups [7]. For a domain wall of a finite thickness, these layer groups define the maximal symmetry of the domain wall.

In principle, internal domain wall structure can have a lower than the maximal symmetry. In the past decade, there have been lots of evidence that within certain types of domain walls of well-known perovskite ferroelectrics like $PbTiO_3$ or $BaTiO_3$, one can find an additional component of polarization. These circumstances imply that the symmetry of such walls is lower than the maximal symmetry compatible with the domain pair and the given permissible domain boundary orientation. By analogy with ferromagnetism, these domain walls are now called Bloch or Néel walls, depending on whether the additional, symmetry-breaking polarization component is oriented along the domain wall normal or perpendicular to it. In fact, the additional polarization components have magnitudes quite comparable to the bulk domain spontaneous values. Since the thickness of these domain walls spans across several atomic planes and polarization varies rather continuously across the domain wall, continuous models are rather relevant in description of these domain walls. Clearly, this nanoscale polarization rotation is largely analogous to the domain wall profiles of ferromagnetic domain walls. The evidence is so far mostly theoretical or rather indirect but still it gives additional confidence in the analogies between ferroelectricity and ferromagnetism. In either case, these findings give hope for future discoveries of ferroelectric analogues of other already well established ferromagnetic topological defects [8–13].

2. Ferroelectric line defects

While ferroelectric domain boundaries are inevitable components of ferroelectric domain structures, one-dimensional anomalies in a ferroelectric

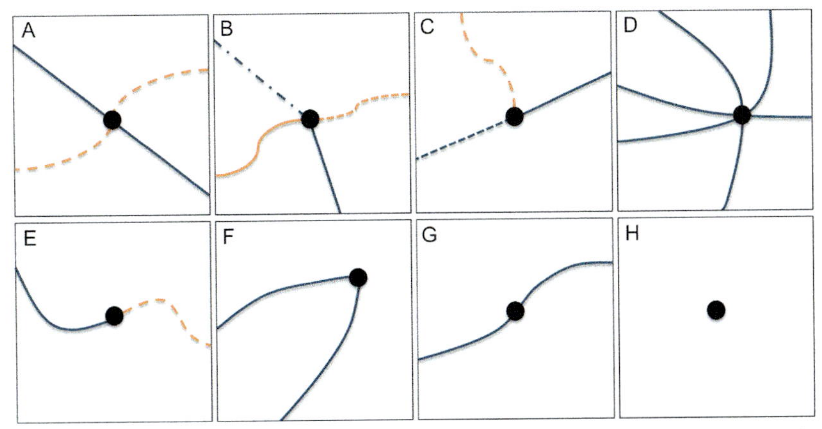

Fig. 1 Line defects in ferroelectric structures and their relations to domain walls: (A) line at the intersection of two domain walls, (B) line junction of four domain walls (C) line junction of three domain walls (D) line junction of six domain walls (E) line separating domain wall areas with qualitatively different internal properties (F) fold line marking a discontinuous change of the domain wall normal on a wedge domain (G) a line defect within an isolated domain wall allowing continuity of domain wall properties (H) isolated, stand-alone line defect. Walls and lines are represented as their intersection with the plane of the figure.

polarization field are encountered less frequently and, in principle, could be even completely absent in some domain structures. Ferroelectric lines can be formed by intersections or junctions of domain walls or independently of it. Typical cases are summarized in Fig. 1.

Probably the most frequently encountered case in ferroelectric perovskite materials is a line defect at the four-domain state junction formed by intersection of non-ferroelastic and ferroelastic domain boundary (Fig. 1A). Otherwise, quadruplets of four distinct ferroelectric domain states rarely fulfill a perfect electromechanical compatibility at a single junction line. They could be nevertheless locally stable as long as the pairs of the opposite domain states cannot be joined by a compatible domain wall instead. Depending on the particular case, the four domain boundaries involved in such domain quadruplets could be also of the same type or, on the contrary, all of them could be of different type (Fig. 1B). A similar situation can happen when three domain walls meet a single line, or when one domain wall ends at the surface of another one (Fig. 1C).

In general, a quadruplet with identical domain states on its opposite sides is unstable because such domains could simply merge together. This is also true for higher order junctions. Thus, observation of multiple junctions allows appreciation of the number of domain states in the material. In the case of improper or partial ferroelectrics, ferroelectric lines obviously can join multiple domains in which one or more pairs have the same polarization direction. Well known are, for example, ferroelectric line defects at the cores of domain sextuplets in hexagonal manganite oxides [14–16]. At the core, six ferroelectric domain walls meet as in Fig. 1D, even though the whole structure contains only two antiparallel ferroelectric states. These ferroelectric lines are very stable because they correspond to topologically protected defects of a continuously varying microscopic primary order parameter.

A ferroelectric line defect can be also associated with a junction of two domain walls. Alternatively, such a line defect can be thought of as located within a single domain wall. One may distinguish interesting cases when the internal nature of the wall on the opposite sides of the line is changing (Fig. 1E), when the domain wall is the same but the domain wall normal is changing abruptly (Fig. 1F) or when domain wall is identical on both sides (Fig. 1G). The situation of (Fig. 1E) corresponds to the case of the Ising line [17] separating parts of the Bloch wall showing the opposite helicity, while the fold line of (Fig. 1F) is typical for all propagating ferroelastic wedge domains. The line defect within a ferroelectric wall as shown in Fig. 1G can arise also in some partial or improper ferroelectrics at the intersection of a ferroelectric wall with a non-ferroelectric one and in this case a more careful consideration is needed as to whether the geometry in Fig. 1E is more appropriate or not. In principle, however, a domain wall of final thickness might host an intrinsic line defect which is related solely to its internal degrees of freedom rather to a hidden domain wall intersection.

Finally, we would like to consider a finite diameter defect of ferroelectric polarization that extends along 1D line in the material, but it stands alone within a ferroelectric domain rather than sticking to a particular domain wall plane or domain wall intersection (Fig. 1H). The polarization distribution within the defect may be, for example, topologically equivalent to that of magnetization in ferromagnetic skyrmions, as invoked in the seminal papers by Bogdanov and Yablonskii in 1989 [18].

3. Second "domainization": Formation of superdomains

Electromechanical requirements on the orientation of ferroelectric domain walls and energy costs of the ferroelastic wall intersections imply that multiaxial ferroelectrics naturally tend to form lamellar twin domain structures (see Fig. 2).

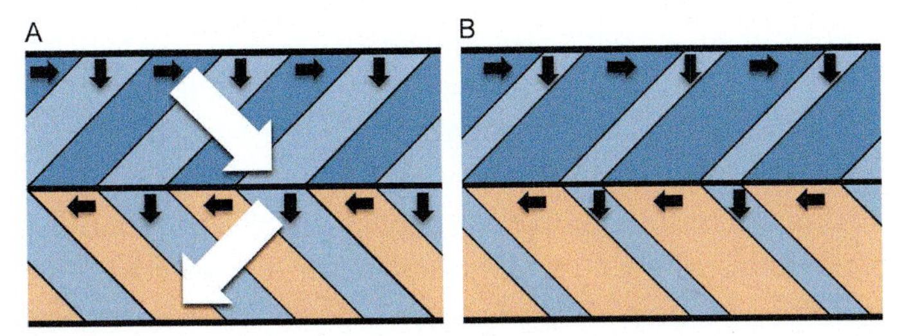

Fig. 2 Superdomains in typical lamellar domain structures in ferroelectrics. The domain volume of different states is equal (A) or largely unequal (B) within superdomains. The black arrows indicate polarization states in domains whereas the white arrows the overall polarization in superdomains.

The overall energy of the long-range electric and elastic interactions is achieved by domain miniaturization and equalization of domain sizes within the twin. Some average macroscopic properties of the twinned area, like its overall strain or the average electric polarization, depend on the volume ratios of the two alternating domain states but are often largely insensitive to the size of the individual domains. In this case, it is actually convenient to coarse-grain the view and to consider the twinned area with one type of lamellar twinning as a secondary domain or a superdomain. It can be often assumed that the optimal geometry of a free standing twin has domain volumes equal to each other (see Fig. 2A). In this case, such a macroscopic symmetry of the twin is the point group symmetry of the domain walls by which the twin is formed. The conditions of electromechanical compatibility of superdomains can be used to derive crystallographic orientations of permissible superdomain walls. If the ratio of domain volumes is largely unequal (see Fig. 2B), we have adaptive superdomains in which the macroscopic properties like strain or average polarization can vary considerably from place to place. The response to external influence may become largely nonlinear and the spontaneous quantities might form a continuous manifold rather than distinct values and the differences between concepts of domains and superdomains may become much more substantial.

4. Nanoscale domains

The small thickness of ferroelectric domain walls implies that in principle, the size of a ferroelectric domain can be of the order of a few nanometers only. They may form dense systems of domain walls or rather rare objects, limited, for example, to the vicinity of surface or phase front. Dense systems of nanoscale domains are often encountered in ferroelectrics with nanostructured morphology, such as in thin films, superlattices and fine grain ceramics, in materials with competing ferroelectric phases, such as those near the so-called morphotropic phase boundaries, and in materials with a nanoscale quenched disorder, as for example, in ferroelectric relaxor materials. Individual nanoscale domains within a bulk macroscopic domain are readily created under an atomically sharp electrode tip or a needle indentor or in the initial phases of a fast polarization switching process occurring at incomplete compensating charge screening conditions. Domain morphology of nanoscale domains does not obey the electromechanical conditions very strictly because the small scale and curvature limit the importance of the long-range electromechanical interactions. Typical shapes are lamellae, wedge domains and needle domains (Fig. 3A), but more complicated dendritic habitus reminiscent of fractals or snowflakes has been also reported [19].

In ferroelectric thin films under a short-circuit condition, domains of antiparallel polarization and a needle shape elongated perpendicularly to the film are typical for uniaxial ferroelectrics as well as for perovskite ferroelectric films under a compressive in-plane strain. These needle domains

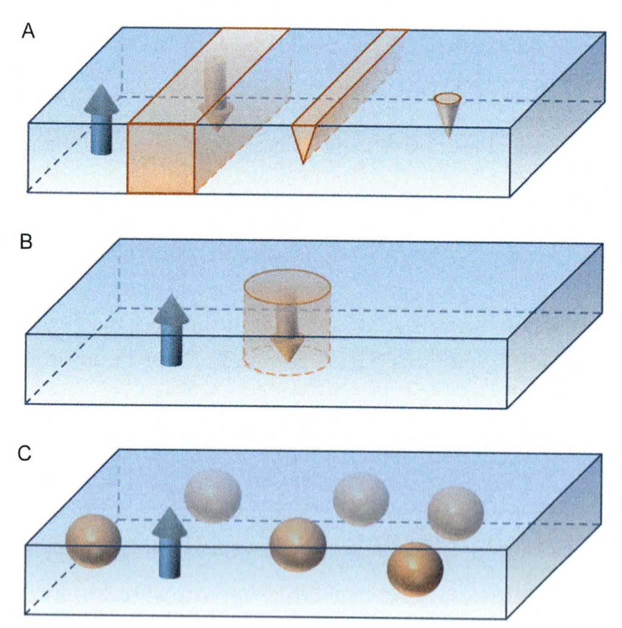

Fig. 3 Typical shapes of nanoscale domains in ferroelectrics: (A) ordinary domains, wedges, and needles, (B) small domains of cylindrical shape, and (C) individual three dimensional bubbles.

would typically grow up into small domains of cylindrical shape (Fig. 3B). Such morphology is equivalent to the so-called bubble domains well known from ferromagnetic garnet, hexaferrite and orthoferrite thin films with high out-of-plane easy axis magnetic anisotropy. These magnetic bubble domains have been very intensively studied by the magnetic community as a prospective memory media [20]. Somewhat unfortunately, the ferroelectric community recently started to use the term bubble domains for laterally confined domains of a nanoscale ball shape, which nevertheless more literally resembles the shape of individual three dimensional soap bubbles (Fig. 3C). These ball shaped bubbles have been demonstrated both in computer simulations [21] and in experiments [22,23]. Interestingly, the term Skyrmion has been probably first invoked in the context of ferroelectricity when referring to creations of cylindrical or ball-like nanodomain precursors in front of a turbulent switching front in ferroelectric lead germanite [24]. While this phenomenon is interesting and its essence is clarified in the above paper, there seems to be no particular reason to expect a deep link between these nanodomains and the ferromagnetic Skyrmions of the Bogadanov-Yablonskii type [18]. The ball-shaped ferroelectric domains sharply violate all macroscopic electromechanical compatibility conditions and therefore, for example, the sub-10 nm diameter of the ball bubble domains reported in perovskite ferroelectrics indicates the scale at which the traditional concept of electromechanically permissible domain walls ceases to have much sense.

5. Heterointerface engineering and closure domains

Domain structures in ferroelectric thin films and artificially designed ferroelectric heterostructures are considerably influenced by epitaxial interfaces between ferroelelectric and non-ferroelectric materials. Electrostriction is one of the dominating interactions. Among others, it was shown experimentally that epitaxial strain can shift the ferroelectric phase transitions by several hundreds of Kelvin. The electric dipole–dipole interaction in proper ferroelectric perovskites is another key player, an open-circuit interface of a few nm sized ferroelectric ball particles normally suppresses the ferroelectricity completely. In a more frequent thin film geometry, the dielectric interface or an open-circuit surface acts on the polarization in a way to make it parallel to the interface. In perovskite films with no other charge screening mechanism, one usually encounters formation of the flux closure domains in the films with thicknesses of the order of 100 nm or more, while in films with thicknesses of the order of 10 nm, a fully in-plane polarization is usually preferred. The size of the domains scales with the thickness of the film and has been shown to follow well the phenomenological Kittel's law through almost five orders of magnitude in the thickness of a ferroelectric slab. The competition of the mechanical and electrical compatibility at the interfaces between ferroelectric and adjacent materials, together with various charge-compensating mechanisms and multiple options for the size and geometrical design thus provide additional opportunities to tailor the ferroelectric nanodomain sizes and orientations that are not encountered in bulk ferroelectric materials. In particular, epitaxial thin films favor formation of superdomains, intersection of ferroelastic walls, and intersections of superdomain walls, such as in Fig. 4 [24a].

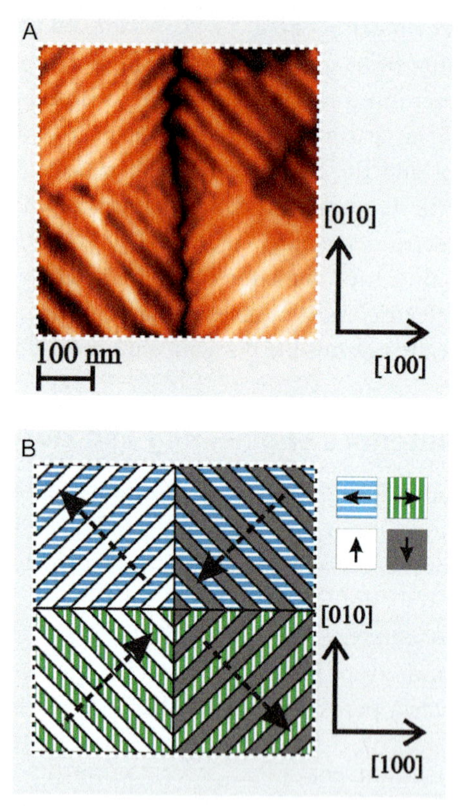

Fig. 4 Quadrant-quadruple domain pattern in a $PbTiO_3$ film on $SmScO_3$ substrate: (A) intersection of two mesoscopic domain walls oriented along [100] and [010] directions measured by lateral PFM. (B) Schematic suggestion of domain structure assignment. Black dashed arrows in (B) correspond to macroscopic polarization directions. *Reprinted from Borodavka, F., Gregora, I., Bartasyte, A., Margueron, S., Plausinaitiene, V., Abrutis, A., Hlinka, J. Ferroelectric nanodomains in epitaxial* $PbTiO_3$ *films grown on* $SmScO_3$ *and* $TbScO_3$ *substrates, J. Appl. Phys. 113 (2013) 187216, with the permission of AIP Publishing.*

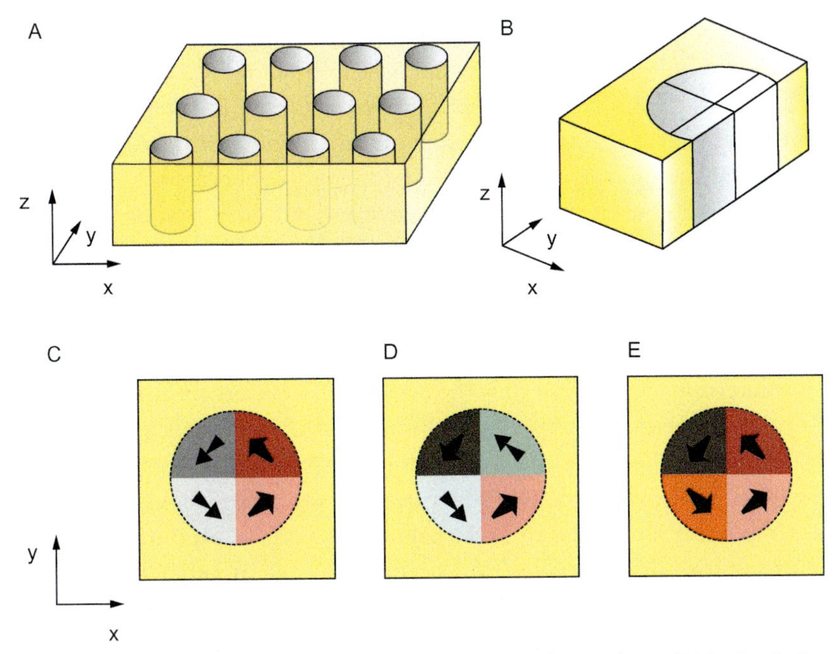

Fig. 5 Heterostructure and its nanodomain states from phase-field simulations: (A) brush-like arrangement of the $BaTiO_3$ nanorods embedded epitaxially in the [001] oriented $SrTiO_3$ epitaxial film; (B) domain boundaries within the 40-nm nanorod; (C) stable domain configuration in the 40-nm nanorod, z components of polarization arranged in a up-up–down-down manner; (D) another stable domain structure, with *z* components of polarization arranged in a up-down–up-down manner; (E) still another stable arrangement, with *z* components of polarization arranged in a up-up–up-up manner. All calculations assumed 3D periodic boundary conditions. *Reprinted figure with permission from Stepkova, V., Marton, P., Setter, N., Hlinka, J. Closed-circuit domain quadruplets in* $BaTiO_3$ *nanorods embedded in a* $SrTiO_3$ *film, Phys. Rev. B 89 (2014) 060101 (R) Copyright (2014) by the American Physical Society.*

In the case of ferroelectric nanodots, nanoscale ferroelectric cylinders, disks, balls, or cubes, one can usually find a typical size of the order of 10 nm, when the ferroelectric nanodots contain only a quadruplet of flux closure domains, with a single ferroelectric line in the center (see Fig. 5 and references [12,25,26]). Obviously, the ferroelectric line is that of Fig. 1A, but confinement often allows one to tolerate the electromechanical incompatibilities here. Since the overall polarization in symmetric nanodots is zero, a more convenient measure of the ferroelectric ordering seems to be some polarization-derived moments of the polarization distribution within the nanodot [27]. The most natural choice is either the integral electric toroidal moment [28] evaluated with respect to the geometrical center of the nanodot particle, or the pseudoscalar index obtained as an integral of the scalar product of polarization **P** with its differential vorticity rot **P** over the nanodot volume [29]. Complementary information can be obtained by calculation surface integrals of the Pontryagin density in the plane perpendicular to the electric toroidal moment of the nanodot but as far as we know, the integral values characteristic for the topology of the skyrmionic lines of Bogdanov-Yablonskii type were not encountered there yet [26].

6. Breakdown of the ferroelectric domain wall picture

In ultrathin films, ultrashort period superlattices and smallest nanodots, heterointerfaces control not only the domain wall density, domain wall positions and orientations, but they can also increase their thickness. At the same time, the heterointerfaces are influencing the magnitude and orientation of the polarization within domains in a strongly inhomogeneous manner and the same holds for the other order parameters such as a spontaneous strain tensor. At the dimensions of the order of 10 nm, it often happens that domain walls become as thick as the nanodomains and the distinction between already inhomogenous domains and significantly broadened domain walls fades out. If there are only two domain states melting together, or several domain states but all having nominal polarization perpendicular to a given axis, for example, the thin film normal, then the area of melted domains and domain walls does not form a superdomain but rather a novel object, which can be called an in-plane polarized sector, where the polarization is smoothly varying from place to place but remaining within a two-dimensional order parameter manifold (normal to the axis). Obviously the rest of the volume can form, for example, a nonpolar sector or an out-of-plane sector, which has no more any symmetry relation to the in-plane sector, but one can still trace residuals of boundaries between the in-plane and out-of-plane domains. In an ultimate case, all domain walls melt out and we have only a polar sector with an arbitrary polarization orientation but no domain walls. In this case, the texture can only be described in terms of differential geometry and topology such as those briefly mentioned in the above paragraph. Fortunately, the topologically essential low-dimensional defects, such as polarization lines, can be still identified. For example, the core of the ferroelectric flux closure domain quadruplet can be identified by the singularity of the differential vorticity rot **P** even if domain walls melt out in an in-plane polarized sector. This crossover somehow mimics the crossover from the situation in Fig. 1A–H, except that the residual ferroelectric line defect is not embedded in a ferroelectric domain, but rather in a ferroelectric sector. Interestingly, domain wall melting can be enhanced also by other influences such as disorder, vicinity of the ferroelectric critical point or by high proximity of domain walls themselves. The concept of polar and nonpolar sectors as opposed to domains or superdomains could thus possibly have some use also in description of other conceptually exceptional situations, such as in ferroelectric relaxors.

7. Geometry of ferromagnetic skyrmions

Properties of ferromagnetic skyrmions have been considered many times in the recent literature and at several places within this book. Emphasis on its topological properties somehow puts aside the original requirements of its line defect character as opposed to Bloch or Néel wall bubble domains on the one hand and considerations about the importance of the stabilization mechanism on the other hand. While some scientists still argue about these conceptual differences, others argue that skyrmion line defects and skyrmion bubble domains might be equally useful for some practical applications. Which of the aspects will be most important for ferroelectric skyrmionics is far from being obvious. In either case, clarity of definitions and concepts is always useful for organizing novel findings.

For the purpose of this brief chapter, it is sufficient to refer to the idealized single Bloch skyrmion texture with its axially symmetric cylindrical symmetry with a cross-section indicated in Fig. 1C of Ref. [30] (reproduced in Fig. 6). It is well understood now that it can be obtained by a stereographic projection of a combed hedgehog's vector "hair" decoration at an auxiliary sphere. Here the sphere is representing the available horizontal real space which is in a unique way full mapped to full order parameter space of all possible directions of the constant magnitude magnetic moment. Such an order parameter space can be thought of as that of saturated magnetization relevant to a ferromagnetic crystal far away from the ferromagnetic phase transition; the magnitude of magnetization is almost constant. Nevertheless, the constant magnitude of the magnetization is not the essential feature of the skyrmion, as long as it is energetically preferable to avoid the zero magnetization everywhere in the profile of the whole defect.

Ferromagnetic Néel skyrmions obtained by stereographic projections of the centrosymmetric hedgehog sphere are actually also known, for example, in ferroelectric substances like GaV_4S_8. There the ferromagnetic Néel skyrmion is naturally decorated by a small improper ferroelectric component, creating thus an interesting multiferroic composite skyrmion object.

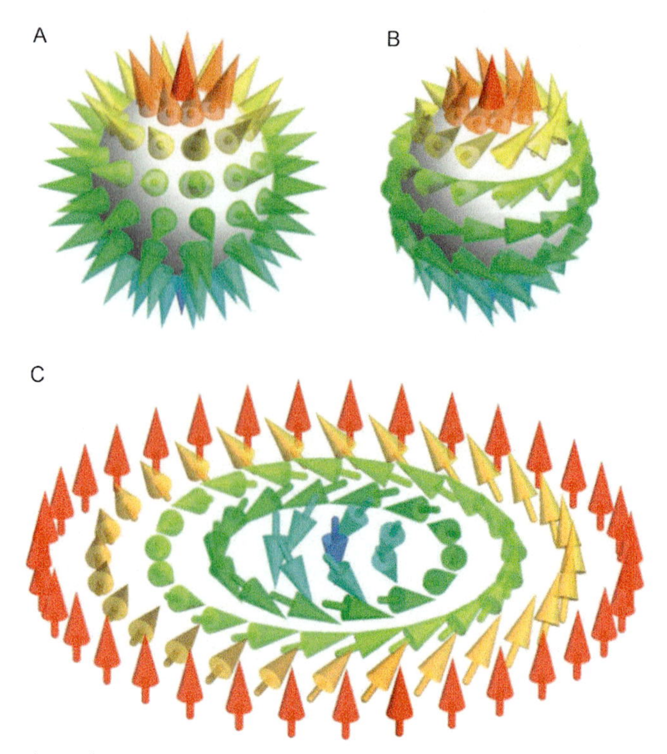

Fig. 6 Topological equivalence of a hairy sphere and a vortex. After combing a hairy sphere (A and B) a stereographic projection onto the plane results in a vortex (C). The hairy sphere and the vortex are topologically equivalent; both are skyrmions. Image courtesy of Achim Rosch. *Reprinted by permission from Springer Nature: Pfleiderer, C. Magnetic order: surfaces get hairy, Nat. Phys. 7 (2011) 673–674. Copyright (2011).*

When it comes to proper ferroelectric skyrmions, the ferroelectric Néel skyrmion would nevertheless require a considerable electric charge needed to compensate for the divergence of the radial gradients of the electric polarization. Therefore, in the case of proper ferroelectric skyrmions, the Bloch skyrmion geometry seems to be the only favorable option.

In order to collect a full integer topological charge, the ferromagnetic skyrmion should in principle include all directions of the magnetization, in particular, its axis should be polarized in an opposite direction to the surrounding domain state (Fig. 7) [30a]. Obviously, the possibility of this complete embedding in a homogeneous single domain state of the defect is a prerequisite for its most compelling aspects, such as possibility to be created, moved or destroyed. To make this ideal story complete, it is fair to invoke that in a thin film topology, the top and bottom of the Bloch skyrmion line is likely

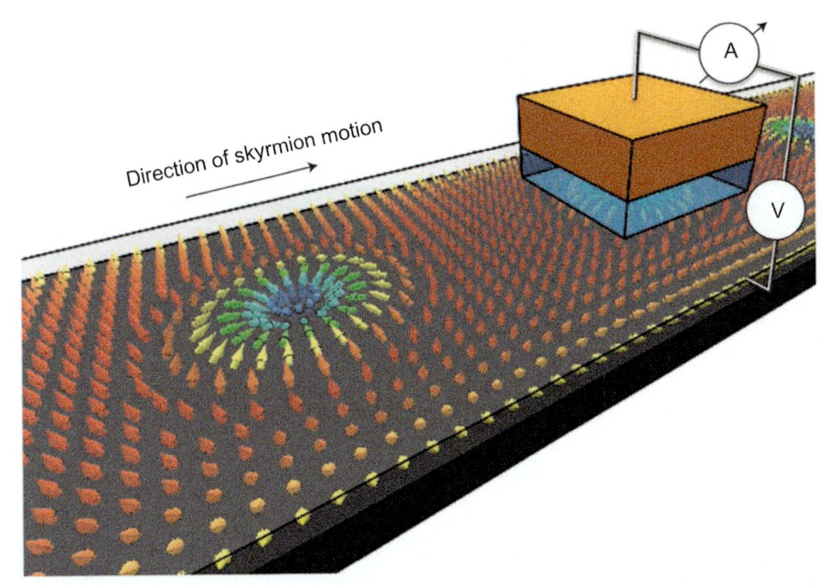

Fig. 7 Skyrmions and their electrical detection in a racetrack memory. A representation of a skyrmion racetrack memory with a tunnel junction read-out device (blue and orange boxes). The measured tunneling current would change as skyrmions are pushed underneath the device with an electrical current. *Reprinted by permission from Springer Nature: Monchesky, T. L. Detection with unpolarized currents, Nat. Nanotechnol. 10 (2015) 1008–1009 Copyright (2011).*

to close up as a Néel skyrmion, as demonstrated experimentally, for example, in the case of magnetic skyrmion lattice shown in Fig. 8 [31]. On the other hand, in a purely bulk form, the skyrmion lattices were predicted to be stable only in a bias magnetic field, and limited to finite areas of the phase diagram due the competing helical or cycloidal phases in Fig. 9 [32].

8. Ferroelectric textures with Bloch skyrmion symmetry

The geometry of Bloch magnetic skyrmions superposes vorticity of the in-plane component of the magnetization field with the chirality of the full radial skyrmion profile. By analogy, in the last few years the ferroelectric community has been searching for domain structures and ferroelectric line defects that would combine vorticity and chirality at a time. It can be easily verified, that symmetry of a ferroelectric skyrmion is the one which simultaneously allows three parallel quantities—polar and axial vector and chiral bidirector [28].

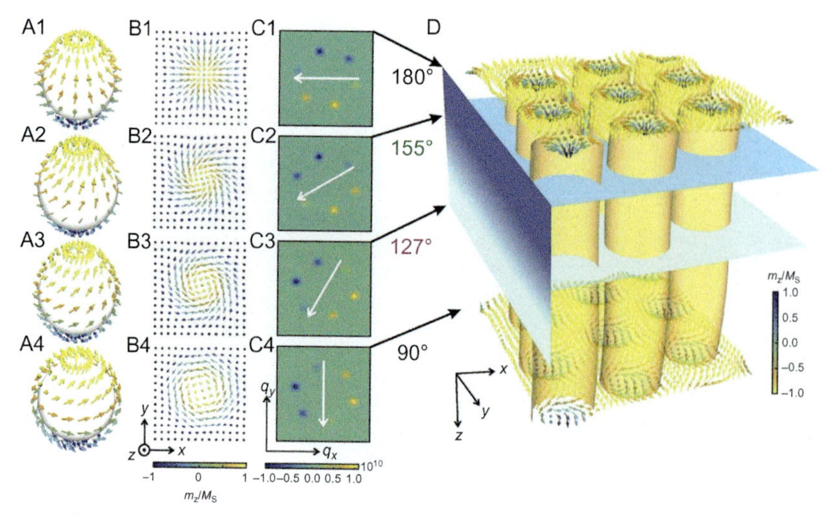

Fig. 8 Illustration of skyrmion order ranging from Néel- to Bloch-twisting with increasing depth below a surface. (A1–A4) Hedgehog spin configuration on the surface of a sphere for skyrmions with winding number $N=1$ varying between pure Néel-twisting (A1) and pure Bloch-twisting (A4). (B1–B4) Real space planar spin configurations. A stereographic projection connects the planar patterns shown in B1–B4 with the hedgehog configurations shown in A1–A4, respectively. (C1–C4) Calculated diffraction pattern associated with spin configurations shown in B1–B4. (D) Schematic illustration of the change from Néel- to Bloch-twisting with increasing depth below the surface. *Reprinted from Zhang, S., van der Laan, G., Müller, J., Heinen, L., Garst, M., Bauer, A., Berger, H., Pfleiderer, C., Hesjedal, T. Reciprocal space tomography of 3D skyrmion lattice order in a chiral magnet, Proc. Natl. Acad. Sci. U. S. A. 115 (2018) 6386–6391.*

The most remarkable progress was brought in series of papers devoted to ferroelectric-dielectric superlattices of $PbTiO_3$-$SrTiO_3$ [33–38]. When $PbTiO_3$ and $SrTiO_3$ layers are about 10 perovskite unit cells thick, the domain structure of each $PbTiO_3$ layer is largely independent and splits in a roughly square-like blocks of flux closure domain quadruplets. Adjacent closure domain quadruplet cores are located in about 10 perovskite unit cell distances and they have opposite electric toroidal moments—vorticity (see Fig. 10). High resolution electron microscopy images and phase-field simulations in Fig. 11 suggest that domain walls are actually very broad. When the almost melted out domain walls are disregarded, the only residual essential features are the ferroelectric line defects at the cores of the vortices. The ferroelectric slabs of the superlattice can be then considered as a compact vortex sector with an array of more or less freely floating vortex cores.

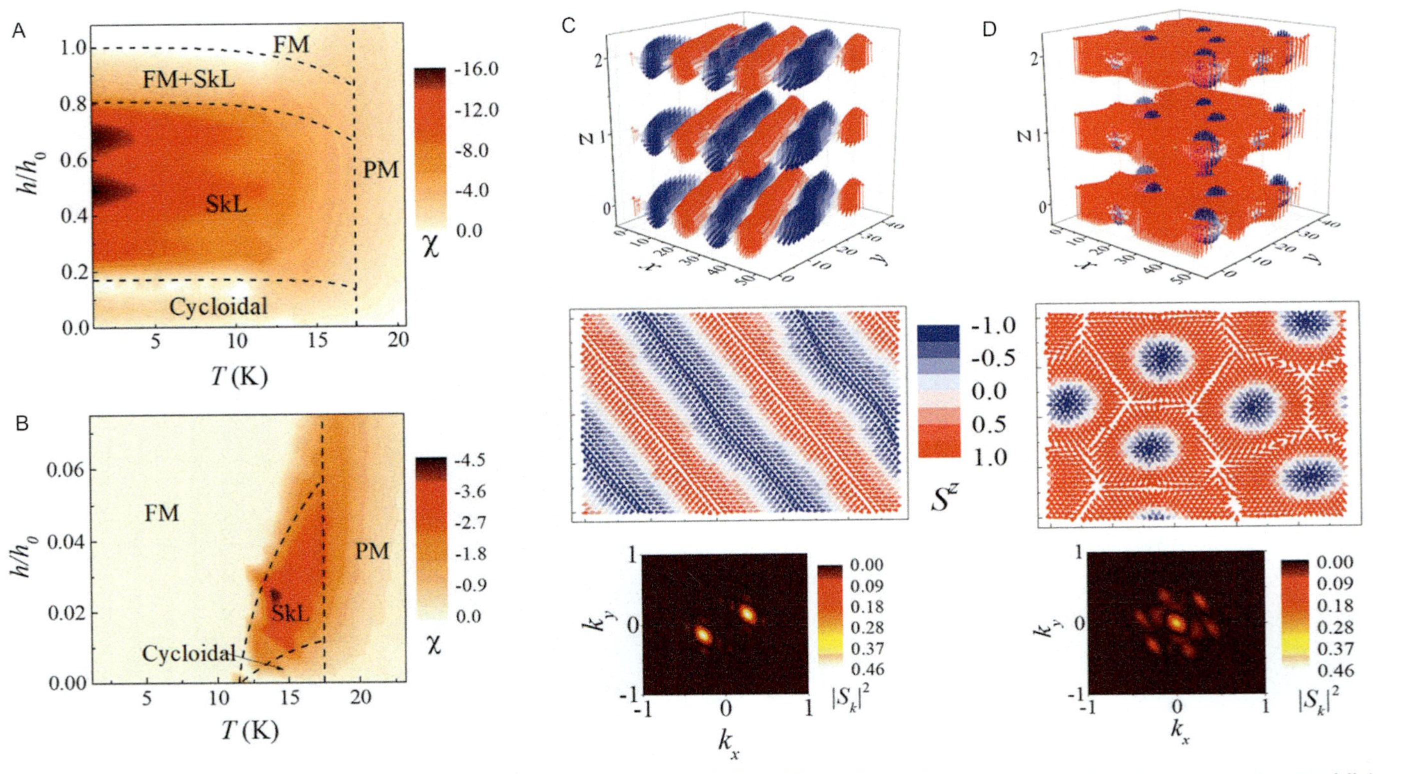

Fig. 9 (A and B) Monte Carlo phase diagrams with skyrmion phases. (C) (Top) Three dimensional snapshot of cycloid spin order. (Middle) In-plane spin structure factor. (Bottom) In-plane spin structure factors. (D) Same as (C) but for the skyrmion lattice. h_0 denotes the saturation magnetization at low temperature. *Reprinted figure with permission from Zhang, H.-M., Chen, J., Barone, P., Yamauchi, K., Dong, S., Picozzi, S. Possible emergence of a skyrmion phase in ferroelectric GaMo4S8, Phys. Rev. B 99 (2019) 214427 Copyright (2019) by the American Physical Society.*

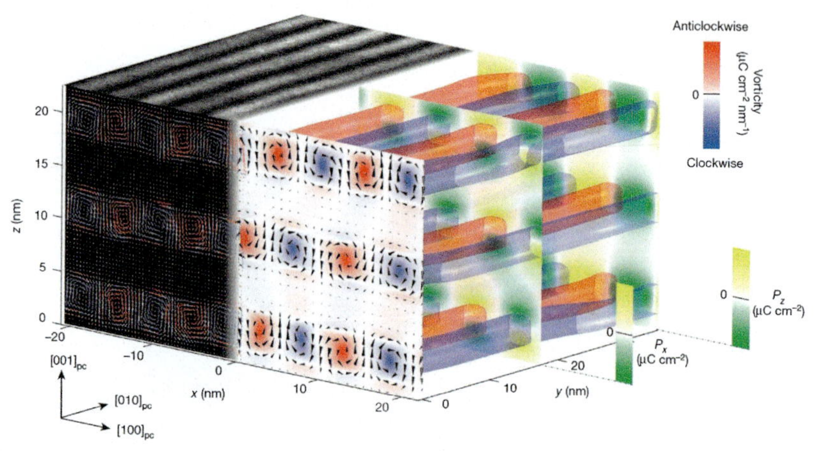

Fig. 10 Three dimensional phase-field simulation for a $(SrTiO_3)_{10}/(PbTiO_3)_{10}$ superlattice. The 3D geometry of the vortex–antivortex array from phase-field simulation of a $(SrTiO_3)_{10}/(PbTiO_3)_{10}$ superlattice is shown on the right. The front cross-section of the model shows the polarization vector **P** ordered into clockwise (blue) and anticlockwise (red) vortex–antivortex states which extend along the $[010]_{pc}$. For comparison, on the left is shown a cross-sectional HR-STEM image overlaid with a polar displacement vector map and a planar-view DF-TEM image projected onto the front and top planes of the axes, respectively. Red/blue color scales correspond to the curl of the polarization extracted from the phase-field model and the HR-STEM polar displacement map. *Reprinted by permission from Springer Nature: Yadav, A.K., Nelson, C. T., Hsu, S.L., Hong, Z., Clarkson, J.D., Schlepütz, C.M., Damodaran, A.R., Shafer, P., Arenholz, E., Dedon, L.R., Chen, D., Vishwanath, A., Minor, A. M., Chen, L.Q., Scott, J.F., Martin, L.W., Ramesh, R. Observation of polar vortices in oxide superlattices, Nature 530 (2016) 198–201 Copyright (2016).*

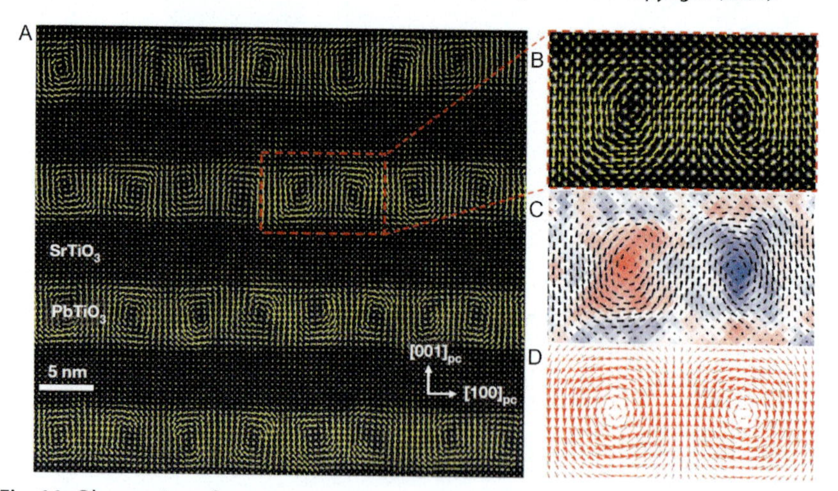

Fig. 11 Observation of vortex–antivortex structures: (A) Cross-sectional HR-STEM image with an overlay of the polar displacement vectors (indicated by yellow arrows) for a $(SrTiO_3)_{10}/(PbTiO_3)_{10}$ superlattice, showing that an array of vortex–antivortex pairs is present in each $PbTiO_3$ layer, (B) a magnified image of a single vortex–antivortex pair, (C) the curl of the polar displacement for the same vortex–antivortex pair reveals the alternating rotation directions of the structures, (D) polarization vectors from a phase-field simulation of the same $(SrTiO_3)_{10}/(PbTiO_3)_{10}$ superlattice. *Reprinted by permission from Springer Nature: Yadav, A.K., Nelson, C. T., Hsu, S.L., Hong, Z., Clarkson, J.D., Schlepütz, C.M., Damodaran, A.R., Shafer, P., Arenholz, E., Dedon, L.R., Chen, D., Vishwanath, A., Minor, A. M., Chen, L.Q., Scott, J.F., Martin, L.W., Ramesh, R. Observation of polar vortices in oxide superlattices, Nature 530 (2016) 198–201 Copyright (2016).*

Later work on superlattices with somewhat thinner $PbTiO_3$ ferroelectric layers has indicated that within the same sample, the vortex sector can coexist with the usual superdomains of a1/a2 in-plane nanodomains (Fig. 12) [34]. These superdomains are essentially the same type of a1/a2 superdomains with electromechanically compatible 90-degree ferroelectric domain walls as those known from tensile strained thin films (Fig. 4).

Detailed inspection of these superlattices by piezoresponse force microscopy suggested, that the vortex sector has also an in-plane polarization, possibly with the polarity oriented along the vortex cores (see Fig. 12F–H). Interestingly, the antipolar and antiferroaxial arrangement (with alternating sense of the polar and axial vector order parameters of the vortex lines) implies same handedness of the adjacent vortices. Thus, the whole vortex sector is of a homochiral nature. This proposition seems to be justified by the X-ray circular dichroism, also reported in a subsequent paper focused on this issue only [36]. It should be stressed that unlike in the chiral magnets, the $PbTiO_3/SrTiO_3$ superlattices are in principle achiral so that the above mentioned observations have to be interpreted as a spontaneous chirality breaking. Indeed, inspection of the sample with the X-ray circular dichroism technique revealed a sector with opposite chirality in a different part of the same sample [36]. It is worth admitting at this point that even though the whole sector as well as individual vortex cores were shown to be chiral, the polarization texture of an individual vortex does not need and apparently does not cover all vector directions as in the true Bloch skyrmion case shown in Fig. 6. Also, the eventual creation or annihilation of new vortices in the vortex sector of these $PbTiO_3/SrTiO_3$ superlattices could probably be only realized in pairs. Still, until the year 2019, these chiral vortex sectors [34,36] were definitely the closest experimentally available textures to the ultimately desired ideal chiral skyrmion phases.

9. Ferroelectric textures with skyrmion topology

The most recent and really surprising results for the whole ferroelectric community were reported in 2019 [37]. When rather similar $PbTiO_3/SrTiO_3$ superlattices were grown on a bulk $SrTiO_3$ crystal substrate instead of on a $DyScO_3$ substrate, the domain morphology has completely changed. Instead of in-plane domains and vortex sectors originating from melted flux closure nanodomains, researchers have observed bubble domains. The most simple trilayer superlattices with a single ferroelectric layer inside have shown cylindrical ferroelectric nanodomain bubbles, elongated in the cross-section

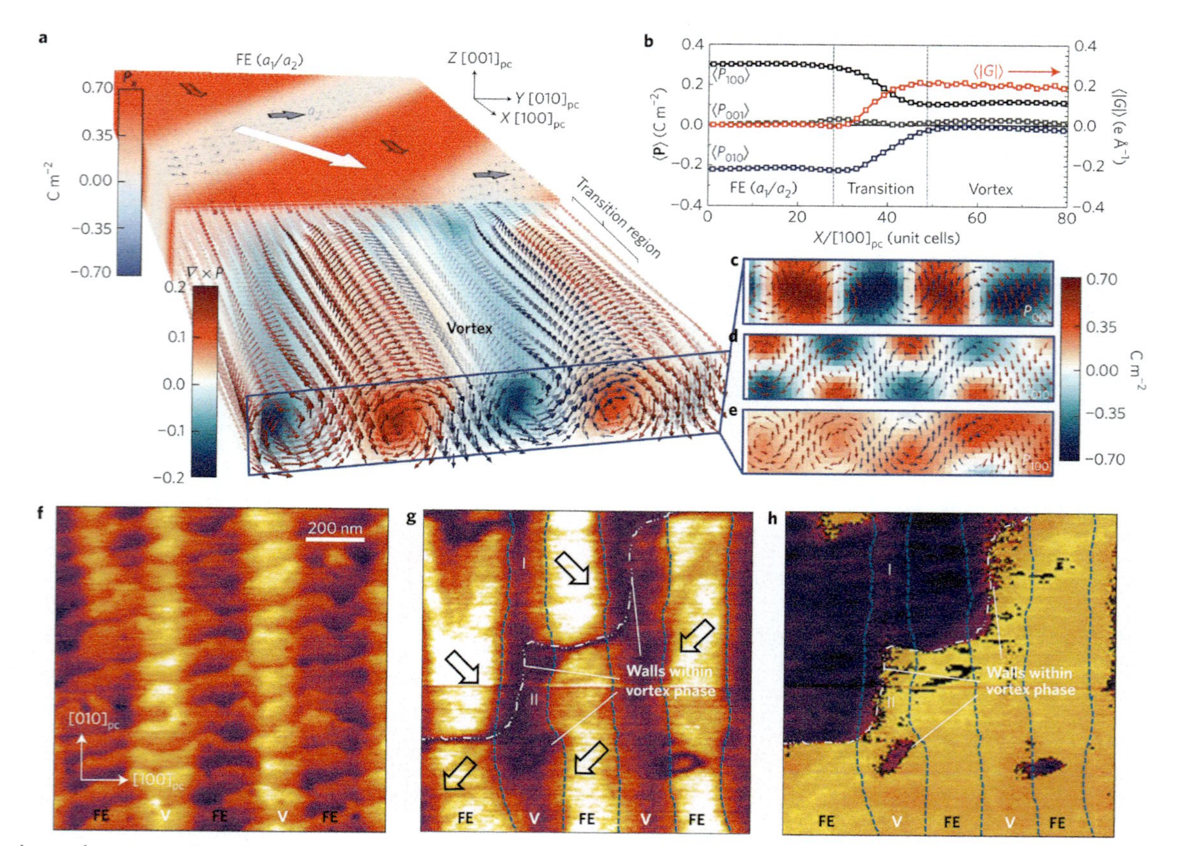

Fig. 12 See legend on opposite page.

(Fig. 13A and C). Superlattices with eight repeats of $PbTiO_3/SrTiO_3$ have shown similar cylindrical ferroelectric nanodomain bubbles but more circular in the cross-section (Fig. 13B and D). State of the art high resolution microscope images revealed that these bubble domains have Bloch walls. Moreover, all inspected bubbles have shown the same vorticity.

The results were supported by phase-field simulations which also indicated formation of bubble domains of the same vorticity, and thus with the same chirality (see Fig. 14). This observation is even more surprising since the Landau potential functional used in phase-field simulations has no chiral interactions involved and the geometry of the simulation box has no chirality by itself. For comparison, our own tentative phase-field simulations of a regular square lattice of Bloch-like ferroelectric bubble domains of similar distances and similar diameters as in the reported experiments are shown in Figs. 15 and 16. Our simulation suggests that a racemic mixture of bubble domains (with about equal amount of right-handed and left-handed bubble domains), resulting from a small random noise in the initial in-plane polarization, is actually fairly stable in the sense that selected handedness of the bubble is preserved. This suggests that, in fact, if one finds a way how to create each individual Bloch skyrmion bubble with a desirable handedness, the handedness could be stored there as a bit of information. At the same time, small energy differences between homochiral and the racemic arrangements are then hardly the sole reason for the reported emergent long-range chirality.

Fig. 12 Exploring the phase boundary between a_1/a_2 and vortex phases: (A) A zoom-in of the phase-field calculation focusing on a single $PbTiO_3$ layer near the phase boundary and revealing that the ferroelectric a_1/a_2 superdomain, color scale showing polarization component along $[100]_{pc}$ or the X-direction) smoothly transitions to the vortex phase comprising of alternating clockwise and anticlockwise polarization vortices, (B) plot showing the spatially averaged ferroelectric polarization components and the averaged magnitude of the axial electrical toroidal moment, (C–E) are color-maps showing the spatial distribution of the (C) P_{001}, (D) P_{010} and (E) P_{100} components for a cross-section, (F–H) Lateral piezoresponse force studies showing (F) topography, (G) lateral amplitude, and (H) lateral phase revealing the nanoscale distribution of a_1/a_2 and vortex phases, and highlighting the presence of domain-wall-like features within the vortex phase that are indicative of axial polarization components. *Reprinted by permission from Springer Nature: Damodaran, A. R., Clarkson, J. D., Hong, Z., Liu, H., Yadav, A. K., Nelson, C. T., Hsu, S.-L., McCarter, M. R., Park, K.-D., Kravtsov, V., Farhan, A., Dong, Y., Cai, Z., Zhou, H., Aguado-Puente, P., García-Fernández, P., Íñiguez, J., Junquera, J., Scholl, A., Raschke, M.B., Chen, L.-Q., Fong, D.D., Ramesh, R., Martin, L.W. Phase coexistence and electric-field control of toroidal order in oxide superlattices, Nat. Mater. 16 (2017) 1003–1009 Copyright (2017).*

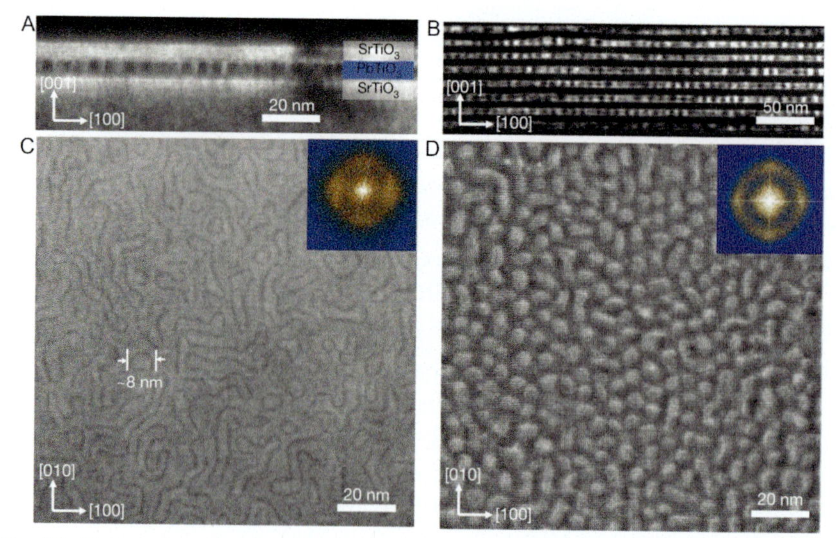

Fig. 13 Observation of ordered polar structure: (A and B) Cross-sectional dark-field TEM images of a $(SrTiO_3)_{16}/(PbTiO_3)_{16}/(SrTiO_3)_{16}$ trilayer (A) and a $[(SrTiO_3)_{16}/(PbTiO_3)_{16}]_8$ superlattice (B), revealing a regular in-plane modulation of about 8nm, (C and D) planar-view dark-field STEM imaging shows the widespread occurrence of nanometer-size round and elongated features in a $(SrTiO_3)_4/(PbTiO_3)_{11}/(SrTiO_3)_{11}$ trilayer (C) and only circular features in a $[(SrTiO_3)_{16}/(PbTiO_3)_{16}]_8$ superlattice (D) along the [100] and [010] directions. Insets, FFT of the images in (C) and (D) show a ring-like distribution with stronger intensities along the cubic directions. *Reprinted by permission from Springer Nature: Das, S., Tang, Y. L., Ramesh, R. Observation of room-temperature polar skyrmions, Nature 568 (2019) 368–372. Copyright (2019).*

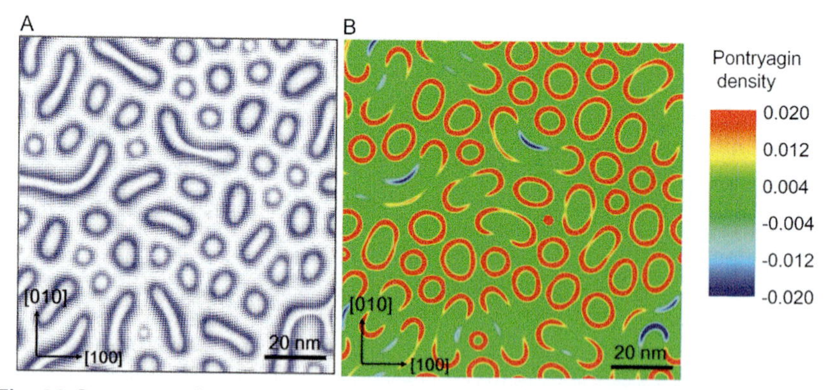

Fig. 14 Pontryagin density distribution from phase-field simulations of skyrmion bubbles and elongated skyrmions for the $[(PbTiO_3)_{16}/(SrTiO_3)_{16}]_3$ superlattice, (A) Planar view of the $[(PbTiO_3)_{16}/(SrTiO_3)_{16}]_3$ superlattice, showing coexistence of skyrmion bubbles with elongated-skyrmion features, (B) The corresponding Pontryagin density calculation. Skyrmion bubbles have a circular shape, indicating a radial distribution of the polar texture, whereas the elongated-skyrmion features have larger density in the two ends and both positive and negative values in the middle. The surface integration gives a topological charge value of +1 for both the skyrmion bubbles and the elongated skyrmions, so the elongated-skyrmion features and the skyrmion bubbles are topologically equivalent structures. *Reprinted by permission from Springer Nature: Das, S., Tang, Y. L., Ramesh, R. Observation of room-temperature polar skyrmions, Nature 568 (2019) 368–372. Copyright (2019).*

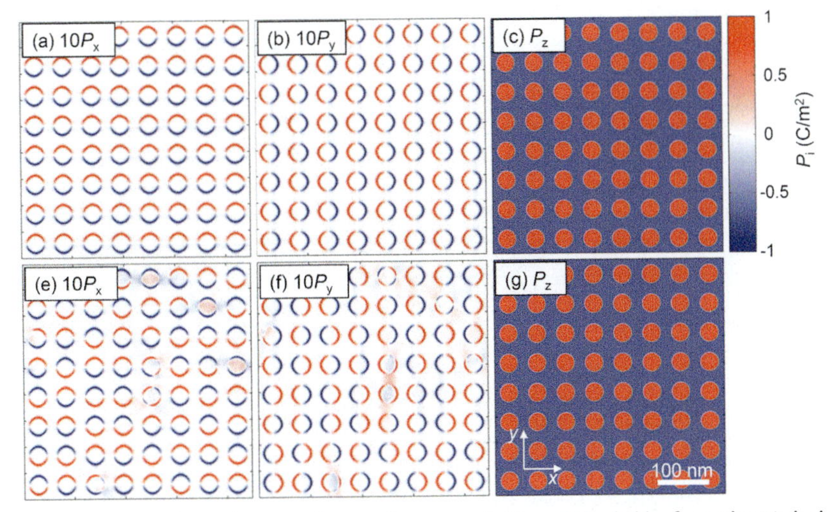

Fig. 15 Phase-field simulations of a regular square lattice of Bloch-like ferroelectric bubble domains. Polarization components of (A–C) homochiral and (E–G) racemic configurations. Parameters of the Giznburg-Landau-Devonshire model were taken from the model III of $PbZr_{1-x}Ti_xO_3$ at $x=0.6$ and $T=0$ K [39].

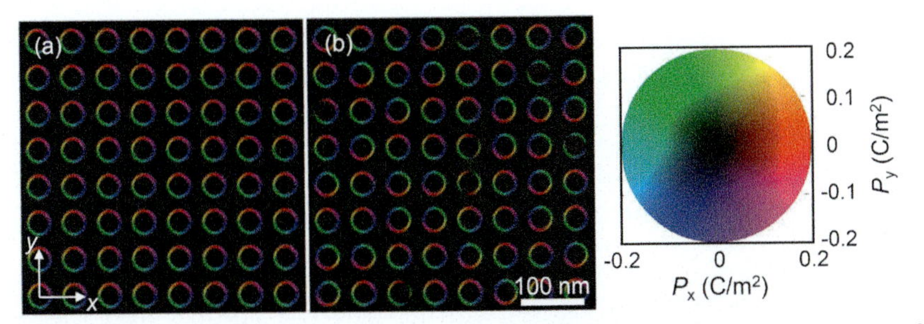

Fig. 16 Bubble domains as in Fig. 15 with P_x and P_y polarization components of (A) homochiral and (B) racemic configurations plotted in polar representation, showing clearly the sense of rotation of the Bloch component.

We speculate that one possible reason could be that there are many tiny chiral bubbles in the early stage of the domain structure formation and during the domain coarsening the more prevalent vorticity somehow proliferates until the numerical excess of the skyrmion charge matches the number of the residual coarse domains. Another possible scenario is that all the bubbles are formed from a common, very long and randomly dispersed single Bloch line that has become homochiral well before it was broken into the separate bubble domains. In fact, the latter scenario seems to be related

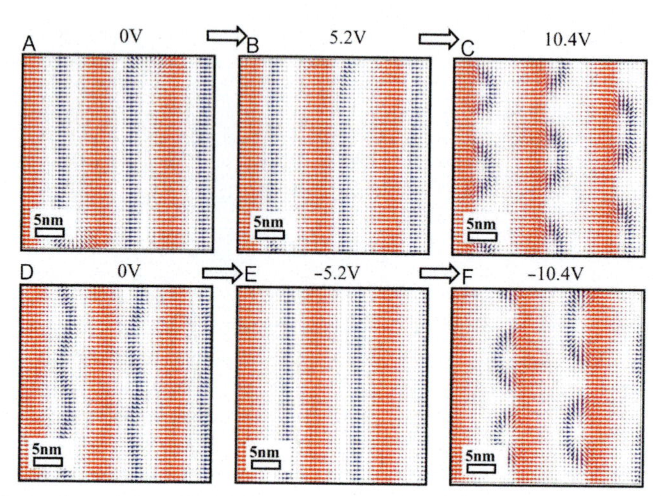

Fig. 17 Planar view of the polarization pattern in a $(PbTiO_3)_{16}/(SrTiO_3)_{16}$ superlattice and its evolution under applied bias: (A)–(C) positive bias and (D)–(F) negative bias one. The melting of the vortex line into "bubble" structure is shown. *Reprinted from Hong, Z., Chen, L.-Q. Blowing polar skyrmion bubbles in oxide superlattices, Acta Mater. 152 (2018) 155–161 Copyright (2018), with permission from Elsevier.*

to a peculiar process of vortex line melting, discovered in an independent phase-field simulation study of field induced effects in superlattices [38] (see Fig. 17).

10. Conclusion

In conclusion, there has been a tremendous progress in experimental studies of ferroelectric structures with ferroelectric vortices over the last 3 years. In particular, it has been reported that in perovskite ferroelectric nanostructures, chiral polarization textures can propagate at length scales considerably exceeding the corresponding superlattice periods. This occurs in two seemingly unrelated cases. First, the originally achiral flux closure superdomains in $PbTiO_3/SrTiO_3$ superlattices grown on scandium oxide substrates were shown to melt into globally chiral vortex sectors supporting a sequence of vortex cores with alternating vorticity when the thickness of $PbTiO_3$ is below about 10 nm. Second, homochiral irregular domain texture formed by cylindrical bubble domains of about 10 nm size was observed in $PbTiO_3/SrTiO_3$ superlattices grown on $SrTiO_3$ substrates. A ferroelectric polarization profile across such a cylindrical bubble domain is topologically equivalent to that of a magnetization distribution within a magnetic Bloch skyrmion. In the context of these results, it seems likely that in future ferroelectric cylindrical bubble domains can be found also in ferroelectrics with chiral paraelectric phases. In this case the vorticity of the ferroelectric cylindrical bubble domains could be uniquely defined by the crystal or superlattice design. This gives hope for observation of intrinsic ferroelectric skyrmion line defects and other interesting phenomena, either analogous to those known from magnetic skyrmion phases or perhaps even being unique for ferroelectric skyrmion phases only. It is also possible that ferroelectric community will exploit the possibility that in the so far available ferroelectric Bloch skyrmion bubble host materials, bubbles of both handedness can be formed and used, for example, to store information.

References

[1] S. Mühlbauer, B. Binz, F. Jonietz, C. Pfleiderer, A. Rosch, A. Neubauer, R. Georgii, P. Böni, Skyrmion lattice in a chiral magnet, Science 323 (2009) 915–919.
[2] V. Dvořák, Improper ferroelectrics, Ferroelectrics 7 (1974) 1–9.
[3] J. Hlinka, J. Privratska, P. Ondrejkovic, V. Janovec, Symmetry guide to ferroaxial transitions, Phys. Rev. Lett. 116 (2016) 177602.
[4] J. Fousek, V. Janovec, The orientation of domain walls in twinned ferroelectric crystals, J. Appl. Phys. 40 (1969) 135–142.
[5] V. Janovec, A symmetry approach to domain structures, Ferroelectrics 12 (1976) 43–53.
[6] A.K. Tagantsev, L.E. Cross, J. Fousek, Domains in Ferroic Crystals and Thin Films, Springer, New York, Dordrecht Heidelberg London, 2010.
[7] V. Janovec, J. Privratska, Domain structures, in: A. Authier (Ed.), International Tables for Crystallography, vol. D, International Union of Crystallography, 2003 (Chapter 3.4).
[8] D. Lee, R.K. Behera, P. Wu, H. Xu, Y.L. Li, S.B. Sinnott, S.R. Phillpot, L.Q. Chen, V. Gopalan, Mixed Bloch-Néel-Ising character of 180° ferroelectric domain walls, Phys. Rev. B 80 (2009) 060102 (R).
[9] J. Hlinka, V. Stepkova, P. Marton, I. Rychetsky, V. Janovec, P. Ondrejkovic, Phase–field modelling of 180-degree "Bloch walls" in rhombohedral $BaTiO_3$, Phase Transit. 84 (2011) 738–746.
[10] M. Taherinejad, D. Vanderbilt, P. Marton, V. Stepkova, J. Hlinka, Bloch-type domain walls in rhombohedral $BaTiO_3$, Phys. Rev. B 86 (2012) 155138.
[11] J.C. Wojdeł, J. Íñiguez, Ferroelectric transitions at ferroelectric domain walls found from first principles, Phys. Rev. Lett. 112 (2014) 247603.
[12] V. Stepkova, P. Marton, N. Setter, J. Hlinka, Closed-circuit domain quadruplets in $BaTiO_3$ nanorods embedded in a $SrTiO_3$ film, Phys. Rev. B 89 (2014) 060101 (R).
[13] V. Stepkova, P. Marton, J. Hlinka, Ising lines: natural topological defects within ferroelectric Bloch walls, Phys. Rev. B 92 (2015) 094106.
[14] M. Šafránková, J. Fousek, S.A. Kižaev, Domains in ferroelectric $YMnO_3$, Czech. J. Phys. B 17 (1967) 559–560.
[15] T. Jungk, Á. Hoffmann, M. Fiebig, E. Soergel, Electrostatic topology of ferroelectric domains in $YMnO_3$, Appl. Phys. Lett. 97 (2010) 012904.
[16] G. Catalan, J. Seidel, R. Ramesh, J.F. Scott, Domain wall nanoelectronics, Rev. Mod. Phys. 84 (2012) 119.
[17] V. Stepkova, J. Hlinka, On the possible internal structure of the ferroelectric Ising lines in $BaTiO_3$, Phase Transit. 90 (2017) 11–16.

[18] A.N. Bogdanov, D.A. Yablonskii, Thermodynamically stable "vortices" in magnetically ordered crystals. The mixed state of magnets, Sov. Phys. JETP 68 (1989) 101–103. Zh. Eksp. Teor. Fiz. 95 (1989) 178–182.

[19] V.Y. Shur, M.S. Kosobokov, E.A. Mingaliev, D.K. Kuznetsov, P.S. Zelenovskiy, Formation of snowflake domains during fast cooling of lithium tantalate crystals, J. Appl. Phys. 119 (2016) 144101.

[20] A.H. Eshenfelder, Magnetic Bubble Technology, SSS14, Springer, 1981.

[21] B.-K. Lai, I. Ponomareva, I.I. Naumov, I. Kornev, H. Fu, L. Bellaiche, G.J. Salamo, Electric-field-induced domain evolution in ferroelectric ultrathin films, Phys. Rev. Lett. 96 (2006) 137602.

[22] Q. Zhang, L. Xie, G. Liu, S. Prokhorenko, Y. Nahas, X. Pan, L. Bellaiche, A. Gruverman, N. Valanoor, Nanoscale bubble domains and topological transitions in ultrathin ferroelectric films, Adv.Mater. 29 (2017) 1702375.

[23] Q. Zhang, S. Prokhorenko, Y. Nahas, L. Xie, L. Bellaiche, A. Gruverman, N. Valanoor, Deterministic switching of ferroelectric bubble nanodomains, Adv. Funct. Mater. 29 (2019) 1808573.

[24] M. Dawber, A. Gruverman, J.F. Scott, Skyrmion model of nano-domain nucleation in ferroelectrics and ferromagnets, J. Phys. Condens. Matter 18 (2006) L71–L79. (a) F. Borodavka, I. Gregora, A. Bartasyte, S. Margueron, V. Plausinaitiene, A. Abrutis, J. Hlinka, Ferroelectric nanodomains in epitaxial $PbTiO_3$ films grown on $SmScO_3$ and $TbScO_3$ substrates, J. Appl. Phys. 113 (2013) 187216.

[25] I. Naumov, A.M. Bratkovsky, Unusual polarization patterns in flat epitaxial ferroelectric nanoparticles, Phys. Rev. Lett. 101 (2008) 107601.

[26] Y. Nahas, S. Prokhorenko, L. Louis, Z. Gui, I. Kornev, L. Bellaiche, Discovery of stable skyrmionic state in ferroelectric nanocomposites, Nat. Commun. 6 (2015) 8542.

[27] S. Prosandeev, L. Bellaiche, Hypertoroidal moment in complex dipolar structures, J. Mater. Sci. 44 (2009) 5235–5248.

[28] J. Hlinka, Eight types of symmetrically distinct vectorlike physical quantities, Phys. Rev. Lett. 113 (2014) 165502.

[29] J. Mangeri, Y. Espinal, A. Jokisaari, S.P. Alpay, S. Nakhmanson, O. Heinonen, Topological phase transformations and intrinsic size effects in ferroelectric nanoparticles, Nanoscale 9 (2017) 1616–1624.

[30] C. Pfleiderer, Magnetic order: surfaces get hairy, Nat. Phys. 7 (2011) 673–674. (a) T.L. Monchesky, Detection with unpolarized currents, Nat. Nanotechnol. 10 (2015) 1008–1009.

[31] S. Zhang, G. van der Laan, J. Müller, L. Heinen, M. Garst, A. Bauer, H. Berger, C. Pfleiderer, T. Hesjedal, Reciprocal space tomography of 3D skyrmion lattice order in a chiral magnet, Proc. Natl. Acad. Sci. U. S. A. 115 (2018) 6386–6391.

[32] H.-M. Zhang, J. Chen, P. Barone, K. Yamauchi, S. Dong, S. Picozzi, Possible emergence of a skyrmion phase in ferroelectric GaMo4S8, Phys. Rev. B 99 (2019) 214427.

[33] A.K. Yadav, C.T. Nelson, S.L. Hsu, Z. Hong, J.D. Clarkson, C.M. Schlepütz, A.R. Damodaran, P. Shafer, E. Arenholz, L.R. Dedon, D. Chen, A. Vishwanath, A.M. Minor, L.Q. Chen, J.F. Scott, L.W. Martin, R. Ramesh, Observation of polar vortices in oxide superlattices, Nature 530 (2016) 198–201.

[34] A.R. Damodaran, J.D. Clarkson, Z. Hong, H. Liu, A.K. Yadav, C.T. Nelson, S.-L. Hsu, M.R. McCarter, K.-D. Park, V. Kravtsov, A. Farhan, Y. Dong, Z. Cai, H. Zhou, P. Aguado-Puente, P. García-Fernández, J. Íñiguez, J. Junquera, A. Scholl, M.B. Raschke, L.-Q. Chen, D.D. Fong, R. Ramesh, L.W. Martin, Phase coexistence and electric-field control of toroidal order in oxide superlattices, Nat. Mater. 16 (2017) 1003–1009.

[35] M.A.P. Gonçalves, C. Escorihuela-Sayalero, P. Garca-Fernández, J. Junquera, J. Íñiguez, Theoretical guidelines to create and tune electric skyrmion bubbles, Sci. Adv. 5 (2) (2019) eaau7023.

[36] P. Shafer, P. García-Fernández, P. Aguado-Puente, A.R. Damodaran, A.K. Yadav, C.T. Nelson, S.-L. Hsu, J.C. Wojdeł, J. Íñiguez, L.W. Martin, E. Arenholz, J. Junquera, R. Ramesh, Emergent chirality in the electric polarization texture of titanate superlattices, Proc. Natl. Acad. Sci. U. S. A. 115 (2018) 915–920.
[37] S. Das, Y.L. Tang, R. Ramesh, Observation of room-temperature polar skyrmions, Nature 568 (2019) 368–372.
[38] Z. Hong, L.-Q. Chen, Blowing polar skyrmion bubbles in oxide superlattices, Acta Mater. 152 (2018) 155e161.
[39] P. Ondrejkovic, P. Marton, V. Stepkova, J. Hlinka, (2020) Domain walls: from fundamental properties to nanotechnology concepts, Chapter: Fundamental Properties of Ferroelectric Domain Walls from Ginzburg-Landau Models, Oxford University Press (in production).

CHAPTER FIVE

Dynamics in artificial spin ice and magnetic metamaterials

Joseph Sklenar[a], Sergi Lendinez[b], M. Benjamin Jungfleisch[b,*]
[a]Department of Physics and Astronomy, Wayne State University, Detroit, MI, United States
[b]Department of Physics and Astronomy, University of Delaware, Newark, DE, United States
*Corresponding author: e-mail address: mbj@udel.edu

Contents

1. Background and introduction

Water ice has many interesting properties that stem from the covalent bond between hydrogen and oxygen ions, see Fig. 1. In its liquid form there is a bond between each oxygen atom and two hydrogen atoms in water. When water freezes it forms a tetragonal structure, in which the oxygen is located at a vertex that has four nearest-neighbor oxygen atoms connected via a shared proton [1]. The proton is not in the center, but closer to one or the other oxygen ion [2]. The "ice rule" describes this lowest energy state that favors an arrangement that has two protons situated close to one oxygen atom and two protons positioned close to a neighboring oxygen atom. This arrangement is also known as "two-in–two-out" arrangement and it can be

Solid State Physics, Volume 70
ISSN 0081-1947
https://doi.org/10.1016/bs.ssp.2019.09.006

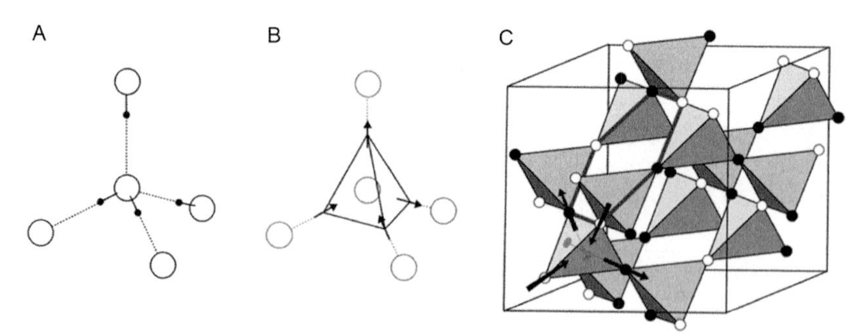

Fig. 1 (A) Local proton arrangement in water ice. Oxygen ions are shown as large white circles and protons (hydrogen ions) are shown as small black circles. The low-energy configurations obey the so-called ice rule. (B) Arrangement is identical to (A), but the position of protons is represented by displacement vectors. (C) Structure of crystalline spin ice such as $Tb_2Ti_2O_7$ and $Ho_2Ti_2O_7$ where two spins are pointing inward and two spins are pointing outward. *Adapted from S.T. Bramwell, M.J. Gingras, Spin ice state in frustrated magnetic pyrochlore materials, Science 294 (2001) 1495–501. https://doi.org/10.1126/science.1064761.*

found in the spin system of pyrochlore materials such as $Tb_2Ti_2O_7$ and $Ho_2Ti_2O_7$ [1,3–5], see Fig. 1. Six possible configurations can be found for each tetrahedron that obey the ice rule. This means that the lowest energy state is macroscopically degenerate [6]. Excitations in real spin ice materials are violations of the ice rule resulting in residual monopoles in the lattice [7,8]. Artificial spin-ice (ASI) lattices are artificially created structures consisting of 2D arrays of interacting nanomagnets, that were originally envisioned to mimic the properties of crystalline spin ice [1]. As the field evolved, ASI became an independent research topic in its own right. Currently, the name "artificial spin ice" broadly refers to a class of magnetic metamaterials where magnetic domains in nanomagnet assemblies can be mapped onto a spin-lattice model.

The field of artificial spin ice (ASI) was largely enabled by the wide availability of two technologies: Electron beam lithography [9,10] and magnetic force microscopy (MFM) [11]. Electron beam lithography is the primary technology used to create the nanoscale magnetic elements found in an ASI. Typically, ASI samples are fabricated from the bottom up. First, a substrate is coated with an electron beam resist (a polymer film), that when exposed to an electron beam is susceptible to being dissolved with a proper solvent. The ASI pattern is made by only exposing selected regions of the resist to the beam. In an ASI, these selected regions are usually elongated "island" shapes which are of the order 100 nm wide and long. The electron beam can be programmed to "expose" these islands repeatedly across the surface to form a lattice constituting the ASI. The key feature of this technology is that the placement of islands is completely controlled by the experimentalist, and there is total creative freedom to explore any configurations that tile the substrate. After the regions of the resist are dissolved by a solvent, in a process called development, the sample substrate can be coated by depositing a thin magnetic film through an evaporative or sputtering process. Once deposition is complete, the sample is placed in another solvent which dissolves the remaining resist in a process called "lift-off." During the lift-off process, regions of the original substrate which are still coated by resist are stripped back down to the original substrate. The final result is that a magnetic film is left behind only on regions that were exposed to the electron beam. Furthermore, top-down approaches can be used for fabricating ASI that involve milling continuous films into an ASI. We note that there are other top-down approaches for creating magnetic nanostructures that include self-assembling structures using polystyrene nanospheres, e.g., Refs. [12–15].

The prototypical artificial spin ice element is an elongated island, for example, in Fig. 2A the individual islands are 80 nm wide and 240 nm long.

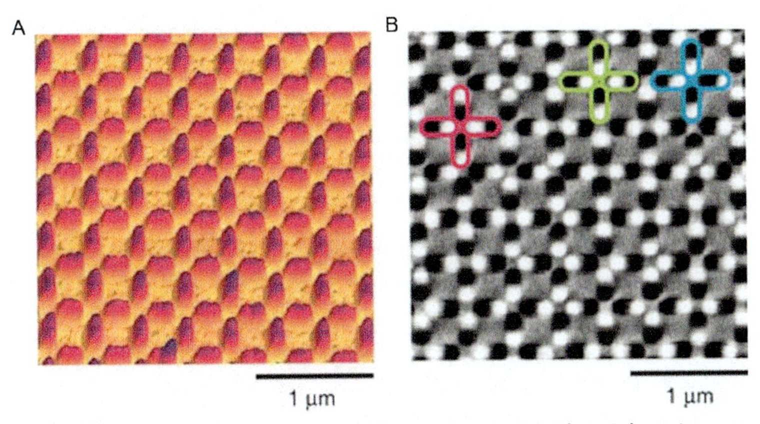

Fig. 2 An AFM image of the first square ASI is shown in (A). The pink regions are permalloy islands that are approximately 25 nm thick, 80 nm wide, and 240 nm long. A MFM image is shown in (B). At the end of the long portion of every island, there is distinct bright and dark magnetic contrast which corresponds to the north and south pole of the magnet. Thus, each island is in a single magnetic domain state where the magnetization lies along the length of the island and each island can be treated as a giant macrospin. *Adapted from R.F. Wang, C. Nisoli, R.S. Freitas, J. Li, W. McConville, B.J. Cooley, M.S. Lund, N. Samarth, C. Leighton, V.H. Crespi, P. Schiffer, Artificial `spin ice' in a geometrically frustrated lattice of nanoscale ferromagnetic islands, Nature 439 (2006) 303–306. https://doi.org/10.1038/nature04447.*

The magnetic materials that are deposited are usually polycrystalline films with negligible magnetocrystalline anisotropy. The most popular material is an alloy of nickel and iron called permalloy (Py, $Ni_{80}Fe_{20}$). With negligible magnetocrystalline anisotropy, the shape anisotropy of the island accounts for the easy magnetic axis [16]. For a range of aspect ratios and thicknesses, it will not be energetically favorable for a domain wall or vortex to form within an individual island and the magnetization will align along the long axis. In this limit, every island is in a single magnetic domain and can be thought of as a "macrospin" that is constrained to point in one of two directions—just as in the Ising model. A simple way to image this Ising "macrospin" employs the second technology which helped the field of ASI grow and proliferate, magnetic force microscopy (MFM). Because ASI islands are often too small to image with magneto-optical techniques such as the magneto-optic Kerr effect, MFM was the first technology used to uncover the exact microstate of an ASI [17]. MFM is a scanning probe microscopy technique very similar to atomic force microscopy (AFM). In conventional MFM, a standard AFM tip is coated with a magnetic layer. When imaging the sample, the scanning probe tip is rastered back and forth across the sample and, via AFM, a height profile of the sample surface is obtained. The scan then repeats except that the tip is lifted vertically, for example by 20nm. When the tip is lifted a fixed height over the surface, the microscope is sensitive to long-range magnetic interactions between the tip and surface of the sample. Specifically, standard MFM is sensitive to a gradient in magnetic field expelled out of the plane of the sample. When imaging disconnected ASI lattices, the MFM technique can resolve the microstate of the system. As can be seen in Fig. 2B, every individual island will show either bright or dark magnetic contrast at the head and tail of the island corresponding to either a north or south pole. In a connected ASI, magnetic contrast will only show up where the islands intersect, and not within individual islands. Consequently, it is impossible for the true and unique microstate of the connected ASI to be identified with MFM. Solutions to the problem of resolving the microstate rely on alternative electron microscopy based technologies, which we will discuss throughout this chapter.

1.1 Frustration in artificial spin ice

In this Section we provide a historical overview of artificial spin ice. We focus our narrative on ASI mainly with a perspective on the square ASI. We will discuss how other lattices and geometries are chosen so that "ice rules" are followed in some manner within the ASI. In parallel, we discuss the magnetic ground state of the square ASI because it tells the story of how progress was made in developing experimental tools and techniques which can thermally equilibrate ASI systems. This will set the stage for our discussion of recent results studying the ground states and slow spin dynamics of ASI in Sections 2.1–2.2.

1.1.1 Birth of artificial spin ice: The square and honeycomb lattices

Wang et al. fabricated the first ASI in 2006 [17], and this square lattice ASI is shown in Fig. 2. The original square lattice consists of an array of "disconnected" magnetic islands, but it is useful to consider vertices, which are the intersection points of the lattice if the islands were connected. With a vertex picture in mind, the connection between the square lattice and real spin ice materials can be made clear. The first microstates observed in the square lattice favored moment configurations where two moments point-in and two point-out of every vertex [17]. Thus, the low energy microstates of the square lattice follow ice rules analogous to real spin ice materials. A common way to characterize the microstate of the square ASI is to count the population of the energetically equivalent vertex types that are enumerated in Fig. 3. Type 1 and Type 2 vertices follow "ice rules," while Type 3 and Type 4 vertices violate the rules. We note that the higher energy Type 3 and 4 vertices draw comparison to the so-called magnetic charges or

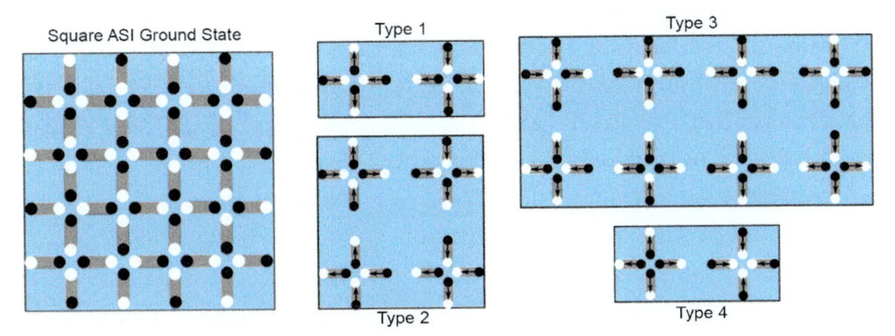

Fig. 3 The left panel shows how the square spin ice can be configured into a long range ordered ground state where every vertex is in a low-energy Type 1 state. The four right panels show all possible vertex configurations of the square lattice from lowest energy (Type 1) to highest energy (Type 4). Note that the two lowest energy configurations (Type 1 and 2) follow the so-called ice rules where two moments point-in and two point-out of every given vertex. We have also illustrated what these vertices look like when imaged with MFM by illustrating black and white poles at the ends of each individual island.

"magnetic monopoles" that are of interest in real spin ice materials. In reality, the "magnetic charge" is an excess of north or south poles, from different islands, adjacent to the vertex. With this picture in mind, it is found that Type 3(4) vertices have an excess of two (four) poles. Throughout this Chapter, we will discuss the role these "charges" play within experimental inquires of ASI.

At roughly the same time the square ASI was reported, Tanaka et al. studied a connected honeycomb network of magnetic islands with magnetotransport and MFM [18]. In 2008, Qi et al. identified the connected honeycomb network as a type of artificial spin ice, and used Lorentz transmission electron microscopy (LTEM) to unambiguously image the microstate of the honeycomb network [19]. After the sample was demagnetized, the microstate of the system was such that the moments at every vertex were either in a two-in one-out or two-out one-in configuration. Vertex configurations where three moments pointed in (or out) of any vertex were not observed. This work demonstrated that low energy microstates of the honeycomb lattice obeyed a different set of "ice rules," compared to the square ASI, where two-in and two-out energetically preferred vertex states are replaced with two-in and one-out (or vice versa). An appealing aspect of the low energy "ice rules" in the Kagome lattice is the fact that all of the "ice rule" vertex types have the same energy. This is in contrast to the square ASI where Type 1 vertex types have a slightly lower energy compared with Type 2 vertices.

1.1.2 Ground state considerations in ASI and thermalization strategies

Early on, it was identified that the original square ASI supports a low-temperature ground state that is not a frustrated spin ice phase [20]. It is more correct to say that any given island in the square ASI experiences a local quasi-frustration the nearest-neighbor interaction directly competes against the next-nearest-neighbor interaction. However, the quasi-frustrated interaction of the square lattice is not a sufficient condition to guarantee a low-temperature ground state analogous to the spin ice phase of real materials. This is because, in the original square ASI, the nearest-neighbor interaction is slightly stronger than the next-nearest-neighbor. Consequently the lowest energy state of the system is able to satisfy every nearest-neighbor interaction at the expense of the next-nearest-neighbor interaction. This microstate corresponds to a configuration where every vertex is Type 1, and across the square lattice the two Type 1 options alternate in a checkerboard like pattern. The alternating pattern of Type 1 vertices represents a doubly degenerate long range ordered phase, and it is referred to in the literature as an "antiferromagnetic" ground state [20]. An illustration of the ground state of the square ASI is shown in Fig. 3.

The predicted long-range ordered phase of the square lattice was not observed in the earliest set of experiments [17,21–24]. The absence of the long-range ordered ground state originates from the fact that the islands are athermal at room temperature where earliest experimental efforts took place. Athermal samples are prevalent in ASI studies largely because of the use of MFM to image the microstates. When using MFM, it is important to have ASI islands thick enough so that the magnetic moment from the tip does not induce a spin-flip in the islands it scans over. This places a lower constraint on the thickness of ASI islands such that most samples are around 20 nm thick. For these thicknesses, at room temperature a permalloy ASI island will have a shape anisotropy and magnetostatic interaction energy within an island corresponding to a temperature of approximately 10^5 K [1]. This temperature scale essentially restricts the ASI from having any thermally induced spin-flips at room temperature, and the ASI is effectively "blocked." In light of this, the first experimental strategies to place ASI samples into magnetic ground states involved demagnetizing the sample [21–24].

For ASI, demagnetization involves rotating the sample in an external field while the field is ramped back and forth from positive to negative values. In each successive ramp back and forth, the upper and lower limit of the external field is reduced until the ramp ends at zero field. The earliest attempts to demagnetize square ASI into the ground states had mixed success, and while demagnetization would place ASI into a microstate with a low energy (relative to a magnetized configuration) the ground state was not achieved. Theoretically, Nisoli et al. developed a model to relate a given demagnetization protocol to an effective temperature [25], and it was concluded that rotational demagnetization should facilitate ordering into the ground state. A discussion on the early failures to demagnetize square ASI into the ground state is beyond the scope of this Chapter and the interested reader will find that this issue has been reviewed in detail [1]. However, we note that in the last three years demagnetization strategies have remarkably improved in terms of achieving the magnetic ground state of the square ASI [26,27]. Ultimately, demagnetization protocols would not be the first experimental method to convincingly demonstrate magnetic ordering in ASI.

To work around the issue of the thermally "blocked" ASI, Morgan et al. found that when the microstate of the square ASI was imaged directly after the permalloy islands were deposited via electron beam evaporation, large

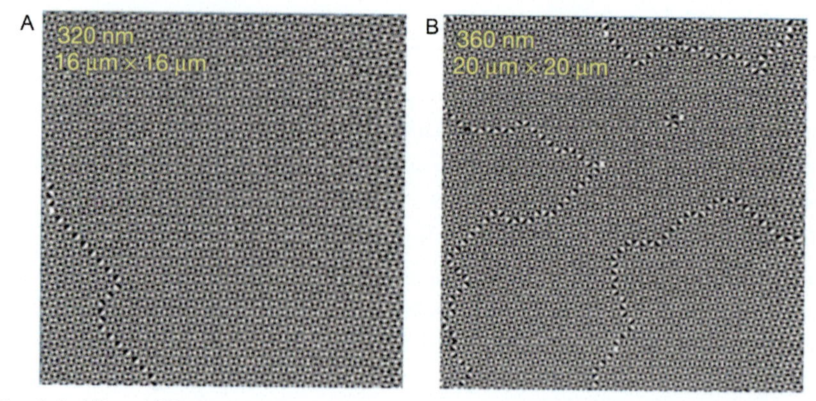

Fig. 4 In (A) and (B) MFM images of the square ASI are shown after the system was controllably "annealed" into the long-range ordered ground state. Individual islands are 220 nm long, 80 nm wide, and 25 nm thick. The lattice constant between islands in (A) and (B) is 320 nm and 360 nm, respectively. Domain walls consisting of Type 2 and Type 3 vertices are present in both images, but it is clear that for the shorter lattice constant the ordered domains of Type 1 vertices are larger. *Adapted from S. Zhang, I. Gilbert, C. Nisoli, G.-W. Chern, M.J. Erickson, L. OBrien, C. Leighton, P.E. Lammert, V.H. Crespi, P. Schiffer, Crystallites of magnetic charges in artificial spin ice, Nature 500 (2013) 553–557. https://doi.org/10.1038/nature12399.*

swathes of the system were in the ordered ground state [28]. This observation was attributed to the islands being thermally active and interacting *as they were growing*. In 2013, Zhang et al. showed that the ground state could be achieved by heating the ASI near the Curie temperature of permalloy, and then slowly cooling the sample down [29]. MFM images of these thermally "annealed" ground states are shown in Fig. 4. This thermal annealing protocol was an important milestone because it represented an experimental method that could be used to reproducibly reconfigure the ASI into the ground state. For example, thermal annealing can reset the ASI into a ground state configuration even after the ASI ground state is "erased" when the ASI is magnetized after being placed in an external field. Further progress along these lines was made by considering different ferromagnetic materials. The most common ASI material, permalloy, has a T_C of approximately 540 °C. Other researchers have demonstrated that by fabricating ASI out of other materials such as FePd alloys or $La_{0.7}Sr_{0.3}MnO_3$, T_C can be lowered so that thermal annealing of ASI is more feasible [30–32].

An alternative approach to thermalize ASI is to deposit thinner films (<3 nm) so that the individual islands are thermally active near room temperature. A thermally active ASI at room temperature can no longer be easily measured with MFM because of the destructive interaction between the moment of the microscope tip and the islands. As a solution to this problem, Farhan et al. first reported the direct observation of thermalization of square ASI having 3 nm thick islands using photoemission electron microscopy (PEEM) combined with X-ray circular magnetic dichroism (XMCD) [33]. PEEM-XMCD is a photon-in, electron-out microscopy technique which had previously been used to image ASI in a non-destructive manner [34–37]. Typically, a sample is illuminated with right- and left-handed circular polarized X-rays. For a common material such as permalloy, the X-ray energy can be tuned for absorption at the L3 edge of Fe. For a given magnetization orientation, the photoemission process will be enhanced or suppressed depending on the circular polarization of the light. After imaging the ASI with both polarizations, the two corresponding images can either be divided by (or subtracted from) one another to create magnetic contrast. The resulting images can then be used to reconstruct the microstate of the ASI. There are other useful advantages to studying ASI with PEEM-XMCD: The time to acquire an image of the ASI microstate takes only a few seconds compared to the many minutes it takes to image the same sample area with scanning probe techniques. In two seminal experiments, this unique advantage of PEEM-XMCD was used by Farhan et al. to study real-time spin dynamics in ASI systems that were controlled with temperature [33,38]. Indeed, when used to study the thermalization of a polarized square ASI into the ordered ground state, PEEM-XMCD allowed for direct real-time observation of the thermalization process. These real-time experiments monitored how a square lattice, magnetized so that every vertex was in a Type 2 state, thermalized into an ordered array of alternating Type 1 vertices. This led to the direct observation of Type 3 monopole pairs in the Type 2 background which separated, propagated, and created a Type 1 string connecting the monopole pairs.

Before continuing, we must note that we have narrowed our focus in this introduction on the square ASI at the expense of the honeycomb ASI. This is entirely for the sake of space and narrative, but we must mention that this geometry has been studied extensively over the years for very important reasons. The interested reader may learn that this was the earliest system where field-driven monopole excitations, in the athermal limit, were observed [35,39]. There is also a rich temperature-dependent phase diagram that the honeycomb ASI is known for [40]. As temperature is lowered, the honeycomb lattice evolves from a phase where all vertices follow ice rules, to a phase where there is a so-called charge order. This is where for a given vertex, it is found that nearest-neighbor vertices have the opposite net magnetic pole on them. As the temperature is further lowered there is an elusive long-range ordered phase where the moments on individual islands exhibit long-range order along with the charge order. Issues with observing the final long-range ordered state of the honeycomb lattice typically arise because the blocking temperature of the system lies above the thermodynamic ordering temperature. There have been recent efforts by Anghinolfi et al. to further miniaturize islands in a bid to lower the blocking temperature, but alternative magnetic probes are needed to study smaller islands. In particular, muon spin resonance experiments have been used to provide strong evidence for the long range ordered phase in honeycomb ASI [41]. For a history of ASI which puts the square and honeycomb lattice on more equal footing, the reader may turn to a recent review on ASI [1]. The honeycomb lattice will be reviewed in more detail in Sections 2.3 and 2.4 since there is a wider range of research utilizing magnetic transport and magnetization dynamics in this lattice.

1.1.3 Restoring macroscopic degeneracy in artificial spin ice

We have now described how the original square ASI has a long-range ordered, low-temperature ground state which eluded researchers until the development of powerful thermalization techniques. With this issue resolved, the original question of how to engineer a low-temperature spin ice phase (with extensive ground state degeneracy) could be properly addressed in modified ASI geometries. A logically straightforward way to recover the extensive ground state degeneracy of the square ASI was proposed as early as 2006 [20], where the square lattice is modified such that Type 1 and Type 2 vertex configurations are energetically degenerate. This can be done by using the third, out-of-plane dimension, and effectively layering an ASI so that there are two lattices with a height offset between them. This will increase the distance between the original nearest-neighbors of the square ASI. Chern et al. theoretically constructed a phase diagram that demonstrated a low-temperature spin ice phase is present as the height is adjusted [42]. In 2016, Perrin et al. experimentally realized a square ASI with extensive ground state degeneracy using this exact strategy [26], see Fig. 5. Operationally, the square ASI was modified so that half of the islands were offset in height by being patterned on a non-magnetic spacer layer. This spacer layer reduces the nearest neighbor interaction while maintaining the next-nearest neighbor interaction. By restoring vertex energy degeneracy in the square lattice, all of these research endeavors have enabled the first clear glimpse

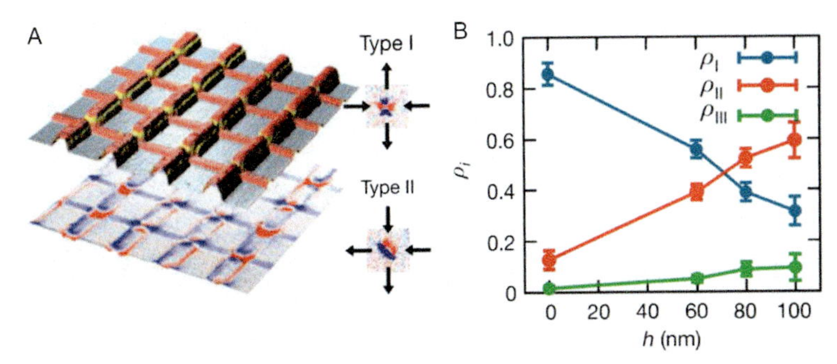

Fig. 5 In the left panel an AFM and MFM image of a square spin ice is shown. Half of the islands, and all of their next-nearest-neighbors are patterned on a non-magnetic spacer layer of a given height h. The effect of the space layer is to diminish the nearest-neighbor interaction so that it is equivalent to the next-nearest. In the right panel the concentration of Type 1, Type 2, and Type 3 vertices are plotted as a function of spacer layer thickness after the sample has been demagnetized. The spacer layer clearly suppresses the Type 1 concentration relative to Type 2 as the vertex energies between these two populations becomes equivalent. *Adapted from Y. Perrin, B. Canals, N. Rougemaille, Extensive degeneracy, Coulomb phase and magnetic monopoles in artificial square ice, Nature 540 (2016) 410.*

into the so-called Coulomb phase of the ASI. Excitations in the spin ice phase (Coulomb phase) are interacting monopoles which can move from vertex to vertex with no discernible string of flipped spins connecting monopole pairs. We discuss further studies of the square ASI with restored degeneracy in Section 2.2.1.

Because artificial spin-ice systems can be patterned into essentially any configuration experimentally desired, there are other approaches to create highly frustrated systems outside the scope of the square lattice. One group of highly frustrated lattices are so-called vertex-frustrated artificial spin ice systems, which were proposed by Morrison et al. in 2013 [43]. With some known exceptions, vertex-frustrated ASI are square lattices that have islands decimated (removed) in a regular manner. These new types of highly frustrated lattices exploit the fact that ASI can be designed with mixed vertex coordination number. The coordination number refers to the number of islands that point into a given vertex. In Fig. 6, we illustrate the first experimentally identified vertex-frustrated ASI, the Shatki lattice [44]. The Shatki lattice is comprised of two-, three-, and four-island vertex coordination numbers. It can be broken down into plaquettes, as illustrated in Fig. 6, which are regions encompassing one two-island vertex, with a perimeter of four three-island and four four-island vertices. With this configuration

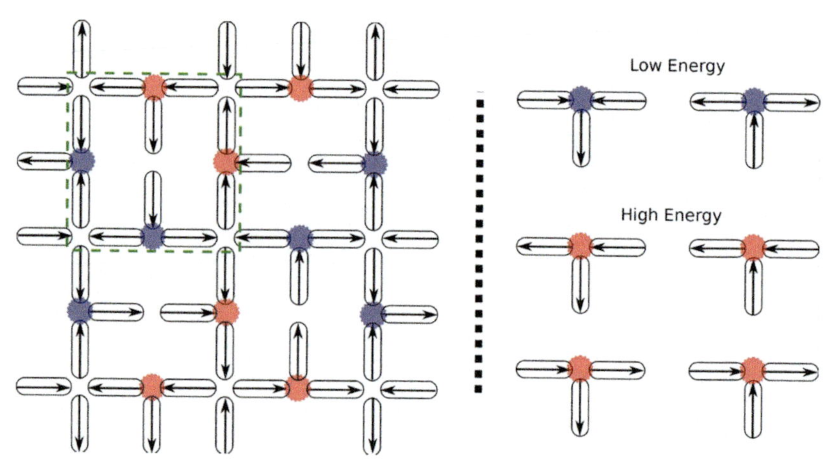

Fig. 6 The left portion of the image is the Shatki ASI geometry. The arrows within every individual island denote the moment configuration, and the system is illustrated in the ground state configuration. One can note that in the ground state, every four-island vertex is in a Type 1 configuration and every two-island vertex is in a low energy state where the moments are aligned head to tail. The topology of the Shatki lattice prohibits every three-island vertex from being placed into the ground state. Red-color shaded three-island vertices represent high energy states while blue-color shaded three-island vertices represent low-energy states. The right portion of the image denotes the possible low and high energy three-island moment configurations. In the left portion of the image a green-dashed line denotes a single plaquette of the Shatki lattice which houses four four-island, four three-island, and one two-island vertices. Note that within the plaquette two of the three-island vertices are in the low energy configuration while the other two are in a high energy configuration. Consequently, the emergent set of ice rules for the Shatki lattice is that the ground state contains two high and two low energy vertices per plaquette. This is somewhat analogous to the two-in and two-out moment configurations for the square lattice.

of islands, it is impossible to place every three-island vertex in the lowest energy vertex configuration. The ground state of the system is such that, for every plaquette there are two three-island vertices in the ground state and two in excited states. From this perspective, a new set of ice rules emerges where every plaquette has two-low and two-high energy vertex types.

In 2013, Bhat et al. reported the first fabrication and study of an artificial magnetic quasicrystal [45]. The original interest of these quasicrystals was motivated by studying the magnon mode spectrum of the aperiodic system [46,47], yet, it was quickly realized that these geometries were remarkably similar to connected ASI systems. The shift in interest raised important questions regarding the ground state of the quasicrystal system. Discerning the ground state through inspection is difficult because quasicrystals are made up of many different types of vertices with mixed coordination numbers, and because there is a lack of translational symmetry in the placement of the magnetic islands. Currently, there are separate reports concerning the ground state of connected [48,49] and disconnected [50] quasicrystals. Thermal annealing techniques, where the quasicrystal samples are heated near the Curie temperature, have been the primary way low energy states have been created in these systems. For the case of disconnected islands, an artificial quasicrystal can be fabricated such that every island size and shape are identical; in this case, recent experimental evidence suggests that the ground state of the quasicrystal is a vertex-frustrated state [50].

1.2 Magnetic transport measurements in nanostructures

The interaction between an electric current and the magnetization in a given structure produces changes in the resistance that can be experimentally observed. This effect is generally known as magnetotransport. The change in resistance can have different origins as we will describe in detail below. Magnetotransport is a topic that is at the basis of the study of magnetization, since the analysis of voltage and resistance measurements reveals the physics of the underlying interactions. Direct access to the magnetic configuration is not always possible. Therefore, exploiting and optimizing the magnetic transport properties is essential in many situations, especially for applications. Studying and understanding the magnetotransport properties is one of the cornerstones of spintronics. For instance, the discovery of the giant magnetoresistance (GMR) effect [51,52], which exploits the different transport characteristics depending on the magnetization direction, is considered to be the birth of spintronics or spin-electronics.

Anisotropic magnetoresistance (AMR) is the change in electric resistance as the orientation of the magnetization with respect to the current direction is changed [53]. Despite being an effect known for decades (it was discovered in the mid-19th Century by Lord Kelvin [54]), it is still an essential mechanism for the detailed investigation of the magnetization of micro- and nanostructures. The electric resistivity can be expressed as:

$$\rho = \rho_0 + \Delta\rho\left(\vec{m} \cdot \vec{j}\right)^2 = \rho_0 + \Delta\rho cos^2\theta$$

with ρ_0 an isotropic resistivity, $\Delta\rho$ the anisotropic resistivity change, $\vec{m}$ a unit vector pointing in the direction of the magnetization, $\vec{j}$ the unit vector pointing along the current direction and θ the angle between current and

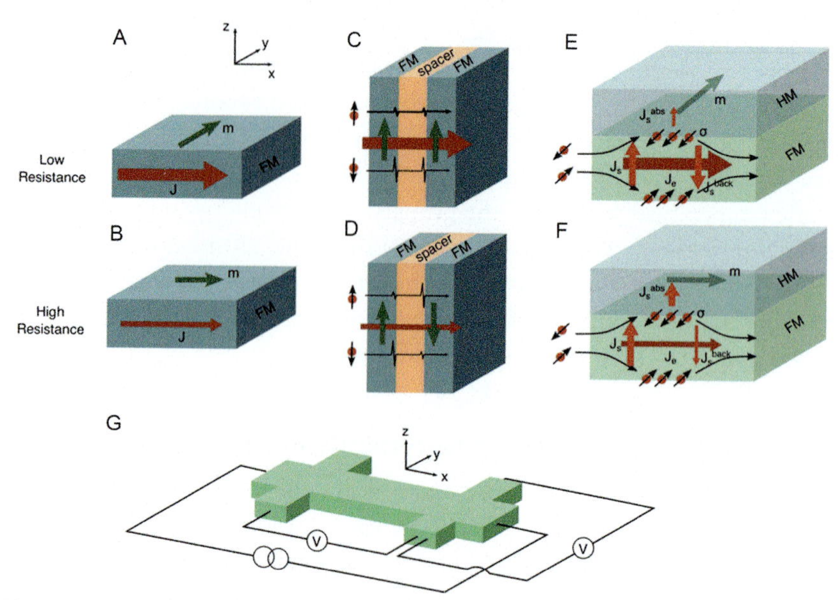

Fig. 7 Low- and high-resistance states for various types of magnetoresistance. In anisotropic magnetoresistance (AMR), the resistance is lower when the current (red arrow) flows perpendicular to the magnetization (green arrow) in a metallic ferromagnet (A). If the current is parallel to the magnetization, the resistance is higher (B). In giant magnetoresistance (GMR), a parallel orientation of two ferromagnetic layers results in low resistance (C), whereas an antiparallel orientation results in high resistance (D). The low-resistance state in spin Hall magnetoresistance (SMR) occurs when a high spin current is backscattered at the interface as a result of the alignment between the spin polarization and the magnetic moment of the ferromagnetic layer (E). An orthogonal alignment produces a lower backscattered spin current, resulting in a high-resistance state (F). (G) Example of the Hall bar configuration to measure the transport properties.

magnetization. From this equation, it is evident that the resistivity is minimum when the magnetization and the current are perpendicular, see Fig. 7A, and maximum when they are in the same direction, see Fig. 7B.

Since the magnetization and the current can have a spatially non-uniform distribution, the total resistance is then a compound of the local resistances. This is particularly important for magnetic micro- and nanostructures. Combining the experimentally acquired magnetoresistance data with the results of micromagnetic simulations, it is oftentimes possible to reconstruct the underlying magnetic configuration, even in structures where the magnetization configuration is not directly accessible otherwise. For this purpose, it is important to determine the current distribution since the angle between the local current direction and the local magnetic moment determines the contribution to the total resistance. This can be achieved by solving Maxwell's equations, usually using a finite element method (some broadly used software distributions are HFSS and Comsol) [55]. It is also possible to assume a certain current distribution that depends on the geometry [56]. The magnetization configuration, on the other hand, can be found by minimizing the total energy of the studied system after taking into account all relevant magnetic interactions. Software like OOMMF [57] and MuMax3 [54] are commonly used to find the magnetic configuration as a function of the external magnetic field. Comparing the computed magnetoresistance and the experimental magnetoresistance results can help discern the underlying magnetic configuration in the experimental system. While this method usually yields a reasonably good agreement with experimental data, it is important to note that different magnetic configurations can produce the same magnetoresistance. Hence, it is important to take the results with caution and to test them against other techniques when possible.

Typically, AMR in magnetic metals is a relatively small effect. The change in resistance between current parallel and perpendicular to magnetization is often less than 5%. In order to have a larger effect, other types of magnetoresistance can be utilized. In GMR the current flows through two magnetic layers separated by a spacer layer. Since the electron scattering depends on the direction of the electron spin with respect to the layer magnetization, the total resistance will vary depending on the relative orientation of the two magnetic layers, see Fig. 7C and D. The conduction band of majority spins in ferromagnets differs from the minority spins. Therefore, majority spin electrons scatter less than minority electrons. When the magnetic moment of the layers are parallel, there is a low-resistivity channel that results in a total lower resistance (see Fig. 7A). On the contrary, if the layers are antiparallel, high scattering is produced in both ferromagnetic layers (see Fig. 7B), which results in a higher resistance state. A common strategy to exploit the GMR effect consists of a layer stack with one of the layers acting as a reference, in which the magnetization remains pointing in the same direction (the so-called fixed layer), and another layer acts as a free layer that can flip its magnetization direction. This strategy is known as spin valve. There are different methods to pin the magnetization of the reference layer, such as using a different material with a higher coercivity [58] or using a third layer made of an antiferromagnet to introduce exchange bias [59]. GMR values up to about 200% can be obtained at low temperature [57].

The discovery of GMR was a big advancement for the computer industry since it allowed a much more sensitive detection of smaller magnetic structures, and hence helped increase the density in computer memories that kept pushing Moore's law [60]. In 2007, Peter Grünberg and Albert Fert received the Nobel Prize in Physics for the discovery of GMR. With the implementation of GMR in hard disk drives and in computer electronics, much research effort was put into further miniaturizing these structures. In order to obtain an even higher effect at higher densities, an insulating barrier was introduced between the two metallic ferromagnets. This is known as tunneling magnetoresistance (TMR) because electrons tunnel through the insulator. The layer stack is called a magnetic tunnel junction (MTJ), and is one of the pillars in current memory development. In this case, the current flows through the stack—unlike in GMR, where it is possible to have an effect even if the current flows along the film. TMR can have values of up to about 1000% at low temperature [61].

The effects described so far in this Section rely on transport in magnetic metals. However, under certain circumstances it can be desirable to utilize magnetic insulators, such as yttrium iron garnet (YIG). YIG exhibits an extremely low magnetic damping, which is important for studies on magnetization dynamics and magnonics. In order to electrically probe a magnetic insulator we can exploit the spin Hall effect (SHE) [62–66]. This effect describes the generation of a transverse spin-polarized electron current ($\vec{J}_s$) from an electric charge current that flows through a heavy metal ($\vec{J}_e$), where electrons with opposite spins scatter in opposite directions due to spin-orbit interaction. As a result a spin accumulation with spin polarization ($\vec{\sigma}$) is built up at the surfaces of the conductor orthogonal to both the charge current and the spin-polarized current [67–69], see Fig. 7E and F. If the heavy metal layer is in contact with a ferromagnet, the spin-polarized current ($\vec{J}_s^{\,abs}$) exerts a torque on the magnetization. At the same time, there is some reflected spins at the surfaces, resulting in a spin current in the opposite direction ($\vec{J}_s^{\,back}$). The efficiency of the spin diffusion depends on the relative orientation between the magnetization in the material and the spin polarization, being higher when they are orthogonal, as there is higher transfer of spin angular momentum, see Fig. 7E. The amount of absorbed spin current in the magnetic material is hence inversely proportional to the spin current backscattered into the heavy metal. As the current is backscattered, the spins pointing in opposite directions re-merge in the heavy metal producing a charge current in the original current direction based on the inverse spin Hall effect (ISHE). Since the backscattered current in the heavy metal depends on the orientation of the magnetization in the magnetic layer (Fig. 7F) it produces a measurable change in resistance, hence this effect is called spin Hall magnetoresistance (SMR) [70].

The experimental configuration to perform magnetotransport measurements depends on the kind of effect that is being explored, as effects such as TMR, and sometimes GMR, require that the current flows perpendicular to the layer stack, see Fig. 7C and D, making the fabrication process more complex. Other magnetoresistance effects allow for a fabrication scheme in a planar structure, which is oftentimes done using microlithography techniques [71]. In this case, structures that allow longitudinal and transverse measurements in the same pattern are usually designed. Fig. 7G shows a typical "Hall bar" design for transport measurements. Using this geometry, it is possible to measure longitudinal resistivity (ρ_{xx}) and transverse resistivity (ρ_{xy}).

1.3 Resonant magnetization dynamics

In stark contrast to the slow dynamics in the spin system associated with temperature effects as discussed previously in Section 1.1, we here briefly introduce resonant magnetization dynamics.[a]

The precessional motion of the magnetization $\vec{m}$ under the influence of an effective magnetic field H_{eff} can be described in the classical limit by the Landau-Lifshitz-Gilbert equation that takes into account the magnetic damping α.

$$\frac{d\vec{m}}{dt} = -\gamma \vec{m} \times \vec{H}_{eff} + \alpha \vec{m} \times \frac{d\vec{m}}{dt}$$

Here, γ is the gyromagnetic ratio, and α is the dimensionless Gilbert damping parameter, see also Fig. 8. The Gilbert damping is considered to be viscous, e.g., α increases with an increasing precession frequency, e.g., $\frac{d\vec{m}}{dt}$. α describes the effective damping of the ferromagnet that includes all possible relaxation channels. The precessional motion of the magnetization is shown Fig. 8.

Due to the rich physics that govern magnetization dynamics in ferromagnets, there has been intensive research over the past decades. Advancements

[a] Oftentimes, the terms "spin dynamics" and "magnetization dynamics" are used synonymously. However, we do want to point out that the meaning of the words varies in different research fields. In magnonics, the field that studies spin waves, *spin dynamics* and *magnetization dynamics* refer to the resonant GHz—or even THz—dynamics in the spin system, whereas in the artificial spin-ice community, *spin dynamics* is understood as effects related to slow spin flip processes and avalanches of the nanomagnets.

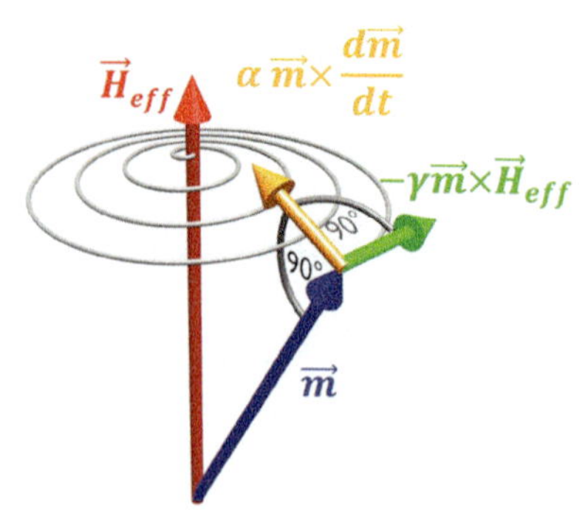

Fig. 8 Illustration of magnetization dynamics governed by the Landau–Lifshitz–Gilbert equation. The magnetization M (blue) precesses under the influence of an effective magnetic field H_{eff} (red). The field-like torque is shown in green and the Gilbert damping torque in orange.

in modern lithography techniques enable the exploration of magnetization dynamics in magnetic micro- and nanostructures. The research field that studies magnetization dynamics in patterned ferromagnets is oftentimes referred to as magnonics. The advent of spin-Hall effect related phenomena invigorated magnonics and established a new direction in magnonics that is called "magnon spintronics" [72]. A detailed introduction in magnonics and the complete derivation of all important equations and formulas is beyond the scope of this Chapter. The interested reader is referred to specialized books [73,74] and review articles [72,75–78] on magnetization dynamics in general and magnonics in particular. Here, we give a brief overview of the theoretical background.

In the following we qualitatively outline the derivation of ferromagnetic resonance in thin films as this is also relevant for the understanding of magnetization dynamics in artificial spin ice and arrays of nanomagnets.

By combining magnetostatic Maxwell's equations and the field dynamics, one can derive the Walker equation, which is the basic equation for magnetostatic modes in homogeneous media [73]. There are two solutions of the Walker equation that describe either the propagation of magnetostatic spin waves, where the specific geometries can be accounted for by a proper choice of the boundary conditions, or the uniform precession. The uniform precession is usually referred to as ferromagnetic resonance and this particular solution of the Walker equation is known as the Kittel equation [79]:

$$\omega_{FMR} = \sqrt{\omega_0(\omega_0 + \omega_M)}$$

where $\omega_0 = -\gamma\mu_0 H_0$ and $\omega_M = -\gamma\mu_0 M_s$.

Spin waves are the magnetic analog to lattice waves in magnetically ordered solids. Similar to quantized lattice waves—phonons—we refer to quantized spin waves as magnons. Important length scales in magnonic devices are determined by the interaction between spins in the magnetically ordered material. The exchange interaction dictates the behavior on the short range, whereas dipolar interactions are responsible for the long-range properties. The competition between both exchange and dipolar interactions leads to an anisotropic spin-wave dispersion, which can be utilized in magnonic devices and applications. Depending on the dominant magnetic interaction in the material, spin waves are classified either as dipolar dominated or exchange dominated. For dipolar dominated spin waves, the displacement between neighboring spins is small, or in other words the phase difference between them is vanishingly small. Therefore, the energy of dipolar dominated spin waves is determined only by long-range dipolar interaction. In contrast, exchange dominated spin waves are characterized by a larger displacement between adjacent spins and their wavelength is therefore short. Here, the exchange interaction dominates the spin-wave energy and the spin-wave frequency is proportional to the quadratic wavevector: $\omega \sim k^2$. Other important quantities that determine the resonant magnetization dynamics include the saturation magnetization M_S, the spin-wave wavevector $\vec{k}$, as well as the relative orientation between $\vec{k}$ and the direction of the magnetization $\vec{m}$, and the spin-wave stiffness D [76].

In patterned media, not only the anisotropic spin-wave propagation, but also the boundaries of the structures lead to distinct dynamic characteristics. As we will see, this is particularly important for the understanding of magnetization dynamics in ASI networks. It is possible to derive the dispersion relation of dipolar spin waves in a thin film by taking into account the boundary conditions. In particular, for the case of an infinitely extended film we can analytically calculate the dispersion relation and distinguish between two important cases if the magnetization lies in plane: (1) Backward volume magnetostatic spin waves, where the dynamic magnetization is extended over the entire film thickness. These waves are characterized by a monotonically decreasing dispersion, i.e., their group velocity $v_g = d\omega/dt$ is negative. (2) Magnetostatic surface waves, which are characterized by a positive group velocity and non-reciprocal propagation direction. We note that for the case of an out-of-plane magnetized film, we can observe so-called forward magnetostatic spin waves, which are less often studied, especially in the case of ASI. Moreover, Kalinikos and Slavin presented an analytical expression for the dispersion characteristics of spin waves in ferromagnetic films taking into account both dipolar and exchange interactions [80].

So far we have only discussed the dispersion characteristics for infinitely extended, thin ferromagnetic layers. However, if we consider magnetic micro- and nanostructures, it is necessary to take into account quantization effects, which may alter the spin-wave dispersion. Considering the propagation of a spin wave with wavevector $k^2 = k_x^2 + k_y^2$ in a longitudinally magnetized wire extending along x-direction, the spin-wave wavevector in y-direction is quantized, i.e., $k_y = \pi/w'$, where w' is the effective width of the wire. The effective width is larger than the geometric width of the wire, which means that the spin-wave precession extends the wire boundaries [81]. In a transversely magnetized wire, demagnetization fields reduce the internal field at the edges and thus the internal field becomes spatially inhomogeneous. As a result, a spin-wave "well" is formed that acts as a spin-wave waveguide along the edges of the wire [82].

In arrays of nanomagnets and artificial spin-ice networks it is rarely possible to derive an analytical or semi-analytical solution of the collective response of the lattice [83,84]. The method of choice in this case is micromagnetic simulations, which we briefly discuss in the following.

1.3.1 Micromagnetic simulations

Numerical simulations have been an indispensable tool to interpret experimental data and to design the properties and behavior of magnonic crystals and devices. As outlined above, the Landau-Lifshitz-Gilbert equation describes the time evolution of the magnetization. We can extend the Landau-Lifshitz-Gilbert (LLG) equation and add spin-torque effects, which gives a modified version of the Landau-Lifshitz-Gilbert equation:

$$\frac{d\vec{m}}{dt} = -\gamma \vec{m} \times \vec{H}_{eff} + \alpha \vec{m} \times \frac{d\vec{m}}{dt} + \frac{\gamma}{\mu_0 M_S}\tau,$$

where M_S is the saturation magnetization, μ_0 is the permeability and τ is the effective torque term that includes both damping-like and field-like torque terms. Since in most cases finding an analytical solution of the modified Landau-Lifshitz-Gilbert equation is impossible, micromagnetic simulations rely on a numerical approach. For this purpose, the space is divided into a grid with cells that are smaller than the exchange length. Each cell houses a single moment that interacts with the rest of the grid. The interaction between the moments in each cell are incorporated into the effective field, which contains the exchange field, magnetostatic energy, anisotropy field, etc. In order to solve the equation and all the interactions between each cell in the grid, different approaches can be used, such as energy minimization or time evolution of the magnetization. Today, various packages such as OOMMF [57] and MuMax3 [85], as well as custom-built codes are available for micromagnetics, and they use different algorithms to solve the problem.

When modeling the dynamics of ASI, micromagnetic simulations have been used frequently to understand the experimentally acquired data. In order to obtain the eigenmodes of the system, the time evolution is simulated over typically 10 ns after a temporally short field pulse is uniformly applied. The magnetization at each time step is then recorded and used to compute the frequency response with a fast Fourier transform algorithm. Among the first reports of ASI eigenmodes was a simulation work by Gliga et al. [86], that explored dynamics of topological defects in the magnetization, such as Dirac monopoles and Dirac strings in ASI structures. It was shown that the presence of topological defects result in a one-to-one correspondence with the spectral features; see Fig. 9. More recently, particular symmetries of ASI networks and the chirality of magnetic defects have been studied by micromagnetics, where a combination of global magnetic fields and local nano-island stray fields due to the microstate were found to significantly alter the high-frequency response of the ASI system [87].

Beyond micromagnetic approaches, other theoretical approaches include analytical formalisms, e.g. [83] and computational methods, e.g. [88].

1.3.2 Experimental techniques

A wide variety of different experimental techniques have been used to investigate magnetization dynamics at the micro- and nanoscale, e.g. [72,75,76,78,89–97]. They range from microwave-based techniques such as ferromagnetic resonance, optical techniques such as magnetic optical Kerr effect and Brillouin light scattering, as well as pump-probe techniques with femtosecond lasers and X-ray microscopy. In the following we review standard techniques and methods that are used to experimentally investigate the dynamics in magnetic micro- and nanostructures in general and artificial spin ice in particular.

1.3.2.1 Ferromagnetic resonance and spin-wave spectroscopy

Owing to their high-sensitivity and relatively easy implementation microwave-based characterization techniques are oftentimes advantageous over other—more time-intensive—measurement methods. The basic idea

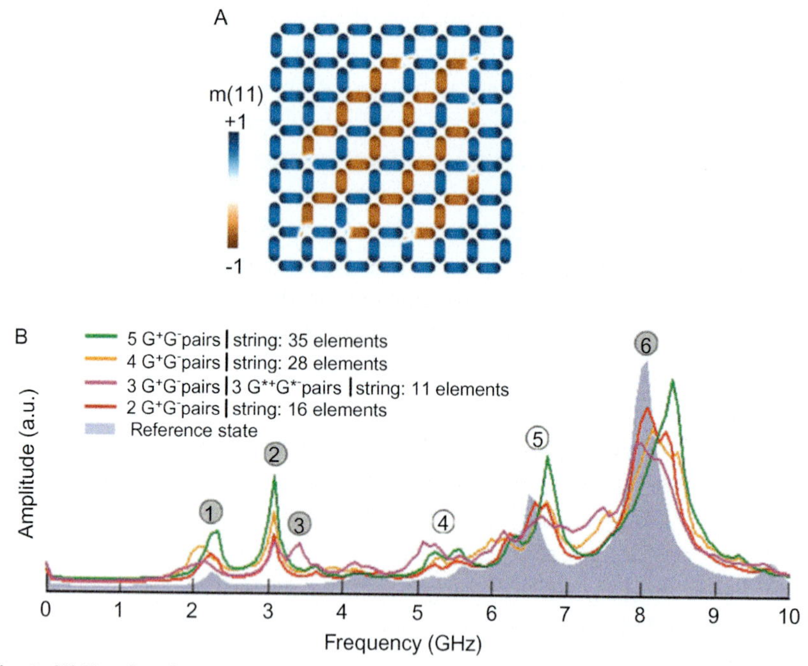

Fig. 9 (A) Simulated spin-ice network with 112 magnetic islands elements containing four monopole-antimonopole pairs connected by Dirac strings. (B) Simulated magnetization dynamics spectra for different string lengths and numbers of monopole-antimononpole pairs (denoted by G^+ and G^-). The reference state is shown in gray. *Adapted from S. Gliga, A. Kákay, R. Hertel, O.G. Heinonen, Spectral Analysis of Topological Defects in an Artificial Spin-Ice Lattice, Phys. Rev. Lett. 110 (2013) 117205. https://doi.org/10.1103/PhysRevLett.110.117205.*

is as follows: a sample is placed in the vicinity of a coplanar waveguide, microwave cavity or microwave antenna. The Oersted field accompanying the microwave signal exerts a torque on the magnetic moments in the sample, which leads to the onset of dynamics if the condition for resonance is fulfilled [98]. The power absorbed by the magnetic sample is proportional to the excited magnetization dynamics, i.e., in the quasiparticle picture the number of magnons. In resonance, the magnetic moments precess at a higher amplitude and, thus, a higher microwave absorption is observed. By varying the bias magnetic field and excitation frequency, we can map out the spin-wave dispersion and experimentally access the eigenmodes of the system. For this purpose, a vector network analyzer is frequently used. Another approach is to use a microwave source, an appropriate antenna structure or cavity in combination with a microwave diode detector that rectifies the transmitted microwave signal.

More recently, spin-transport based techniques have become increasingly important for studying the underlying dynamics in ferromagnetic micro- and nanostructures. Spin-torque ferromagnetic resonance (ST-FMR) was envisioned as an easy-to-integrate metrology for the determination of the spin-Hall angle of a material [99]. However, it is also a useful method for studying the associated magnetization dynamics of the ferromagnet, e.g. [100,101]. For this purpose, a microwave signal is passed through a heterostructure consisting of a magnetic and a heavy metal layer. The heavy metal layer in this scenario acts as a spin-current source by means of the spin Hall effect. Thus, a material with strong spin-orbit coupling is preferential. As for the magnetic layer, a metallic [99,102–106] or insulating ferromagnet [100,101,107–110] can be employed for the ST-FMR measurements. In the ST-FMR process, magnetization dynamics is simultaneously excited by the Oersted field generated by the microwave current and an oscillating transverse spin current that is created in the heavy metal layer due to the spin Hall effect. This spin current exerts an oscillatory spin torque on the magnetic moments leading to the onset of dynamics if the conditions for resonance are achieved. The resonant oscillation of the ferromagnetic layer then leads to a concomitant variation of the bilayer resistance either due to the anisotropic magnetoresistance in the case of a metallic ferromagnet, or the spin Hall magnetoresistance in the case of an insulating ferromagnet such as YIG. This time-varying resistance leads to a rectified voltage due to the mixing of the resistance with the microwave charge current that can be detected either by a lock-in amplifier or a voltmeter. More information and a detailed review of ST-FMR can be found in Ref. [111].

Moreover, it is even possible to electrically drive magnetization dynamics in ferromagnetic heterostructures utilizing the spin Hall effect. In this scenario, a dc charge current supplied to a ferromagnet/heavy metal layer leads to a charge-to-spin current conversion in the heavy metal. The spin current is subsequently injected into the adjacent ferromagnetic layer, which can either be insulating or metallic. The spin current exerts a torque on the magnetic moments in the magnetic layer which can either impede or favor the onset of magnetization precession depending on the orientation of the spin-polarization vector with respect to the moments. Above a threshold current the damping in the system is compensated and the onset of auto-oscillations of the magnetization can be observed. The resonant oscillation can be detected by microwave techniques such as a spectrum analyzer [112–114] or by optical means [115–118]. Auto-oscillations were even reported to be observable as a dc voltage variation [119], which might be a promising route for the investigation of auto-oscillations in connected ASI networks.

1.3.2.2 Optical techniques

Brillouin light scattering alongside time-resolved magneto-optical Kerr effect measurements have become standard techniques for the investigation of magnetic micro- and nanostructures. In the following we present the basics of both techniques. For more detailed reviews we refer the reader to Ref. [120] for an excellent review on microfocused Brillouin light scattering and to Ref. [121] for the time-resolved magneto-optical Kerr effect technique.

Magneto-optical Kerr Effect (MOKE). The effect is based on magnetic circular dichroism, where exchange interaction and spin-orbit coupling in a magnetic material lead to different absorption for left- and right-circularly polarized light. As a result, the polarization of light is rotated when it is reflected from a magnetic material and by measuring this change in polarization one can access the magnetization state of the sample. In time-resolved MOKE an ultrashort laser (ps) is used to probe the magnetization dynamics in the time-domain. More details can be found in Ref. [121].

Brillouin light scattering (BLS). BLS, on the other hand, probes magnetization dynamics in the frequency domain. It is based on the inelastic scattering of magnons—the elementary quanta of spin waves—and laser photons. The creation and annihilation process of magnons upon scattering laser photons are energy and momentum conserving [120]. Therefore, the inelastically scattered photons carry frequency and momentum information about the probed magnons. Using BLS it is possible to detect either thermally activated magnons or microwave-driven excitations. In the context of ASI, it is important to note that the laser focus of a conventional BLS setup is tens of micrometers in diameter, which renders it as not suitable for studies of individual micro- or even nanoscale magnets. To circumvent this issue, there has been considerable progress in implementing scanning microscopy with BLS, e.g. [120], and near field effects with BLS [122].

1.3.2.3 X-ray-based techniques

As outlined in Section 1.1.2, X-ray-based techniques for magnetic studies offer a high resolution as well as element selectivity and the possibility of time-resolved characterization of the magnetic configuration in real space [123–125].

Photoemission electron microscopy (PEEM) is based on the excitation of electrons in a material by X-ray radiation. Differences in the material properties result in a change in the intensity of the emitted electrons. The photoemitted electrons are then extracted by a large electric field and sent to an electron-to-visible-light detector, which finally generates an image detected by a CCD [126,127]. Magnetic contrast of the samples can be obtained from X-ray magnetic circular dichroism (XMCD): the absorption of the incoming polarized radiation depends on the magnetization direction and, thus, an image of the magnetization can be detected. Typically, a sample is illuminated with right- and left-handed circular polarized X-rays. For a common material such as permalloy, the X-ray energy can be tuned for absorption at the L3 edge of Fe. X-ray radiation generated in synchrotrons is usually pulsed, which enables time-resolved PEEM (TR-PEEM) studies, where spin dynamics can be probed with high spatial resolution. For detailed reviews of X-ray PEEM we refer to Refs. [128–130].

Another X-ray-based technique is scanning transmission X-ray microscopy (STXM), where the transmitted portion of X-rays through a sample is measured using a photodiode. Similar to PEEM, XMCD has been combined with STXM to measure the magnetization and to observe the state in artificial spin ice [131]. Time-resolved measurements can be achieved similar to TR-PEEM [132–135]. For more detailed reviews we refer to Refs. [136, 137].

2. Experimental results

This Section is devoted to recent experimental results in the field of artificial spin ice. Section 2.1 reviews developments seeking the ground state of artificial spin ice including thermal annealing and demagnetization processes (Section 2.1.1) and direct writing of the ground state (Section 2.1.2). In Section 2.1.3 we provide a future outlook on ground state studies, before we discuss slow spin dynamics and PEEM in Section 2.2. This Section also covers studies on tuning the frustration by geometry (Section 2.2.1) and vertex frustration (Section 2.2.2), as well as new directions with PEEM-XMCD (Section 2.2.3). After having laid the foundation and background we discuss magnetotransport in artificial spin ice in Section 2.3 and resonant magnetization dynamics in arrays of nanomagnets and artificial spin ice in Section 2.4. This Subsection covers inductive (Section 2.4.1) and Brillouin light scattering (Section 2.4.2) studies, as well as new directions involving spintronic-based effects (Section 2.4.3).

2.1 Recent developments seeking ground states of artificial spin ice

The intention of Section 2.1 is to focus on recent progress in studying the ground states of ASI system using thermal annealing and demagnetization strategies. We selectively highlight recent work that represents transformative research which builds upon the history we summarized in Section 1.1.

2.1.1 Thermal annealing and demagnetization

Thermal annealing has recently been used to study the effects of topological defects on the ground state of ASI systems. After identifying $FePd_3$ as a useful material for thermal annealing of ASI [31], Drisko et al. deliberately patterned edge defects into a square ASI of the same material [138]. In addition to distorting four-island vertices in the vicinity of a defect, each topological defect leads to the creation of a three-island vertex as illustrated in Fig. 10.

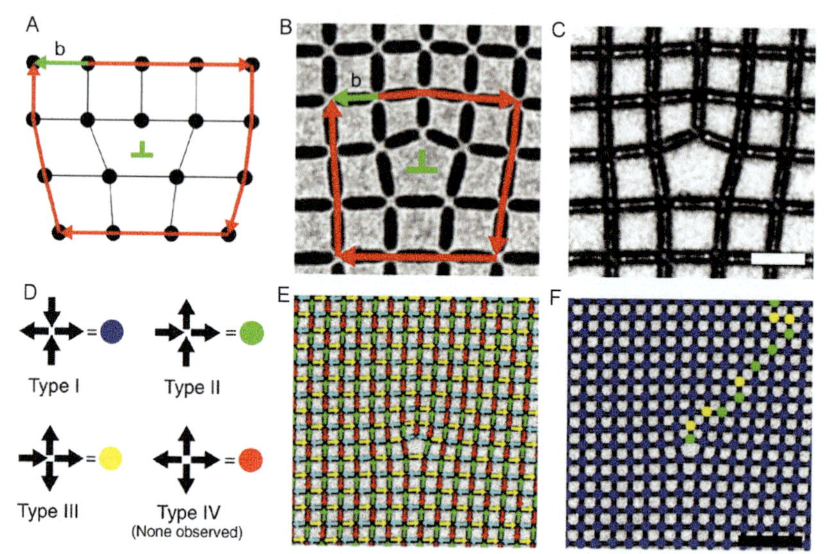

Fig. 10 In (A) an edge defect is illustrated in a square lattice. A Burgers circuit is illustrated in red, and the Burgers vector is drawn with a green arrow. In (B) and (C) a realization of a square ASI with the same edge defect is shown with transmission electron microscopy, and Lorentz transmission microscopy to realize the microstate of the system. (D) shows the typical ASI vertex types of a square lattice. The microstate of the annealed square lattice is shown in (E) and (F) from a spin and vertex mapping, respectively. It is clear from (F) the system prefers to be in an ordered Type 1 ground state except for a domain wall which nucleated from the edge dislocation to the boundary of the system. *Adapted from J. Drisko, T. Marsh, J. Cumings, Topological frustration of artificial spin ice, Nat. Commun. 8 (2017) 14009. https://doi.org/10.1038/ncomms14009.*

After being thermally annealed into a ground state it was found that the system is generally in a long range ordered ground state consisting of Type 1 vertices, except there is a caveat. Extending from every three-island vertex introduced from the topological defect, there is a domain wall consisting of Type 2 and Type 3 vertices growing from the defect. If there is only one defect in the system the domain wall will extend to the edge of the sample and terminate on the boundary of the system. If there are two defects, there can be two domain walls that extend to the edge of the system and terminate, or a domain wall can nucleate off of one defect and terminate on the other. In all of these scenarios there is generally freedom in how the domain wall propagates throughout the square lattice, and although domain walls with short lengths to the edge of the system are energetically favorable they are not typically observed. The authors describe these domain wall configurations as being "entropically driven" as opposed to being energetically driven. Thus, beyond demonstrating that artificial spin systems can be platforms to simulate defect driven effects in magnetic order, Drisko et al. conclude that topological defects may provide an approach to controllably reintroduce frustration into the otherwise unfrustrated square ASI [138].

Thermal annealing has also proven to be an important tool in helping to find ground states in complicated geometries. An excellent example is the disconnected Penrose quasicrystal studied by Shi et al. [50]. Here, the authors propose a ground state by imagining a deconstruction of the quasicrystal into two different decagon units, which house a number of different vertex types. This ground state is shown in Fig. 11. The ground states of the individual decagons are identified, and it is discussed how when the quasicrystal is reconstructed using the individual decagons, it is impossible for all vertices to remain in a ground state configuration. The inability for all vertices to remain in the ground state was subsequently identified as a "topologically induced emergent frustration," or a vertex frustration. During the reconstruction of the quasicrystal from the decagons, Shi et al. identify regions of the quasicrystal termed the "skeleton" which are made up of vertices allowed to be in a doubly degenerate, ordered ground state. The remaining regions are the vertex-frustrated portions of the quasicrystal. Experimentally, the best evidence for the proposed ground state was obtained after thermal annealing (as compared with as-grown and demagnetized samples). The entire "skeleton" region of the quasicrystal was not found to be in a single domain, but subdomains of the two long-range ordered states were experimentally observed.

Other types of modified square lattices have also been the subject of ground state studies which employ refined demagnetization protocols. Parakkat

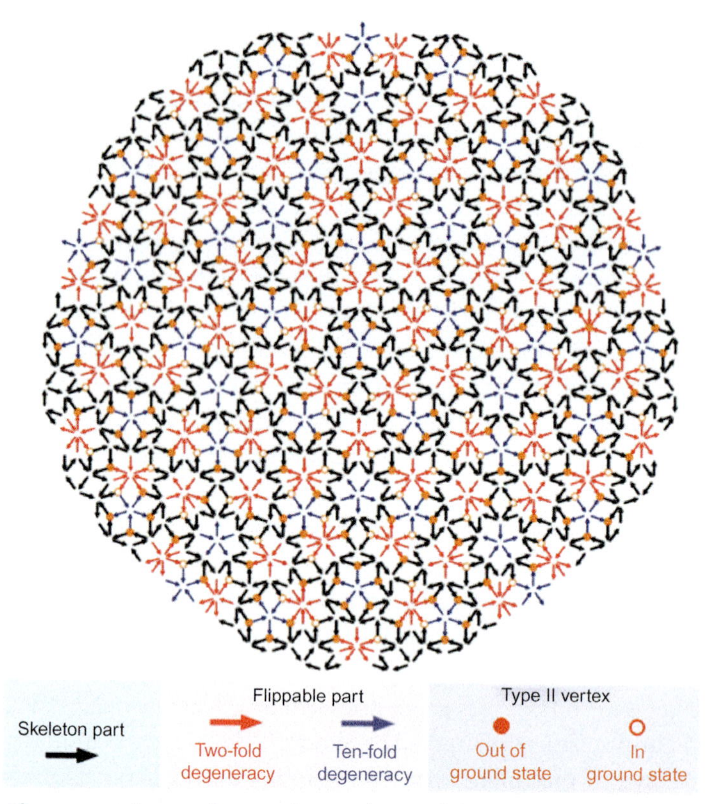

Fig. 11 The proposed vertex-frustrated ground state of the Penrose P2 quasicrystal is illustrated above. The skeleton portion of the quasicrystal is a region where every vertex is configured into a lowest energy configuration. Analogous to the Shatki lattice, these regions are similar to the four-island and two-island vertices which are placed into a ground state. The red and blue colored moments participate in low energy vertices that do not have a unique moment configuration and have twofold and 10-fold degenerate configurations. Consequently, vertex frustration emerges in the three-island vertices marked with orange filled and open circles (high and low energies). The topology of the quasicrystal makes it impossible for every three-island vertex to be placed within its ground state. *Adapted from B.P. Tonner, D. Dunham, T. Droubay, J. Kikuma, J. Denlinger, E. Rotenberg, A. Warwick, The development of electron spectromicroscopy, J. Electron Spectrosc. Relat. Phenom.. 75 (1995) 309–332. https://doi.org/10.1016/0368-2048(95)02523-5.*

et al. studied a square ASI heterostructure, where individual islands in the square lattice can be exchange biased. Individual islands consist of a stacked MgO/Fe/IrMn/Pt heterostructure [27]. Reference samples in this work had a missing IrMn layer so that no exchange bias was present. To fabricate the ASI arrays, a top-down approach was employed where continuous films are grown and milled down into square ASI arrays. With an exchange bias field present between the Fe and antiferromagnetic IrMn, the lattice constant of the square lattice can be adjusted to engineer a situation where two of the Type 2 vertex types are energetically equivalent to the Type 1 vertex types. For missing IrMn layers, or for situations with the smallest lattice constants, the demagnetization protocol allowed the square ASI to develop large crystallites of the long-range ordered phase. When the lattice constant was increased such that exchange bias competed with nearest neighbor interactions, the same demagnetization protocols returned disordered microstates. The message here is clear: the energy hierarchy of all of the vertex configurations within a given ASI geometry can be modified by adopting ideas from spintronics and creating heterostructures.

2.1.2 Direct writing of ground state

Recently there have been remarkable new techniques which allow the experimentalist to directly "write" the ground state, and many other microstates for that matter, into ASI. These types of experiments take place entirely in the athermal limit, and they do not utilize demagnetization strategies. Wang et al. first demonstrated unprecedented re-writability within a disconnected ASI that was named an "artificial magnetic charge ice" [139]. In this work, the lattice chosen can be thought of as a modified square ASI. As seen in Fig. 12A, the modification is that there are equal number of islands

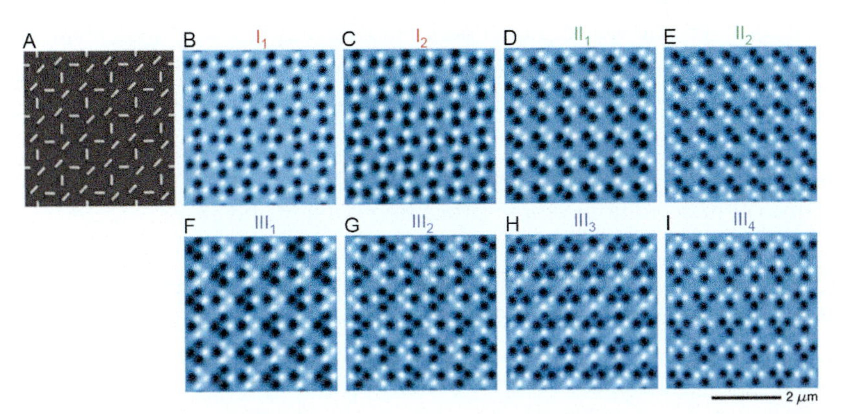

Fig. 12 (A) The geometry of the "artificial magnetic charge ice" is shown. In (B) and (C) the two ground states consisting of only Type 1 vertices were directly written into the system. In (D) and (E) two different excited states consisting of only Type 2 vertices are shown. In (F)-(I) an unusual set of excited states consisting of only Type 3 vertices were directly written into artificial spin system. These states are unusual in that they cannot be obtained through conventional thermal annealing techniques, and they cannot be achieved by applying external magnetic fields. *Adapted from Y.-L. Wang, Z.-L. Xiao, A. Snezhko, J. Xu, L.E. Ocola, R. Divan, J.E. Pearson, G.W. Crabtree, W.-K. Kwok, Rewritable artificial magnetic charge ice, Science 352 (2016) 962–966. https://doi.org/10.1126/science.aad8037.*

that are oriented horizontally, vertically, and also at 45° between horizontal and vertical. Within a typical square ASI every vertex has a four-island coordination number, and the four magnetic "charges" that are located around a vertex come from the four participating islands. In the "magnetic charge ice" every vertex has a three-island coordination number yet there are still four magnetic "charges" that participate in a vertex. This is allowed if one island (in this case the 45° oriented islands) contributes two "charges" to a given vertex. Wang et al. used a homemade MFM station where the ASI sample is placed within a vector magnet as the MFM tip is rastered over a preprogrammed area of the sample. By applying an external field as the MFM tip scans the sample, the differing magnetic anisotropy of the three types of islands (horizontal, vertical, and in-between) can be exploited and the tip can selectively flip one of these island types. The freedom to flip individual sets of islands allows for the ASI to be manually configured into a ground state consisting of only Type 1 vertices. Additionally, because the system is athermal, it is also possible to configure the system into unusual and energetically unfavorable configurations which consist of only Type 3 vertex populations. All of the possible manually obtained, and magnetically ordered configurations are imaged via MFM in Fig. 12B–I. Gartside et al. recently used a similar direct writing approach that involved a programmable MFM configuration [140]. To write, or selectively flip individual elements of the honeycomb lattice, a high moment MFM tip was used to controllably introduce a vortex and domain wall in the center of an individual island. The writing operation allows one edge of the domain wall to propagate along the length of the nanowire until it encounters a vertex, effectively flipping the ASI island. This on demand writing technique was used to manually configure a connected honeycomb lattice into the spin and charge ordered ground state.

Direct writing of the ground state, or any other arbitrary state in ASI research tends to be motivated from applied interests. For example, Wang et al. used direct writing technologies to manually reconfigure the microstate of the "artificial magnetic charge ice" when it was patterned on top of a superconducting film [141]. Depending on the microstate of the artificial magnetic charge ice, flux quanta in the underlying superconducting film can crystallize or become frustrated. This enables the critical current density of the superconducting film to be modulated by the artificial spin system. The manual reconfiguration of ASI has also been identified as a way to control the magnonic or high-frequency magnetic response of ASI (a topic which we expound upon in Section 2.4). In the context of this Section, Haldar et al. used two different types of elongated rhomboid shaped islands in a one-dimensional artificial spin lattice [142]. Using two different rhomboids that are mirror images of one another, a field protocol can be used to switch the magnetic state of one rhomboid with respect to another. Thus, the microstate of the system can be toggled between a uniformly magnetized lattice, versus a lattice where one set of rhomboids is reversed with respect to the partner set. These two microstates can be used to propagate or stop magnetization dynamics in the one-dimensional lattice as described in more detail in Section 2.4.

2.1.3 A future outlook on ground state studies

Thermal annealing and demagnetization protocols are an excellent first line of inquiry when it comes to studying ground states in ASI. The experiments mentioned above do not require any X-ray microscopy tools and can easily be done in house with a heater, electromagnet, and MFM. There is plenty of space in this experimental arena to innovate and use these tools. Currently, emerging studies which employ these techniques focus on the study of ground states in the context of phase changes or phase transitions within ASI systems. Within this Subsection we highlight some of the most exciting developments in this area.

Beyond the traditional ASI lattices, there are new geometries, which support new and/or exotic ground states. Macêdo et al. theoretically proposed a variant of the square ASI where every individual island is rotated 45° about its midpoint so that vertices resemble a "pinwheel" structure [143]. The pinwheel ASI can still be described at the vertex level. It was theoretically found that for a range of angles near 45°, Type 2 vertices will be lower in energy than Type 1 vertices. This dramatically alters the ordered ground state of the system from a state with no net moment to a state with a net moment, or an emergent ferromagnetic phase. Simulations demonstrated that the ground state the pinwheel ASI are sensitive to the finite boundaries of the system and can lead to the formation of a myriad of ferromagnetic domain configurations that are found in real materials. Field-driven effects of the pinwheel ASI have recently taken place, but there is still experimental opportunity for these systems to be thermalized or demagnetized into the various predicted ground state configurations [144].

The vast majority of ASI system are comprised of magnetic elements that take the form of elongated islands. Due to the shape anisotropy, the magnetization of every island, connected or disconnected in a lattice, prefers to be in one of two orientations. For this reason, it is fair to say that individual elements of an ASI system have binary, or Ising degrees of freedom. We note that the Ising-like nature of ASI ice systems has also been studied in an alternative material system and geometry. By patterning arrays of magnetic discs with perpendicular magnetic anisotropy, Ising-like ASI systems have been studied extensively with MFM and MOKE techniques [145–147]. Although there is a traditional focus on Ising degrees of freedom, nanofabrication techniques do allow for other basic elements, aside from elongated islands, to make up an artificial spin system. By seeking inspiration from spin-lattice models that go beyond Ising model physics, new magnetic metamaterials can be designed. These new systems, with more than two-degrees of freedom, can house more exotic phases and phase transitions which can be studied using thermal annealing and demagnetization strategies.

One pioneering, non-binary, artificial spin system is the four-state artificial Potts spin system studied by Louis et al. [148]. As seen in Fig. 13, a four-state artificial Potts lattice can be realized by patterning arrays of nanomagnetic squares in a square lattice. Within a given nanomagnetic square, the magnetization will tend to point in one of four, corner-to-corner, configurations. These four corner-to-corner states represent Potts, rather than Ising internal degrees of freedom. It was shown that the magnetic order of the artificial Potts lattice could be tuned based on whether the nanomagnetic squares on a square lattice were rotated or not. Similar to the pinwheel ASI, by rotating each nanomagnetic Potts element the ground state can be adjusted to be ferromagnetic, antiferromagnetic, or a spin ice phase. Another Potts-like system, shown in Fig. 13, was studied by Sklenar et al., where pairs of islands are placed on a square lattice [149]. The island-pair orientation alternates between being vertically or horizontally aligned. In this lattice, the nearest neighbor interaction between island-pairs dominates all other neighboring interactions and it was found that the quadrupole configuration (or lack thereof) of every pair tends to order. In the absence of an external field, the ground state was found to be ferroquadrupolar, where every pair has the same quadrupole state. When this "quadrupolar" lattice was thermalized in the presence of an external field, polarized states on a plaquette competed with the two quadrupolar states and a phase transition from ferroquadrupolar to antiferroquadrupolar order was observed. The salient details of the phase diagram were captured using a Potts Hamiltonian reminiscent of the Blume-Emery-Griffiths models [150], where a three-state Potts variable was mapped onto the two quadrupole states and the polarized states.

2.2 Slow spin dynamics and PEEM

In this Section we discuss recent experiments which employ PEEM-XMCD to study slow spin dynamics in ASI systems. We first discuss experiments that

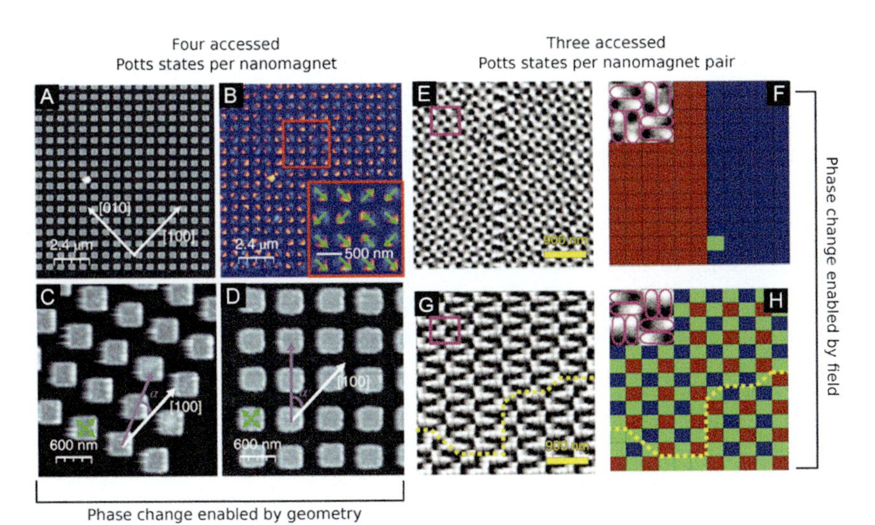

Fig. 13 Two examples of artificial spin systems that realize different types of Potts or Potts-like models are shown. In the left panel (A) is an AFM image of a dipolar four-state Potts lattice. In (B) the corresponding MFM image is shown, and based on the magnetic contrast it is clear that each individual element can be in one of four configurations where the magnetization is pointing from corner-to-corner within the square. In (C) and (D) two additional AFM images, where the square elements have two different angular orientations between elements in the lattice. In (C) a ferromagnetic order between Potts states is favored while in (D) a spin ice phase is favored. In the right panel, (E) and (F) illustrate a ferroquadrupolar ordered phase in an artificial spin system made up of pairs of islands as seen in the inset of (F). Red and blue pixels in (F) correspond to two different quadrupole states for each island pair after the system has been thermalized in a zero-field environment. When thermalized in the presence of a field, a phase change occurs as shown in (G) and (H) to antiferroquadrupolar order. Green pixels in (H) correspond to polarized spin states as opposed to quadrupoles. Images were obtained using MFM. *Adapted from D. Louis, D. Lacour, M. Hehn, V. Lomakin, T. Hauet, F. Montaigne, A tunable magnetic metamaterial based on the dipolar four-state Potts model, Nat. Mater. 17 (2018) 1076–1080. https://doi.org/10.1038/s41563-018-0199-x and J. Sklenar, Y. Lao, A. Albrecht, J.D. Watts, C. Nisoli, G.-W. Chern, P. Schiffer, Field-induced phase coexistence in an artificial spin ice, Nat. Phys. 15 (2019) 191–195. https://doi.org/10.1038/s41567-018-0348-9.*

tune some geometric property of a given lattice to transform an ASI system from having an ordered ground state to a frustrated state. Next, we review progress on vertex-frustrated ASI systems measured with PEEM-XMCD. Although the first two subsections exclusively deal with frustration, this is not representative of all spin dynamic studies in ASI with PEEM-XMCD. We conclude this Section with a variety of innovative experiments that measure spin dynamics of ASI with both applied and fundamental motivations that fall outside of frustration.

2.2.1 Tuning frustration with geometry

As discussed Section 1.1.3 a true spin ice phase was realized in a square ASI by patterning half of the islands on a non-magnetic spacer layer [26]. If the spacer layer is not thick enough (or non-existent), the square ASI will prefer a long-range ordered ground state consisting of only Type 1 vertices. If the spacer layer is too thick, the square ASI will prefer a different long-range ordered state consisting of only Type 2 vertices. A core idea in this work was that a geometric property, e.g., the height, of an ASI can be tuned to a "sweet spot" so that a truly frustrated ground state exists. There have been three recent works which follow this strategy, and exploit height or some other spacing parameter to achieve a maximally frustrated ground state. In all of these studies PEEM-XMCD was the imaging tool of choice because it allows one to observe thermalization effects and low energy excitations simultaneously. We begin with a discussion of the square lattice tuned to be maximally frustrated. Here, thermally excited "monopoles" are closer analogs to monopole excitations in real spin ice materials. Excitation and propagation of such monopoles can now be directly observed with PEEM-XMCD.

As opposed to fabricating ASI islands on a spacer layer, one can utilize etching techniques and fabricate islands inside of trenches. This approach was taken by Farhan et al. to study square ASI with restored degeneracy between Type 1 and Type 2 vertex populations [151]. First, the square ASI was imaged for various height offsets between islands. The magnetic structure factor, a spin-spin correlation function in reciprocal space, was calculated for the thermalized images in a manner similar to previous work [26]. For a magnetically ordered ASI, the structure factor shows sharp Bragg peaks. As the height offset induces a spin-ice phase, the sharp peaks become diffuse indicating increasing disorder within the system, and an extended analysis of the structure factor can be used to estimate monopole densities. By using PEEM-XMCD the motion of Type 3 monopoles was also able to be tracked in real time. The motion of the monopoles was analyzed using Debeye-Hückel theory, which can model the density of correlated monopole charge pairs versus the density of uncorrelated monopole excitations in the ASI. Operationally, this analysis allowed for a calculation of the effective magnetic "charge" of the Type 3 defects. The value of calculated charge can be related back to the saturation magnetization of the magnetic material, which in this case was permalloy. In the Debeye-Hückel framework the measured charge corresponded to a saturation magnetization, which was approximately 60% less than the expected value, but this discrepancy was attributed to blocking temperature effects within individual islands.

Ostman et al. performed an alternative PEEM-XMCD study that realized a maximally frustrated square lattice where magnetic discs were patterned at the center of every vertex (between all four islands) [152]. In general, the presence of a disc favors Type 2 vertex configurations which have a net moment. In these configurations the disc will tend to magnetize along the net moment of the Type 2 vertex, and in this manner the disc equally satisfies the dipolar interaction between all four neighboring islands. For Type 1 vertex configurations there is no net moment and it is impossible for the disc to magnetize in a direction that satisfies all four islands. Consequently, in this experiment the key parameter was the diameter of the disc; increasing the diameter of the disc makes Type 2 vertices energetically more favorable relative to Type 1. By calculating magnetic spin structure maps from the real-space data similar to previous experiments, a similar argument for the Coulomb or spin-ice phase in the modified square lattice was made.

Beyond the square lattice, Farhan et al. recently reported a new type of "trident lattice" that is able to be tuned through island spacing to have a frustrated ground state [153]. Here three islands are placed side-by-side as a plaquette. By tiling a plane with alternating plaquettes, such that the islands alternate to align horizontally or vertically, a geometry is achieved where islands either participate in a four islands vertex or with no vertex at all. The trident lattice is shown in Fig. 14A. In this geometry, two important

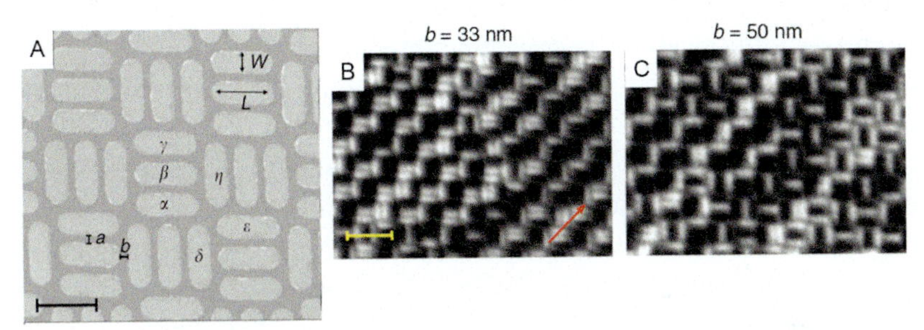

Fig. 14 An electron microscopy image of the trident lattice is shown in (A). The geometry is altered by varying the spacing, B, and in (B) and (C) b = 33 nm and 50 nm, respectively. The ground state in (B) does have a magnetically ordered phase which is evident when the image is analyzed via a magnetic structure factor. In (C), there is no clear magnetic order observed and the spacing has been so that the lattice is maximally frustrated. Note, that unlike MFM images magnetic contrast is present throughout the entire island. The red arrow in (B) indicates the X-ray propagation direction and black islands have a moment along the X-ray direction while white islands have a moment antiparallel against the X-ray direction. *Adapted from A. Farhan, C.F. Petersen, S. Dhuey, L. Anghinolfi, Q.H. Qin, M. Saccone, S. Velten, C. Wuth, S. Gliga, P. Mellado, others, Nanoscale control of competing interactions and geometrical frustration in a dipolar trident lattice, Nat. Commun. 8 (2017) 995.*

parameters are the lateral spacing between islands within a given plaquette and the spacing between plaquettes. The ratio between these two length can be adjusted to control whether or not the system has a long-range ordered ground state or a frustrated ground state, as shown in Fig. 14B and C. Similar to what is observed in the square lattice, when the island spacings in the trident lattice are tuned such that there is a frustrated ground state, the magnetic structure factor of the system evolves from being sharply peaked to diffuse, indicating the presence of a disordered phase.

2.2.2 Vertex frustration

Vertex-frustrated ASI systems have also become a more recent system studied with PEEM-XMCD. By imaging the microstate of the Shatki lattice at temperatures just above the blocking temperature, low-energy thermal fluctuations within the lattice were carefully studied. In the square lattice, excitations above the ground state often are monopole pairs, which are two charged Type 3 vertices in a Type 1 background. Because the Shatki lattice has a mixed vertex coordination number, excitations cannot be well described with effective magnetic charges on individual vertices. Thus, a picture of quasiparticle excitations in vertex-frustrated ASI was not well-established prior to PEEM-XMCD studies. Using PEEM-XMCD, Lao et al. showed that the Shatki lattice can be mapped onto an emergent dimer-cover model [154]. The dimer-cover model allows for experimentally imaged spin maps to be converted into an emergent vector field. This vector field can be used to define topologically protected charges corresponding to long lifetime quasiparticles within the Shatki lattice. It was concluded that the topological charges, which represent excitations above the Shatki's ground state, play an important role in limiting thermal equilibration of vertex-frustrated ASI systems because the movement of the charges is kinetically constrained at low temperature. This is in contrast to the movement of charges in the conventional square ASI, where single spin flips lead to both charge-pair creation and propagation. In another PEEM-XMCD experiment, Stopfel et al. studied thermalization effects between a modified Shatki lattice, where the vertices with two-islands coordination are replaced by a single long island roughly twice the length of a single island, and the original Shatki lattice [155]. A central focus of this work was to study ground state formation when there are two different blocking temperatures within the Shatki lattice. This is possible because the effective blocking temperature of a long magnetic island is significantly higher (nearly 30 K) than the short islands.

While the Shatki lattice has been the most extensively studied vertex-frustrated ASI with PEEM-XMCD, other lattices have been considered as well. The Tetris lattice is one of the original vertex-frustrated lattices identified by Morrison et al. [43]. Similar to the Shatki lattice, the Tetris lattice is comprised of vertices with four-, three-, and two- island coordination numbers. Unlike the Shatki lattice, the Tetris lattice has two types of two-island vertices where the islands are either in-line with each other or are arranged to make a right angle with respect to one another. When studying slow spin dynamics of the Tetris lattice, Gilbert et al. noted that the lattice can be decomposed into one set of diagonal stripes consisting of four-island and two-island vertices, and another set consisting of three-island and two-island vertices [156]. The stripes consisting of four-island and two-island vertices are able to thermalize into ordered ground states while the stripes consisting of three-island and two-island vertices remain disordered and behave as one-dimensional Ising spin chains. Therefore, although the Tetris lattice is a two-dimensional tiling, the dimensionality of the system was effectively reduced along the effective one-dimensional Ising spin chains.

2.2.3 New directions with PEEM-XMCD

Because of its experimental versatility, PEEM-XMCD has been a testbed for new ideas involving ASI. The hallmark of ASI research is creativity in lattice and geometry design, so it is not surprising that new lattices are studied with PEEM-XMCD for a myriad of reasons. One example, was the fabrication of the so-called dipolar-dice lattice by Farhan et al. [157]. This lattice was explicitly designed to study the role of magnetic charge screening effects in the thermalization of an ASI. The dipolar-dice lattice is a lattice with mixed coordination consisting of a collection of six-island vertices surrounded by five three-island vertices. Prior to relaxing into the equilibrated ground state, where no charges remain on the six-island vertices, there is an intermediate state where the charged six-island vertices are screened by charge on the surrounding three-island vertices. This work clearly demonstrated that there are "charge" interactions in ASI and that the appearance of screening effects is not merely a consequence of the lattice geometry. Rather, charge screening is yet another consequence of "monopole" interactions, and in this particular case the screened state was an essential step in the thermalization pathway the dipolar dice.

PEEM-XMCD has also been used to study potential applications of ASI systems. Gliga et al. studied the pinwheel ASI, which acts as a "spin ratchet" [158]. Experimentally it was shown that if the spin ratchet is magnetized so

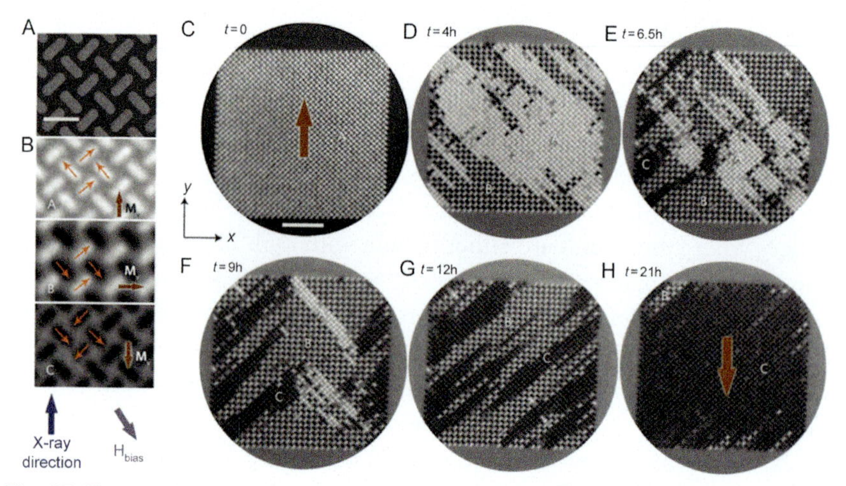

Fig. 15 The pinwheel geometry is shown in (A). The PEEM-XMCD magnetic contrast is shown in (B) when the pinwheel is magnetized along the +y, +x, and −y direction. In (C)-(H) the time evolution of the pinwheel ASI is shown to evolve over the course of 21 h. The magnetization is observed to rotate with a clockwise directional preference. *Adapted from S. Gliga, G. Hrkac, C. Donnelly, J. Büchi, A. Kleibert, J. Cui, A. Farhan, E. Kirk, R. V Chopdekar, Y. Masaki, N.S. Bingham, A. Scholl, R.L. Stamps, L.J. Heyderman, Emergent dynamic chirality in a thermally driven artificial spin ratchet, Nat. Mater. 16 (2017) 1106–1111. https://doi.org/10.1038/nmat5007.*

that the collection of islands have a net moment along the y-axis, and then heated just above the blocking temperature of individual islands, slow spin dynamics will proceed in a manner where the net magnetization rotates with a directional preference as seen in Fig. 15. Although such rotational dynamics can be controlled with an external magnetic field, in this work the chirality of the magnetization rotation was established in the absence of an external field. To explain the origin of the rotational preference micromagnetic simulation were employed. Micromagnetic simulations showed, in a magnetostatic picture, that the magnetic surface charges on the edges of individual islands within the ASI rearrange in a way to promote the emergence of chirality within the ratchet. Previously, it was identified that edge effects representing deviations from ideal Ising behavior in ASI could be useful from the perspective of modulating high-frequency magnetization dynamics in ASI. This work effectively showed how the same magnetostatic edge effects can be used to engineer spin dynamics that may have utility as nanomagnetic motors or sensors [158].

In Section 2.1.3 we discussed how an emerging trend is to study ASI systems that are more direct analogs of spin-lattice models that do not consist of

Ising spins. PEEM-XMCD has been used to study a subset of these models, in particular XY spin systems were fabricated and measured by Streubel et al. [159]. To obtain magnetic elements that resemble XY spins, individual elements are discs with a diameter of approximately 100 nm that are 10 nm thick. Ordered phases of the artificial XY spin system were studied in both a square and honeycomb lattice, and spin dynamics in the long-range ordered magnetic phase of the hexagonal XY lattice were examined. Streubel et al. did note some challenges with the XY spin system that arise from nanofabrication issues. Random edge roughness effects on individual discs can cause random preferred easy axes that vary from disc to disc.

2.3 Magnetotransport in artificial spin ice

As the field of artificial spin ice received more attention and research efforts, other measurements techniques used to explore the state of the ice provided new insights. In particular, the interaction of current and magnetism proved especially useful, since the lithographically designed spin ices produced unique magnetoresistance traces. Being able to control the magnetic state of a lattice and to measure its transport response is of great importance, since it enables the use of ASI as logic gates [160–162].

2.3.1 Magnetotransport results in honeycomb artificial spin ice

The first magnetotransport studies were performed by Tanaka et al. [18] in a honeycomb lattice made out of $Ni_{81}Fe_{19}$, with each wire length $l=400$ nm, width $w=40$ nm and thickness $t=20$ nm. MFM images confirmed that the magnetization of each wire behaves as a single macrospin, without any internal magnetic domain wall. In order to measure the magnetoresistance signal, the lattice needs to be connected to have a continuous path for the current to flow. In their study, they observed steps in the magnetoresistance curve as different wires were reversed at varying magnetic fields depending on the relative angle between the external magnetic field and the lattice direction, see Fig. 16A. Since the honeycomb lattice geometry is threefold, up to three steps in the magnetoresistance are expected for certain in-plane angles. However, only up to two steps were observed. This was produced by the interactions at the vertex, which prohibit the all-in and all-out configurations of the neighboring wires, thus obeying the ice rules. By applying an out-of-plane field, see Fig. 16B, the magnetoresistance traces eventually showed three steps, as shown in Fig. 16C and D, indicating the appearance of all-in and all-out states, as the energy of the systems was modified. These results show that the ground state of the lattice can be manipulated and

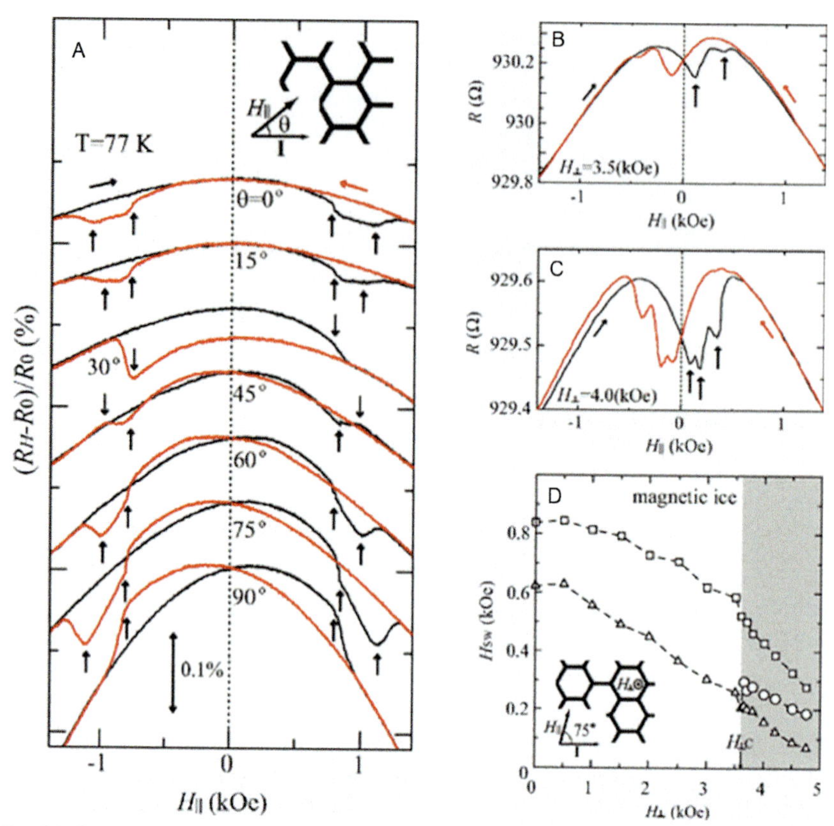

Fig. 16 Experimental magnetoresistance traces of a honeycomb lattice at (A) different in-plane angles, (B)-(C) a fixed in-plane field of 75° and out-of-plane field of (B) 3.5 kOe and (C) −4 kOe. (D) Evolution of the peaks marked in (B) and (C) as a function of the out-of-plane field, with a fixed in-plane field of 75°. At a field above 3.5 kOe, three peaks are observed. *Adapted from M. Tanaka, E. Saitoh, H. Miyajima, T. Yamaoka, Y. Iye, Magnetic interactions in a ferromagnetic honeycomb nanoscale network, Phys. Rev. B. 73 (2006) 052411. https://doi.org/10.1103/PhysRevB.73.052411.*

determined using magnetotransport, and that typical spin ice behavior such as the ice rules can be observed not only by direct imaging, but also from transport measurements.

The effects of the magnetization in the transport properties of a Co honeycomb lattice were also used in a parallel Hall geometry by Branford et al. [163] to observe the appearance of a chiral order as the temperature was decreased. They correlated the Hall transport measurements with the spatial distribution of the magnetization obtained by micromagnetic simulations using OOMMF [57]. The hexagonal components of the honeycomb lattice were characterized by their clockwise or anticlockwise chirality, depending on the direction of the magnetization in the wires. As they swept the magnetic field in the simulations, they observed the appearance of different chiralities in different regions of the simulated lattice at certain magnetic fields, which corresponded to the magnetic fields presenting an asymmetric peak in the transport data. As the temperature was increased in the experiment, the observed peak decreased, indicating the disappearance of the chiral order, and a transition to a short-range order.

Similar asymmetric magnetoresistance traces were observed in a $Ni_{81}Fe_{19}$ honeycomb lattice at temperatures below 20 K [164]. The longitudinal and transverse transport properties of the lattice made out of $Ni_{81}Fe_{19}$ were compared to the properties of lattices with an extra capping layer of alumina. Since the observation of an unusual asymmetry onset at temperatures below 20 K was present in the samples without capping, it was concluded that the formation of antiferromagnetic oxides in the $Ni_{81}Fe_{19}$ was responsible for the asymmetric traces, since these oxides are known to have an antiferromagnetic behavior and to produce an exchange bias in the ferromagnetic $Ni_{81}Fe_{19}$. The addition of an alumina capping layer prevents the oxidation of the $Ni_{81}Fe_{19}$ surface, hence preventing any AFM/FM interactions. These results further showed the importance of the care that must be taken when interpreting MR results, since the response can be produced by different complex mechanisms, especially the Hall contributions, as they can come from the AMR response (planar Hall effect), non-coplanar spin textures producing a Berry phase (anomalous Hall effect), etc.

To gain further understanding on how the magnetic configuration of the lattice affects the measured magnetoresistance it is important to know the current distribution in the lattice as the magnetoresistance curve is calculated by summing the local contributions $V_{AMR} \propto \Sigma(m_i \cdot j_i)$. The work by Chern [165] analyzed the response of a honeycomb lattice modeled as a resistor network, calculating the voltage drop at the vertices and taking into account the current distribution at the vertices, see Fig. 17, and it shows the path to predict the longitudinal and transverse magnetoresistive response of arbitrary-shaped lattices.

The state found by micromagnetic simulations was used in combination with the current distribution to produce a simulated magnetoresistance curve that could be compared to the experiment [56,166]. In the work by Le et al. [56], the current through the honeycomb lattice was modeled as having a magnitude I in the wires parallel to the current direction and

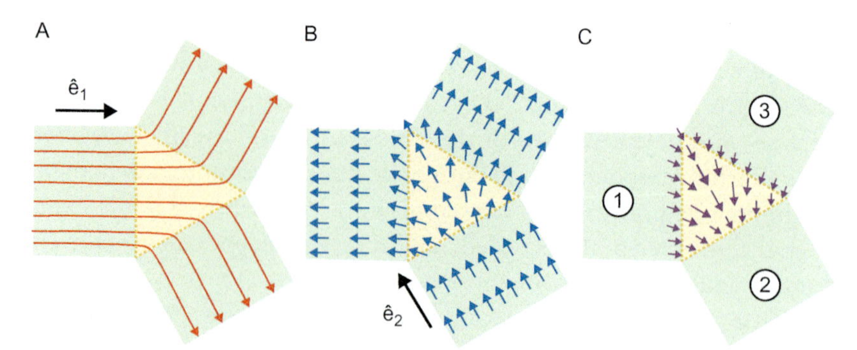

Fig. 17 Vertex of a honeycomb lattice used to compute the magnetoresistance response. (A) Current distribution, (B) magnetization profile and (C) resulting electric field. *Adapted from G.-W. Chern, Magnetotransport in Artificial Kagome Spin Ice, Phys. Rev. Appl. 8 (2017) 064006. https://doi.org/10.1103/PhysRevApplied.8.064006.*

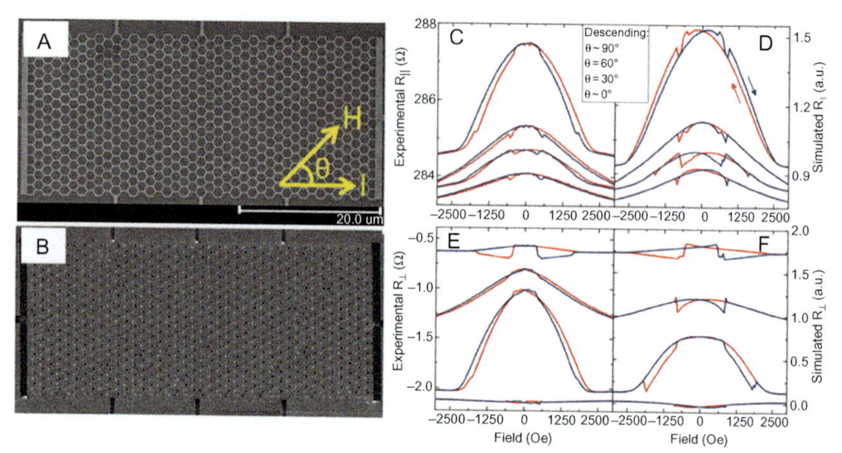

Fig. 18 (A) Connected honeycomb lattice of $Ni_{81}Fe_{19}$ with wire width $w = 75$ nm, length $l = 800$ nm and thickness $t = 25$ nm. (B) MFM image of the lattice. Experimental longitudinal (C) and transverse (E) magnetoresistance curves, and the corresponding longitudinal (D) and transverse (F) simulated data. *Adapted from B.L. Le, J. Park, J. Sklenar, G.-W. Chern, C. Nisoli, J.D. Watts, M. Manno, D.W. Rench, N. Samarth, C. Leighton, P. Schiffer, Understanding magnetotransport signatures in networks of connected permalloy nanowires, Phys. Rev. B. 95 (2017) 060405. https://doi.org/10.1103/PhysRevB.95.060405.*

$I/2$ in the others, with the current direction parallel to the wires. In this manner, the local current is used alongside the magnetization distribution obtained using micromagnetic simulations with MuMax3 [85]. The result of their calculation at different fields is shown in Fig. 18D and F. Fig. 18C and E show the corresponding measured curves, with a remarkable agreement. Since the experimental and simulated curves agree, it is reasonable to state that the magnetic configuration obtained in the simulation is hence similar to the magnetic configuration in the experimentally measured lattice. By analyzing the magnetic configuration from the simulations, it was observed that the vertices of the lattice play an important role in determining the shape and the steps in the magnetoresistance data.

However, as mentioned earlier, this method has to be applied with care, since the magnetic configuration is not uniquely determined from the magnetoresistance, as different magnetic configurations can produce the same resistance value. It is unlikely that different magnetic configurations will evolve with the magnetic field in a way that they produce the same magnetoresistance curve, but it is possible in principle. Experimental techniques such as X-ray PEEM (see Sections 1.3.2 and 2.2) can help determining the actual magnetic configuration.

2.3.2 Magnetotransport results in other artificial spin ices: Brickwork lattice

Magnetoresistive measurements in combination with simulation provided a deeper understanding of the magnetization reversal process in honeycomb lattices. The same technique was also applied to study the behavior of a brickwork lattice [166], Fig. 19A, which shows a higher degree of frustration. In this study, a more precise current density distribution was obtained by solving the continuity equation with a finite element method implemented

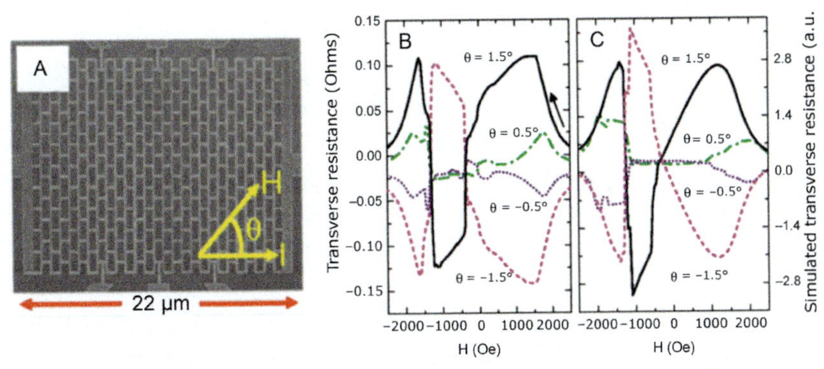

Fig. 19 (A) Connected brickwork lattice of $Ni_{81}Fe_{19}$ with wire width $w = 140$ nm, length $l = 660$ nm and thickness $t = 40$ nm. (B) Measured transverse resistance at four different field angles very close to the direction of the current. (C) Corresponding simulated resistance, showing a good agreement with the experimental curves. *Adapted from J. Park, B.L. Le, J. Sklenar, G.-W. Chern, J.D. Watts, P. Schiffer, Magnetic response of brickwork artificial spin ice, Phys. Rev. B. 96 (2017) 024436. https://doi.org/10.1103/PhysRevB.96.024436.*

by the commercial software Ansys Maxwell [167]. This approach resulted in a more detailed magnetoresistance curve which, in agreement with their previous study in the honeycomb lattice [56], confirmed the greater contribution to the magnetoresistance of the vertices compared to the wires. The work also showed that the lowest energy ground state of the brickwork lattice can be identified from magnetoresistance traces. Furthermore, in the vicinity of the ground state, the magnetoresistance exhibited an extremely sensitive response to the orientation of an applied external magnetic field. The traces were found experimentally, Fig. 19B, and were in good agreement with the computed traces combining micromagnetic simulations and current distribution calculations, Fig. 19C.

2.3.3 Perspectives for magnetotransport-based techniques

Given the common geometry between anisotropic magnetoresistance measurements and dynamic spin-torque ferromagnetic resonance, the two methods were combined to study an artificial spin ice with a square lattice geometry [168], as will be further explained in Section 2.4. The possibility to perform magnetoresistive measurements in combination with dynamic measurements creates new opportunities for experimental studies involving connected spin ices, allowing for correlations to be made between the dynamic behavior and the static state (which is the state measured with magnetotransport).

Magnetotransport measurements in artificial spin ices, especially Hall transport measurements, have shown the sensitivity to explore different effects, from the usual planar Hall effect to anomalous Hall effect. It is hence possible to optimize the material parameters to obtain a higher response from a specific Hall effect. The magnetoresistive signal itself could also be improved by a proper choice of materials and structure, and it would even be possible to implement giant magnetoresistance in artificial spin ice devices to increase the response. Other kinds of magnetoresistance, present in different materials, are also of interest in ASI. For example, spin-Hall magnetoresistance could be exploited in low-damping ASI made of YIG in contact with Pt, or one could even think of using less often used magnetic insulators in combination with different heavy metals.

A restriction of magnetoresistive measurements is that the current requires a continuous path and hence the necessity to fabricate connected lattices. However, it would be interesting to determine the magnetoresistance response of disconnected lattices. This could be achieved by patterning the islands on top of a conductor, and it would open magnetoresistive studies, which have proven to give a great insight in ASI.

2.4 Resonant magnetization dynamics in artificial spin ice

While one-dimensional arrays of nanomagnets and so-called antidot lattices, where periodic arrays of holes are patterned into extended magnetic thin films, have been extensively studied over the past two decades, the investigation of the resonant excitation of magnetization dynamics in ASI started only about 5 years ago. In this Subsection, we review some of those experimental findings that illustrate the intimate relationship between confined magnetization dynamics and static magnetization configuration and states in nanostructures.

2.4.1 Inductive measurements

Among the first experimental studies on magnetization dynamics in artificial spin ice [45,46,84,169–174] were detailed FMR studies of aperiodic artificial quasicrystals, which consist of tilings with long-range order, but lacking translational variance. Bhat et al. [45] reported dc magnetization and ferromagnetic resonance absorption on quasicrystalline Penrose P2 tilings. It was found that the magnetization switching events of individual segments in the quasicrystal is accompanied by a local FMR response of clusters of segments with several magnetic field directions. These results highlight that the orientation of the clusters with respect to the external field strongly affects the static as well as the dynamic magnetization in the high-field range close to saturation [45]. Moreover, angular-dependent spin-wave spectroscopy studies of quasicrystalline magnetic lattices such as Ammann [46] and Penrose P2 and P3 lattices made of large arrays of interconnected Py nanobars revealed distinct resonances with characteristic angular dependencies for applied in-plane magnetic fields [175]. Scanning electron micrographs of the various lattices and the corresponding angular-dependent results are shown in Fig. 20. Using micromagnetic simulations with OOMMF the detected resonances (Fig. 20E–H) were correlated to mode profiles with specific mirror symmetries. In particular, the results indicate that aperiodic quasicrystalline lattices could be used as reprogrammable magnonic devices since the emergence and disappearance of specific modes can be systematically controlled in these lattices.

This controllability of the specific modes was also shown in square artificial spin-ice lattices in a minor loop experiment [84], where the maximum applied field is chosen low enough so that the magnetization is not

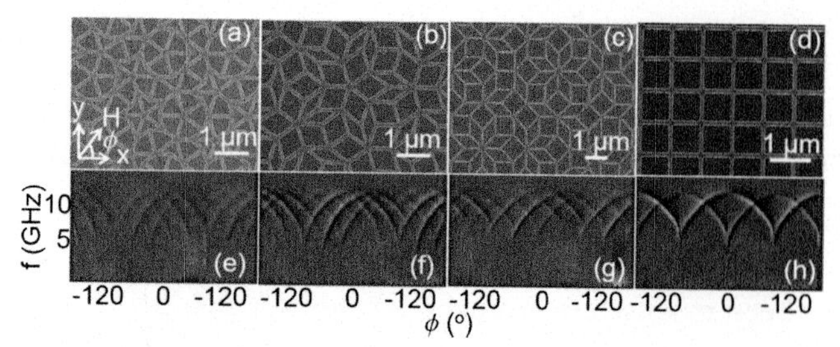

Fig. 20 SEM images of studied (A) Penrose P2 tiling, (B) Penrose P3 tiling, (C) Ammann tiling, and (D) connected square lattice. Bright (dark) regions correspond to Py (substrate). (E)-(H) Grayscale plots summarize spin-wave absorption spectra measured at 1000 Oe as a function of the in-plane field angle φ. *Adapted from V.S. Bhat, D. Grundler, Angle-dependent magnetization dynamics with mirror-symmetric excitations in artificial quasicrystalline nanomagnet lattices, Phys. Rev. B. 98 (2018) 174408. https://doi.org/10.1103/PhysRevB.98.174408.*

completely switched throughout the lattice. Furthermore, a field-history-dependent behavior in the high-field regime was observed in a narrow angular range in a square lattice [169]. This result implies that the local magnetization configuration of nearby islands changes when the field is swept at specific in-plane field angles. Besides the standard square lattice, anti-square spin-ice systems [176] consisting of extended ferromagnetic thin films where only the lattice part is removed, were studied by ferromagnetic resonance [170,177]. These systems are interesting owing to the possibility of creating a periodic structure of magnetic vortices in the center regions.

2.4.2 Brillouin light scattering

While inductive techniques such as ferromagnetic resonance and spin-wave spectroscopy exhibit a high sensitivity, optical methods such as Brillouin light scattering (BLS) have the advantage of obtaining a real-space image of the magnetization dynamics. Furthermore, it is possible to access non-zero wavevector excitations by varying the angle of incidence of the probing laser light. Li et al. reported wavevector-resolved BLS measurements of thermal spin-waves in a square ASI [178]. They studied both Damon-Eshbach and backward-volume magnetostatic waves and recorded the corresponding dispersion curves. It turns out that in the studied lattice, any magnonic dynamic inter-island interaction is negligible as the observed dispersion curves are almost flat [178]. Furthermore, BLS measurements were carried out in anti-spin-ice systems [88], see Fig. 21. Mamica et al. interpreted the wavevector-resolved BLS data using a theory based on

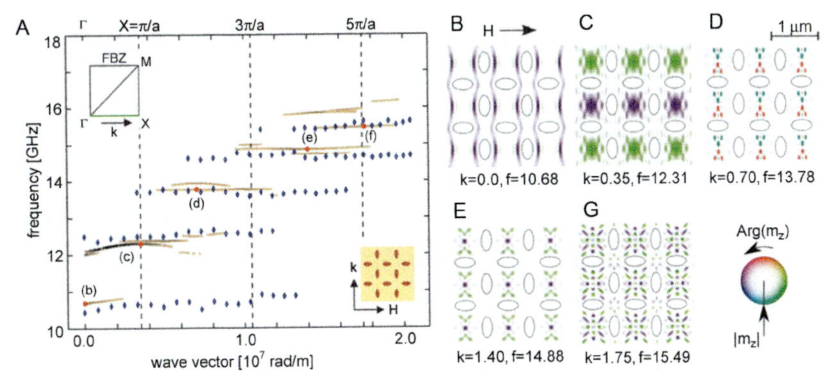

Fig. 21 (A) Dispersion curves of square anti-ASI measured by wavevector-resolved BLS (blue data points) and corresponding theoretical results obtained by the plane-wave method (brown lines). (B)-(G) Simulations of a square anti-ASI with different eigenmodes characterized by their oscillation amplitude and phase. *Adapted from S. Mamica, X. Zhou, A. Adeyeye, M. Krawczyk, G. Gubbiotti, Spin-wave dynamics in artificial anti-spin-ice systems: Experimental and theoretical investigations, Phys. Rev. B. 98 (2018) 054405. https://doi.org/10.1103/PhysRevB.98.054405.*

the plane-wave method. Fig. 21A shows the results of the wavevector-resolved BLS measurements (blue symbols) and a comparison to the calculations (brown lines). As is obvious from the figure, magnonic bandgaps are observed in this configuration. However, when the external magnetic field is rotated by 45°, the spectra transform from predominantly non-propagating spin waves exhibiting flat bands into propagating spin waves in narrow channels. Fig. 21E-G illustrates the corresponding calculated 2D maps of the spin-wave profiles as labeled in the dispersion in Fig. 21A. The lowest lying mode detected in Fig. 21A is confined to a small area along the sides of holes perpendicular to the external field direction. In contrast, the profiles for the higher-frequency modes observed by wavevector-resolved BLS are confined between the holes, as shown in Fig. 21E-G. Due to this lateral confinement spin waves have flat bands and thus propagate at a low group velocity or are non-propagating. When the field is rotated by 45° straight and narrow channels of spin-wave propagation are formed. This results in broadening of the bands and eventually closing of the magnonic bandgaps.

Although most of the optical measurements of ASI by BLS reported to date were done using a conventional BLS system with a laser spot size much larger than individual spin-ice elements [88,178–181], it is possible to adapt a microfocused approach that enables spatially-resolved studies of nanomagnet arrays [182].

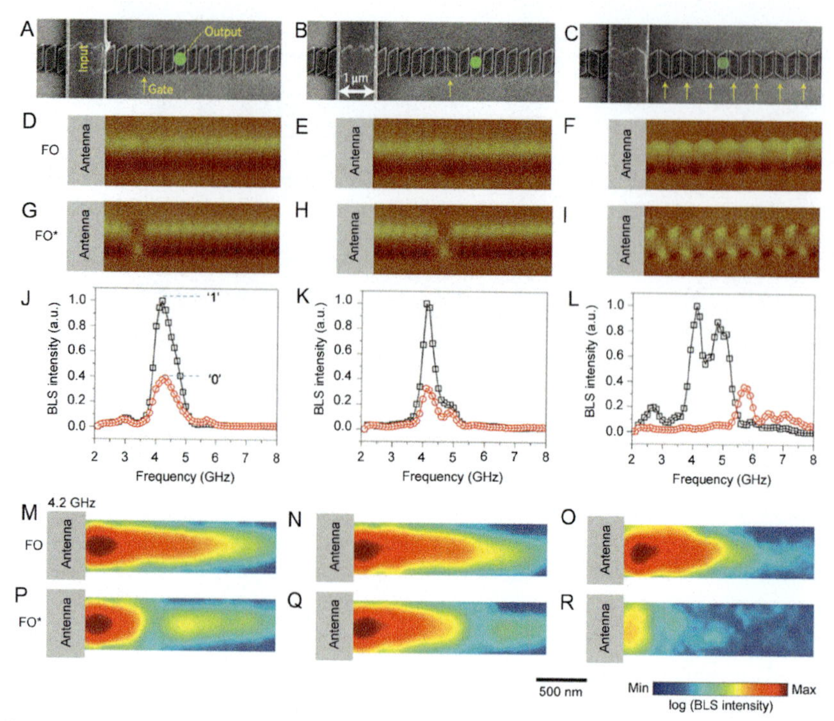

Fig. 22 Gating of spin-wave propagation. (A-C) SEM images of devices with different orientations of the rhomboid shaped nanomagnets in the waveguides. (D-F) Corresponding MFM images at remanence after a biasing field was applied along the long axis of the nanomagnets. (G-I), MFM images at remanence after a biasing field was applied along the short axis of the nanomagnets. (J-L) BLS spectra measured at FO (black squares) and FO* states (red circles). (M-R) Spatially-resolved intensity map for three devices for FO (M-O) and FO* (P-R) states. *Adapted from A. Haldar, D. Kumar, A.O. Adeyeye, A reconfigurable waveguide for energy-efficient transmission and local manipulation of information in a nanomagnetic device, Nat. Nanotechnol. 11 (2016) 437–443. https://doi.org/10.1038/nnano.2015.332.*

Haldar et al. presented an interesting spin-wave waveguide design based on rhomboid shaped nanomagnets that possess specific magnetization directions at remanence due to shape anisotropy, see Fig. 22 [142]. As such, those elements in principle resemble the building blocks of any ASI lattice. Haldar and co-workers demonstrated that this type of waveguide is reconfigurable so that spin waves can be transmitted and locally controlled without any external field [142]. For this purpose, a binary gating approach is used that is based on a simple field initialization routine, see Fig. 22. Fig. 22A-C shows SEM images of the waveguide and the laser spot location for the measurements shown in Fig. 22J-L. The corresponding MFM images are depicted in

Fig. 22D-I. In Fig. 22D-F the initialization field is applied along the long axis and then reduced to zero (denoted as FO state), whereas in Fig. 22G-I the field is applied the short axis and reduced to zero (denoted as FO* state). The difference between FO and FO* states are shown in the spectra Fig. 22J-L. The resulting spin-wave transmission is shown in the bottom panels, Fig. 22M-R. It turns out that the reversed magnetization orientation in the FO* state originates from the predefined shape of the nanomagnets at the gate position as indicated in Fig. 22A-C, [142,183]. Previously, it was shown that transmitting a spin-wave signal through a curved corner can be achieved using the Oersted field created by an underlying current stripe [184,185]. Haldar et al. also demonstrated that the waveguide comprising rhomboid shaped nanomagnets can be used for a spin-wave transmission at an angle without the need of applying any bias magnetic field or an electric current [142].

2.4.3 Magnetization dynamics in nanostructures meets spintronics: Perspectives

More recently, the implementation of electron-carried spin currents, generated by the spin-Hall effect, has given magnonics a new twist. Similar approaches and methods can be used in the context of ASI [168,186]. For instance, it was shown that Oersted field-driven dynamics in a connected ASI network consisting of a metallic ferromagnet (Py)/heavy metal (Pt) bilayer can be studied by means of spin pumping and inverse spin Hall effect [186]. The magnetization dynamics pumps a spin-polarized electron current into the adjacent Pt layer where it is converted into an electronic charge current due to the inverse spin Hall effect. Spin dynamics in the Py layer are modified due to the lateral confinement of the ASI pattern as confirmed by micromagnetic simulations. The spin-pumping results in connected ASI structures suggest that a local detection and excitation of spin dynamics might be possible by covering only selective areas of the lattice with Pt.

While the detection of GHz spin dynamics by dc electrical means in nanostructures is relatively easy, the realization of full-circle conversion of an electronic to a magnonic signal and back is more challenging—even in micrometer-sized devices. Auto-oscillations were reported in micrometer-scaled magnetic heterostructures based on insulators such as yttrium iron garnet, as well as metals such as Py, e.g. [115,117,119,187–190]. A step forward toward the electrical excitation of dynamics in nanopatterned arrays by the spin-Hall effects was recently demonstrated using spin-torque ferromagnetic resonance [168]. As discussed in more detail in Section 1.3, the magnetization dynamics in the bilayer is simultaneously excited by the microwave generated Oersted field and an alternating spin current that diffuses from the heavy metal to the ferromagnet where it exerts a torque on the magnetization. If the ferromagnet is metallic, the precession leads to a time-varying resistance change of the ferromagnet on account of the anisotropic magnetoresistance. Spin-torque ferromagnetic resonance combines these two effects to electrically drive and detect spin dynamics [168]. It was shown that the same concept can be applied to connected ASI networks as shown in Fig. 23. Similar to the spin-pumping results, it is possible to use ST-FMR to detect and identify

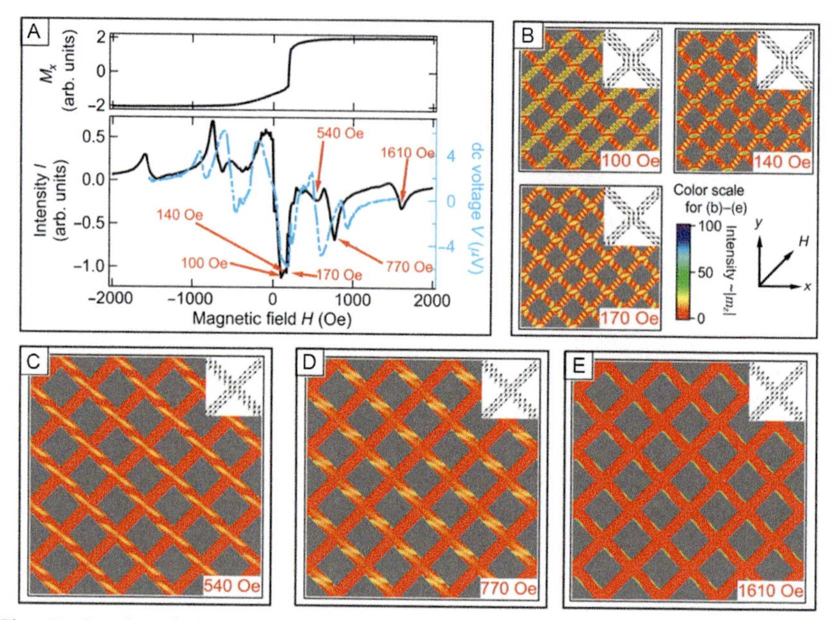

Fig. 23 Results of dynamic micromagnetic simulations in connected square lattice. (A) Top panel: Simulated static magnetization in x-direction as a function of the biasing magnetic field. Bottom panel: Predicted resonance modes (black, straight line) and their comparison to experimental results (blue, dashed line) at 6 GHz. (B)-(E) Spatially resolved maps of the dynamic response at a given magnetic field. The color scale represents the dynamic m_z component of the precession at 6 GHz. The 2D maps illustrate where the particular modes are located. The insets represent the corresponding static magnetization states. The static magnetization configuration in the reversal regime (B) only changes slightly, while the dynamic maps show a big change when the field is swept. *Adapted from M.B. Jungfleisch, J. Sklenar, J. Ding, J. Park, J.E. Pearson, V. Novosad, P. Schiffer, A. Hoffmann, High-Frequency Dynamics Modulated by Collective Magnetization Reversal in Artificial Spin Ice, Phys. Rev. Appl. 8 (2017) 064026. https://doi.org/10.1103/PhysRevApplied.8.064026.*

the various resonant spin-wave modes which are determined by the specific geometry of the lattice, as well as the specific magnetization configuration; see Fig. 23A. The results demonstrate that the collective magnetization behavior in an ASI strongly affects not only the magnetoresistive behavior, but also the dynamic ST-FMR spectra. As shown in Fig. 23, based on micromagnetic simulations a microscopic picture of the collective response can be revealed. To obtain the 2D maps of the dynamic micromagnetic simulations, the approach outlined in Section 1.3.1 is used for each cell. The sharp features observed in the low-field regime occur due to a sudden change in the magnetization configuration when the field is swept. In particular, it turns out that the angular-dependent alignment of the vertex region strongly affects the resistance [56], as well as the resonance condition resulting in an altered ST-FMR response. These results can also be viewed as the first stepping stone to the realization of engineered, reconfigurable microwave oscillators and magnetoresistive devices based on connected networks of nanomagnets.

Looking forward, we expect that the implementation of spintronic effects such as spin-orbit torques, as well as Dzyaloshinskii-Moriya interaction in artificial spin-ice structures will offer entirely new functionalities in magnonics. Indeed, recent work by Luo et al. demonstrated the ability to create engineered magnetic domains in Pt/Co/AlO_x trilayers with alternating in-plane and out-of-plane magnetizations [191]. They used this concept to realize lateral exchange bias and synthetic antiferromagnets, as well as magnetization textures such as skyrmions, and artificial spin ices, see Fig. 24. They also showed that the same platform can be used for field-free current-induced switching paving the way for logic gates and memory devices using this concept.

3. Future directions and outlook

As outlined in this Chapter, the exploration of dynamics in nanopatterned arrays of ferromagnets and artificial spin ice in particular offers a rich playground for studying fundamental physics questions, as well as for the development and realization of novel magnonic applications.

Now that thermalization strategies through either annealing or PEEM-XMCD experiments have demonstrated that ASI can be configured into ground states, the field is well-poised to study thoroughly study phase transitions within these magnetic metamaterials. In particular, we have highlighted how moving beyond Ising-like ASI to Potts [148,149] and XY artificial spin systems [159], more exotic phase transitions are enabled. By combining beyond Ising artificial spin systems with experimental tools

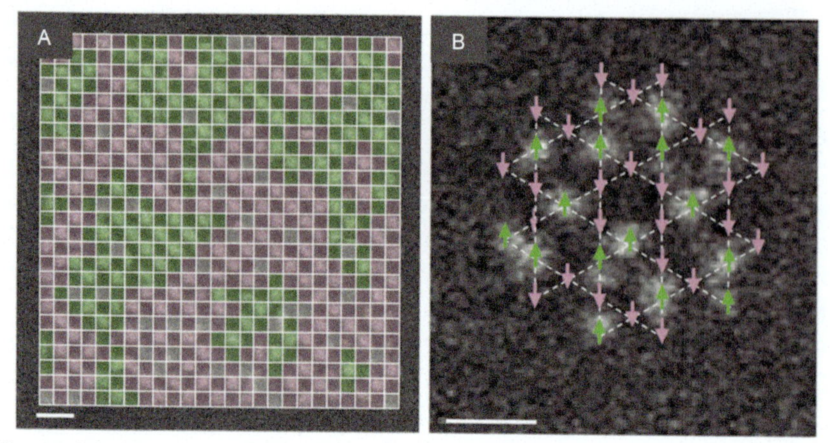

Fig. 24 (A) MFM image of an artificial spin system consisting of out-of-plane elements arranged on a square lattice. Coupling between the Ising-like moments is achieved by means of in-plane spacers. Antiferromagnetic domains are shaded in green and purple. (B) MFM image of an artificial kagome spin system. Here, the green and purple arrows indicate the orientation of the magnetization of the out-of-plane vertices. Scale bars are 500 nm. *Adapted from Z. Luo, T.P. Dao, A. Hrabec, J. Vijayakumar, A. Kleibert, M. Baumgartner, E. Kirk, J. Cui, T. Savchenko, G. Krishnaswamy, L.J. Heyderman, P. Gambardella, Chirally coupled nanomagnets, Science 363 (2019) 1435–1439. https://doi.org/10.1126/science.aau7913.*

that can image samples as both a function of temperature and field, phase diagrams in the temperature-field plane can be extensively mapped out. For these purposes, we believe that temperature controlled PEEM-XMCD experiments, designed to have an integrated *in-situ* applied field, will be able to lead the way in mapping the phase diagrams of artificial spin systems.

Among the earliest works on artificial spin ice were anisotropic magnetoresistance (AMR) measurements on connected ASI lattices [18]. It would be a remarkable step forward to combine other, well-established magnetoresistance effects such as GMR or TMR with ASI. Moreover, since the first AMR results on ASI were reported more than a decade ago, many new and exciting magnetoresistance effects such as unidirectional magnetoresistance [192], spin Hall magnetoresistance [70], or inverse Rashba Edelstein magnetoresistance [193,194] have been discovered. The combination of these effects with ASI might provide new perspectives for memory devices, and computing as well as logic applications, e.g. [160–162]. In particular, magnetic field and/or electric current history-dependent behaviors in ASI could be exploited by these magnetoresistance effects and be utilized for concepts in neuromorphic spintronics [195] and reservoir computing [196–198].

From a resonant magnetization dynamics perspective, exploring the interaction between defect strings connecting monopole-antimonopole pairs and magnetization dynamics is an interesting research area [86,171,199]. Utilizing spin-torque and spin-orbit effects to locally switch the magnetization could potentially be used for a targeted manipulation of defect strings in the 2D lattices and thus modify the dynamics. The recent work by Luo et al., where magnetic domains in Pt/Co/AlO_x with alternating in-plane and out-of-plane magnetization due to interfacial Dzyaloshinskii-Moriya interaction was demonstrated [191], is a first step in this direction. They were not only able to engineer various magnetization textures such as skyrmions and ASI in this material system, but also achieved field-free current-induced switching in these artificially created elements. Similar concepts may enable the realization of engineered, reconfigurable microwave oscillators based on networks of nanomagnets, which can easily be driven by dc currents.

Acknowledgments

We acknowledge support by the National Science Foundation under Grant No. 1833000.

References

[1] C. Nisoli, R. Moessner, P. Schiffer, Colloquium : artificial spin ice: designing and imaging magnetic frustration, Rev. Mod. Phys. 85 (2013) 1473–1490, https://doi.org/10.1103/RevModPhys.85.1473.

[2] L.J. Heyderman, R.L. Stamps, Artificial ferroic systems: novel functionality from structure, interactions and dynamics, J. Phys. Condens. Matter 25 (2013) 363201, https://doi.org/10.1088/0953-8984/25/36/363201.

[3] M.J. Harris, S.T. Bramwell, D.F. McMorrow, T. Zeiske, K.W. Godfrey, Geometrical frustration in the ferromagnetic pyrochlore Ho 2 Ti 2 O 7, Phys. Rev. Lett. 79 (1997) 2554–2557, https://doi.org/10.1103/PhysRevLett.79.2554.

[4] J.S. Gardner, S.R. Dunsiger, B.D. Gaulin, M.J.P. Gingras, J.E. Greedan, R.F. Kiefl, M.D. Lumsden, W.A. MacFarlane, N.P. Raju, J.E. Sonier, I. Swainson, Z. Tun, Cooperative paramagnetism in the geometrically frustrated pyrochlore antiferromagnet Tb 2 Ti 2 O 7, Phys. Rev. Lett. 82 (1999) 1012–1015, https://doi.org/10.1103/PhysRevLett.82.1012.

[5] S.T. Bramwell, M.J. Gingras, Spin ice state in frustrated magnetic pyrochlore materials, Science 294 (2001) 1495–1501, https://doi.org/10.1126/science.1064761.

[6] I. Gilbert, C. Nisoli, P. Schiffer, Frustration by design, Phys. Today 69 (2016) 54–59, https://doi.org/10.1063/PT.3.3237.

[7] C. Castelnovo, R. Moessner, S.L. Sondhi, Magnetic monopoles in spin ice, Nature 451 (2008) 42–45, https://doi.org/10.1038/nature06433.

[8] T. Fennell, P.P. Deen, A.R. Wildes, K. Schmalzl, D. Prabhakaran, A.T. Boothroyd, R.J. Aldus, D.F. McMorrow, S.T. Bramwell, Magnetic coulomb phase in the spin ice Ho2Ti2O7, Science 326 (2009) 415–417, https://doi.org/10.1126/science.1177582.

[9] C. Vieu, F. Carcenac, A. Pepin, Y. Chen, M. Mejias, A. Lebib, L. Manin-Ferlazzo, L. Couraud, H. Launois, Electron beam lithography: resolution limits and applications, Appl, Surf. Sci. 164 (2000) 111–117.

[10] A.A. Tseng, K. Chen, C.D. Chen, K.J. Ma, Electron beam lithography in nanoscale fabrication: recent development, IEEE Trans, Electron. Packag. Manuf. 26 (2003) 141–149.

[11] U. Hartmann, Magnetic force microscopy, Annu. Rev. Mater. Sci. 29 (1999) 53–87, https://doi.org/10.1146/annurev.matsci.29.1.53.

[12] R. Streubel, P. Fischer, F. Kronast, V.P. Kravchuk, D.D. Sheka, Y. Gaididei, O.G. Schmidt, D. Makarov, Magnetism in curved geometries, J. Phys. D Appl. Phys. 49 (2016) 363001, https://doi.org/10.1088/0022-3727/49/36/363001.

[13] D. Makarov, E. Bermúdez-Ureña, O.G. Schmidt, F. Liscio, M. Maret, C. Brombacher, S. Schulze, M. Hietschold, M. Albrecht, Nanopatterned CoPt alloys with perpendicular magnetic anisotropy, Appl. Phys. Lett. 93 (2008) 153112, https://doi.org/10.1063/1.2993334.

[14] M. Albrecht, G. Hu, I.L. Guhr, T.C. Ulbrich, J. Boneberg, P. Leiderer, G. Schatz, Magnetic multilayers on nanospheres, Nat. Mater. 4 (2005) 203–206, https://doi.org/10.1038/nmat1324.

[15] T.C. Ulbrich, D. Makarov, G. Hu, I.L. Guhr, D. Suess, T. Schrefl, M. Albrecht, Magnetization reversal in a novel gradient nanomaterial, Phys. Rev. Lett. 96 (2006) 077202, https://doi.org/10.1103/PhysRevLett.96.077202.

[16] A. Imre, G. Csaba, L. Ji, A. Orlov, G.H. Bernstein, W. Porod, Majority logic gate for magnetic quantum-dot cellular automata, Science 311 (2006) 205–208,

[17] R.F. Wang, C. Nisoli, R.S. Freitas, J. Li, W. McConville, B.J. Cooley, M.S. Lund, N. Samarth, C. Leighton, V.H. Crespi, P. Schiffer, Artificial 'spin ice' in a geometrically frustrated lattice of nanoscale ferromagnetic islands, Nature 439 (2006) 303–306, https://doi.org/10.1038/nature04447.

[18] M. Tanaka, E. Saitoh, H. Miyajima, T. Yamaoka, Y. Iye, Magnetic interactions in a ferromagnetic honeycomb nanoscale network, Phys. Rev. B 73 (2006) 052411, https://doi.org/10.1103/PhysRevB.73.052411.

[19] Y. Qi, T. Brintlinger, J. Cumings, Direct observation of the ice rule in an artificial kagome spin ice, Phys. Rev. B 77 (2008) 094418, https://doi.org/10.1103/PhysRevB.77.094418.

[20] G. Möller, R. Moessner, Artificial square ice and related dipolar nanoarrays, Phys. Rev. Lett. 96 (2006) 237202.

[21] R.F. Wang, J. Li, W. McConville, C. Nisoli, X. Ke, J.W. Freeland, V. Rose, M. Grimsditch, P. Lammert, V.H. Crespi, P. Schiffer, Demagnetization protocols for frustrated interacting nanomagnet arrays, J. Appl. Phys. 101 (2007) 09J104.

[22] X. Ke, J. Li, C. Nisoli, P.E. Lammert, W. McConville, R.F. Wang, V.H. Crespi, P. Schiffer, Energy minimization and ac demagnetization in a nanomagnet array, Phys. Rev. Lett. 101 (2008) 037205, https://doi.org/10.1103/PhysRevLett.101.037205.

[23] J. Li, X. Ke, S. Zhang, D. Garand, C. Nisoli, P. Lammert, V.H. Crespi, P. Schiffer, Comparing artificial frustrated magnets by tuning the symmetry of nanoscale permalloy arrays, Phys. Rev. B 81 (2010) 092406, https://doi.org/10.1103/PhysRevB.81.092406.

[24] Z. Budrikis, J.P. Morgan, J. Akerman, A. Stein, P. Politi, S. Langridge, C.H. Marrows, R.L. Stamps, Disorder strength and field-driven ground state domain formation in artificial spin ice: experiment, simulation, and theory, Phys. Rev. Lett. 109 (2012) 37203.

[25] C. Nisoli, J. Li, X. Ke, D. Garand, P. Schiffer, V.H. Crespi, Effective temperature in an interacting vertex system: theory and experiment on artificial spin ice, Phys. Rev. Lett. 105 (2010) 047205, https://doi.org/10.1103/PhysRevLett.105.047205.

[26] Y. Perrin, B. Canals, N. Rougemaille, Extensive degeneracy, Coulomb phase and magnetic monopoles in artificial square ice, Nature 540 (2016) 410,

[27] V.M. Parakkat, K. Xie, K.M. Krishnan, Tunable ground state in heterostructured artificial spin ice with exchange bias, Phys. Rev. B 99 (2019) 54429.
[28] J.P. Morgan, A. Stein, S. Langridge, C.H. Marrows, Thermal ground-state ordering and elementary excitations in artificial magnetic square ice, Nat. Phys. 7 (2011) 75–79, https://doi.org/10.1038/nphys1853.
[29] S. Zhang, I. Gilbert, C. Nisoli, G.-W. Chern, M.J. Erickson, L. O'Brien, C. Leighton, P.E. Lammert, V.H. Crespi, P. Schiffer, Crystallites of magnetic charges in artificial spin ice, Nature 500 (2013) 553–557, https://doi.org/10.1038/nature12399.
[30] V. Kapaklis, U.B. Arnalds, A. Harman-Clarke, E.T. Papaioannou, M. Karimipour, P. Korelis, A. Taroni, P.C.W. Holdsworth, S.T. Bramwell, B. Hjörvarsson, Melting artificial spin ice, New J. Phys. 14 (2012) 35009.
[31] J. Drisko, S. Daunheimer, J. Cumings, FePd 3 as a material for studying thermally active artificial spin ice systems, Phys. Rev. B 91 (2015) 224406, https://doi.org/10.1103/PhysRevB.91.224406.
[32] R.V. Chopdekar, B. Li, T.A. Wynn, M.S. Lee, Y. Jia, Z.Q. Liu, M.D. Biegalski, S.T. Retterer, A.T. Young, A. Scholl, others, Nanostructured complex oxides as a route towards thermal behavior in artificial spin ice systems, Phys. Rev. Mater. 1 (2017) 24401.
[33] A. Farhan, P.M. Derlet, A. Kleibert, A. Balan, R.V. Chopdekar, M. Wyss, J. Perron, A. Scholl, F. Nolting, L.J. Heyderman, Direct observation of thermal relaxation in artificial spin ice, Phys. Rev. Lett. 111 (2013) 057204, https://doi.org/10.1103/PhysRevLett.111.057204.
[34] E. Mengotti, L.J. Heyderman, A.F. Rodriguez, A. Bisig, L. Le Guyader, F. Nolting, H.B. Braun, Building blocks of an artificial kagome spin ice: photoemission electron microscopy of arrays of ferromagnetic islands, Phys. Rev. B 78 (2008) 144402.
[35] S. Ladak, D.E. Read, G.K. Perkins, L.F. Cohen, W.R. Branford, Direct observation of magnetic monopole defects in an artificial spin-ice system, Nat, Phys. 6 (2010) 359.
[36] N. Rougemaille, F. Montaigne, B. Canals, A. Duluard, D. Lacour, M. Hehn, R. Belkhou, O. Fruchart, S. El Moussaoui, A. Bendounan, others, Artificial kagome arrays of nanomagnets: a frozen dipolar spin ice, Phys. Rev. Lett. 106 (2011) 57209.
[37] U.B. Arnalds, A. Farhan, R.V. Chopdekar, V. Kapaklis, A. Balan, E.T. Papaioannou, M. Ahlberg, F. Nolting, L.J. Heyderman, B. Hjörvarsson, Thermalized ground state of artificial kagome spin ice building blocks, Appl, Phys. Lett. 101 (2012) 112404.
[38] A. Farhan, P.M. Derlet, A. Kleibert, A. Balan, R.V. Chopdekar, M. Wyss, L. Anghinolfi, F. Nolting, L.J. Heyderman, Exploring hyper-cubic energy landscapes in thermally active finite artificial spin-ice systems, Nat. Phys. 9 (2013) 375–382, https://doi.org/10.1038/nphys2613.
[39] E. Mengotti, L.J. Heyderman, A.F. Rodriguez, F. Nolting, R.V. Hügli, H.-B. Braun, Real-space observation of emergent magnetic monopoles and associated Dirac strings in artificial kagome spin ice, Nat, Phys. 7 (2011) 68.
[40] G. Möller, R. Moessner, Magnetic multipole analysis of kagome and artificial spin-ice dipolar arrays, Phys. Rev. B 80 (2009) 140409.
[41] L. Anghinolfi, H. Luetkens, J. Perron, M.G. Flokstra, O. Sendetskyi, A. Suter, T. Prokscha, P.M. Derlet, S.L. Lee, L.J. Heyderman, Thermodynamic phase transitions in a frustrated magnetic metamaterial, Nat, Commun. 6 (2015) 8278.
[42] G.-W. Chern, C. Reichhardt, C. Nisoli, Realizing three-dimensional artificial spin ice by stacking planar nano-arrays, Appl, Phys. Lett. 104 (2014) 13101.
[43] M.J. Morrison, T.R. Nelson, C. Nisoli, Unhappy vertices in artificial spin ice: new degeneracies from vertex frustration, New J. Phys. 15 (2013) 045009, https://doi.org/10.1088/1367-2630/15/4/045009.
[44] I. Gilbert, G.-W. Chern, S. Zhang, L. O'Brien, B. Fore, C. Nisoli, P. Schiffer, Emergent ice rule and magnetic charge screening from vertex frustration in artificial spin ice, Nat. Phys. 10 (2014) 670–675, https://doi.org/10.1038/nphys3037.

[45] V.S. Bhat, J. Sklenar, B. Farmer, J. Woods, J.T. Hastings, S.J. Lee, J.B. Ketterson, L.E. De Long, Controlled magnetic reversal in permalloy films patterned into artificial quasicrystals, Phys. Rev. Lett. 111 (2013) 077201, https://doi.org/10.1103/PhysRevLett.111.077201.
[46] V.S. Bhat, J. Sklenar, B. Farmer, J. Woods, J.B. Ketterson, J.T. Hastings, L.E. De Long, Ferromagnetic resonance study of eightfold artificial ferromagnetic quasicrystals, J. Appl. Phys. 115 (2014) 17C502.
[47] B. Farmer, V.S. Bhat, J. Sklenar, E. Teipel, J. Woods, J.B. Ketterson, J.T. Hastings, L.E. De Long, Magnetic response of aperiodic wire networks based on Fibonacci distortions of square antidot lattices, J. Appl. Phys. 117 (2015) 17B714.
[48] B. Farmer, V.S. Bhat, A. Balk, E. Teipel, N. Smith, J. Unguris, D.J. Keavney, J.T. Hastings, L.E. De Long, Direct imaging of coexisting ordered and frustrated sublattices in artificial ferromagnetic quasicrystals, Phys. Rev. B 93 (2016) 134428, https://doi.org/10.1103/PhysRevB.93.134428.
[49] V. Brajuskovic, F. Barrows, C. Phatak, A.K. Petford-Long, Real-space observation of magnetic excitations and avalanche behavior in artificial quasicrystal lattices, Sci. Rep. 6 (2016) 34384, https://doi.org/10.1038/srep34384.
[50] D. Shi, Z. Budrikis, A. Stein, S.A. Morley, P.D. Olmsted, G. Burnell, C.H. Marrows, Frustration and thermalization in an artificial magnetic quasicrystal, Nat, Phys. 14 (2018) 309.
[51] M.N. Baibich, J.M. Broto, A. Fert, F.N. Van Dau, F. Petroff, Giant magnetoresistance of (001)Fe/(001)Cr magnetic superlattices, Phys. Rev. Lett. 61 (1988) 2472–2475, https://doi.org/10.1103/PhysRevLett.61.2472.
[52] G. Binasch, P. Grünberg, F. Saurenbach, W. Zinn, Enhanced magnetoresistance in layered magnetic structures with antiferromagnetic interlayer exchange, Phys. Rev. B 39 (1989) 4828–4830, https://doi.org/10.1103/PhysRevB.39.4828.
[53] T. McGuire, R. Potter, Anisotropic magnetoresistance in ferromagnetic 3d alloys, IEEE Trans. Magn. 11 (1975) 1018–1038, https://doi.org/10.1109/TMAG.1975.1058782.
[54] W. Thomson, On the electro-dynamic qualities of metals:–effects of magnetization on the electric conductivity of nickel and of iron, Proc. R. Soc. Lond. 8 (1856) 546–550, https://doi.org/10.1098/rspl.1856.0144.
[55] C.C. Wang, A.O. Adeyeye, N. Singh, Y.S. Huang, Y.H. Wu, Magnetoresistance behavior of nanoscale antidot arrays, Phys. Rev. B 72 (2005) 174426, https://doi.org/10.1103/PhysRevB.72.174426.
[56] B.L. Le, J. Park, J. Sklenar, G.-W. Chern, C. Nisoli, J.D. Watts, M. Manno, D.W. Rench, N. Samarth, C. Leighton, P. Schiffer, Understanding magnetotransport signatures in networks of connected permalloy nanowires, Phys. Rev. B 95 (2017) 060405, https://doi.org/10.1103/PhysRevB.95.060405.
[57] M.J. Donahue, D.G. Porter, OOMMF user's guide, Version 1.0, Interagency Report NISTIR 6376, NIST, Gaithersburg, MD, 1999.
[58] T. Shinjo, H. Yamamoto, Large magnetoresistance of field-induced giant ferrimagnetic multilayers, J. Physical Soc. Japan 59 (1990) 3061–3064, https://doi.org/10.1143/JPSJ.59.3061.
[59] J. Nogués, I.K. Schuller, Exchange bias, J. Magn. Magn. Mater. 192 (1999) 203–232, https://doi.org/10.1016/S0304-8853(98)00266-2.
[60] E.Y. Tsymbal, D.G. Pettifor, Perspectives of giant magnetoresistance, in: H. Ehrenreich, F. Spaepen (Eds.), Solid State Physics, vol. 56, Academic Press, 2001, pp. 113–237.
[61] S. Ikeda, J. Hayakawa, Y. Ashizawa, Y.M. Lee, K. Miura, H. Hasegawa, M. Tsunoda, F. Matsukura, H. Ohno, Tunnel magnetoresistance of 604% at 300K by suppression of Ta diffusion in CoFeB / MgO / CoFeB pseudo-spin-valves annealed at high temperature, Appl. Phys. Lett. 93 (2008) 082508, https://doi.org/10.1063/1.2976435.

[62] J.E. Hirsch, Spin Hall effect, Phys. Rev. Lett. 83 (1999) 1834–1837, https://doi.org/10.1103/PhysRevLett.83.1834.
[63] S. Zhang, Spin Hall effect in the presence of spin diffusion, Phys. Rev. Lett. 85 (2000) 393–396, https://doi.org/10.1103/PhysRevLett.85.393.
[64] J. Sinova, D. Culcer, Q. Niu, N.A. Sinitsyn, T. Jungwirth, A.H. MacDonald, Universal intrinsic spin Hall effect, Phys. Rev. Lett. 92 (2004) 126603, https://doi.org/10.1103/PhysRevLett.92.126603.
[65] Y.K. Kato, R.C. Myers, A.C. Gossard, D.D. Awschalom, Observation of the spin Hall effect in semiconductors, Science 306 (2004) 1910–1913, https://doi.org/10.1126/science.1105514.
[66] J. Wunderlich, B. Kaestner, J. Sinova, T. Jungwirth, Experimental observation of the spin-Hall effect in a two-dimensional spin-orbit coupled semiconductor system, Phys. Rev. Lett. 94 (2005) 047204, https://doi.org/10.1103/PhysRevLett.94.047204.
[67] A. Hoffmann, Spin Hall effects in metals, IEEE Trans. Magn. 49 (2013) 5172–5193, https://doi.org/10.1109/TMAG.2013.2262947.
[68] M.B. Jungfleisch, W. Zhang, W. Jiang, A. Hoffmann, New pathways towards efficient metallic spin Hall spintronics, SPIN 05 (2015) 1530005,
[69] M.I. Dyakonov, V.I. Perel, Possibility of orienting electron spins with current, Sov, JETP Lett. 13 (1971) 467.
[70] H. Nakayama, M. Althammer, Y.-T. Chen, K. Uchida, Y. Kajiwara, D. Kikuchi, T. Ohtani, S. Geprägs, M. Opel, S. Takahashi, R. Gross, G.E.W. Bauer, S.T.B. Goennenwein, E. Saitoh, Spin Hall magnetoresistance induced by a nonequilibrium proximity effect, Phys. Rev. Lett. 110 (2013) 206601, https://doi.org/10.1103/PhysRevLett.110.206601.
[71] R.E. Howard, P.F. Liao, W.J. Skocpol, L.D. Jackel, H.G. Craighead, Microfabrication as a scientific tool, Science 221 (1983) 117–121, https://doi.org/10.1126/science.221.4606.117.
[72] A.V. Chumak, V.I. Vasyuchka, A.A. Serga, B. Hillebrands, Magnon spintronics, Nat, Phys. 11 (2015) 453–461.
[73] D.D. Stancil, A. Prabhakar, Spin Waves Theory and Applications, Springer, 2009,
[74] A.G. Gurevich, G.A. Melkov, Magnetization Oscillations and Waves, CRC, New York, 1996.
[75] A.V. Chumak, A.A. Serga, B. Hillebrands, Magnonic crystals for data processing, J. Phys. D Appl. Phys. 50 (2017) 244001.
[76] S. Neusser, D. Grundler, Magnonics: spin waves on the nanoscale, Adv, Mater. 21 (2009) 2927–2932.
[77] V.V. Kruglyak, S.O. Demokritov, D. Grundler, Magnonics, J. Phys. D Appl. Phys. 43 (2010) 264001.
[78] B. Lenk, H. Ulrichs, F. Garbs, M. Münzenberg, The building blocks of magnonics, Phys. Rep. 507 (2011) 107–136.
[79] C. Kittel, On the theory of ferromagnetic resonance absorption, Phys. Rev. 73 (1948) 155–161, https://doi.org/10.1103/PhysRev.73.155.
[80] B.A. Kalinikos, A.N. Slavin, Theory of dipole-exchange spin wave spectrum for ferromagnetic films with mixed exchange boundary conditions, J. Phys. C Solid State Phys. 19 (1986) 7013–7033, https://doi.org/10.1088/0022-3719/19/35/014.
[81] K.Y. Guslienko, S.O. Demokritov, B. Hillebrands, A.N. Slavin, Effective dipolar boundary conditions for dynamic magnetization in thin magnetic stripes, Phys. Rev. B. 66 (2002) 132402, https://doi.org/10.1103/PhysRevB.66.132402.
[82] J. Jorzick, S.O. Demokritov, B. Hillebrands, M. Bailleul, C. Fermon, K.Y. Guslienko, A.N. Slavin, D.V. Berkov, N.L. Gorn, Spin wave wells in nonellipsoidal micrometer size magnetic elements, Phys. Rev. Lett. 88 (2002) 47204.

[83] E. Iacocca, S. Gliga, R.L. Stamps, O. Heinonen, Reconfigurable wave band structure of an artificial square ice, Phys. Rev. B 93 (2016) 134420, https://doi.org/10.1103/PhysRevB.93.134420.
[84] M.B. Jungfleisch, W. Zhang, E. Iacocca, J. Sklenar, J. Ding, W. Jiang, S. Zhang, J.E. Pearson, V. Novosad, J.B. Ketterson, O. Heinonen, A. Hoffmann, Dynamic response of an artificial square spin ice, Phys. Rev. B 93 (2016) 100401, https://doi.org/10.1103/PhysRevB.93.100401.
[85] A. Vansteenkiste, J. Leliaert, M. Dvornik, M. Helsen, F. Garcia-Sanchez, B. Van Waeyenberge, The design and verification of MuMax3, AIP Adv. 4 (2014) 107133, https://doi.org/10.1063/1.4899186.
[86] S. Gliga, A. Kákay, R. Hertel, O.G. Heinonen, Spectral analysis of topological defects in an artificial spin-ice lattice, Phys. Rev. Lett. 110 (2013) 117205, https://doi.org/10.1103/PhysRevLett.110.117205.
[87] D.M. Arroo, J.C. Gartside, W.R. Branford, Sculpting the Spin-Wave Response of Artificial Spin Ice via Microstate Selection, http://arxiv.org/abs/1805.01397, 2018. (accessed August 23, 2019).
[88] S. Mamica, X. Zhou, A. Adeyeye, M. Krawczyk, G. Gubbiotti, Spin-wave dynamics in artificial anti-spin-ice systems: experimental and theoretical investigations. Phys. Rev. B 98 (2018) 054405, https://doi.org/10.1103/PhysRevB.98.054405.
[89] M. Vogel, A.V. Chumak, E.H. Waller, T. Langner, V.I. Vasyuchka, B. Hillebrands, G. von Freymann, Optically reconfigurable magnetic materials, Nat. Phys. 11 (2015) 487.
[90] K. Wagner, A. Kákay, K. Schultheiss, A. Henschke, T. Sebastian, H. Schultheiss, Magnetic domain walls as reconfigurable spin-wave nanochannels. Nat. Nanotechnol. 11 (2016) 432–436, https://doi.org/10.1038/nnano.2015.339.
[91] K. Schultheiss, R. Verba, F. Wehrmann, K. Wagner, L. Körber, T. Hula, T. Hache, A. Kákay, A.A. Awad, V. Tiberkevich, A.N. Slavin, J. Fassbender, H. Schultheiss, Excitation of whispering gallery magnons in a magnetic vortex, Phys. Rev. Lett. 122 (2019) 97202, https://doi.org/10.1103/PhysRevLett.122.097202.
[92] Z.K. Wang, V.L. Zhang, H.S. Lim, S.C. Ng, M.H. Kuok, S. Jain, A.O. Adeyeye, Observation of frequency band gaps in a one-dimensional nanostructured magnonic crystal, Appl, Phys. Lett. 94 (2009) 83112–83114.
[93] A.V. Chumak, A.A. Serga, B. Hillebrands, M.P. Kostylev, Scattering of backward spin waves in a one-dimensional magnonic crystal, Appl, Phys. Lett. 93 (2008) 22508.
[94] H. Qin, G.-J. Both, S.J. Hämäläinen, L. Yao, S. van Dijken, Low-loss YIG-based magnonic crystals with large tunable bandgaps, Nat, Commun. 9 (2018) 5445.
[95] T. Goto, K. Shimada, Y. Nakamura, H. Uchida, M. Inoue, One-dimensional magnonic crystal with Cu stripes for forward volume spin waves, Phys. Rev. Appl. 11 (2019) 14033.
[96] M. Langer, R.A. Gallardo, T. Schneider, S. Stienen, A. Roldán-Molina, Y. Yuan, K. Lenz, J. Lindner, P. Landeros, J. Fassbender, Spin-wave modes in transition from a thin film to a full magnonic crystal, Phys. Rev. B. 99 (2019) 24426.
[97] A.V. Chumak, P. Pirro, A.A. Serga, M.P. Kostylev, R.L. Stamps, H. Schultheiss, K. Vogt, S.J. Hermsdoerfer, B. Laegel, P.A. Beck, B. Hillebrands, Spin-wave propagation in a microstructured magnonic crystal, Appl, Phys. Lett. 95 (2009) 262508.
[98] S.S. Kalarickal, P. Krivosik, M. Wu, C.E. Patton, M.L. Schneider, P. Kabos, T.J. Silva, J.P. Nibarger, Ferromagnetic resonance linewidth in metallic thin films: Comparison of measurement methods, J. Appl. Phys. 99 (2006) 93909.
[99] L. Liu, T. Moriyama, D.C. Ralph, R.A. Buhrman, Spin-torque ferromagnetic resonance induced by the spin Hall effect, Phys. Rev. Lett. 106 (2011) 036601, https://doi.org/10.1103/PhysRevLett.106.036601.

[100] M.B. Jungfleisch, J. Ding, W. Zhang, W. Jiang, J.E. Pearson, V. Novosad, A. Hoffmann, Insulating nanomagnets driven by spin torque, Nano Lett. 17 (2017) 8–14, https://doi.org/10.1021/acs.nanolett.6b02794.
[101] J. Sklenar, W. Zhang, M.B. Jungfleisch, W. Jiang, H. Chang, J.E. Pearson, M. Wu, J.B. Ketterson, A. Hoffmann, Driving and detecting ferromagnetic resonance in insulators with the spin Hall effect, Phys. Rev. B 92 (2015) 174406, https://doi.org/10.1103/PhysRevB.92.174406.
[102] H.L. Wang, C.H. Du, Y. Pu, R. Adur, P.C. Hammel, F.Y. Yang, Scaling of spin Hall angle in 3d, 4d, and 5d metals from Y3Fe5O12/Metal spin pumping, Phys. Rev. Lett. 112 (2014) 197201.
[103] L. Liu, C.-F. Pai, Y. Li, H.W. Tseng, D.C. Ralph, R.A. Buhrman, Spin-torque switching with the giant spin Hall effect of tantalum, Science 336 (2012) 555–558, https://doi.org/10.1126/science.1218197.
[104] W. Zhang, W. Han, S.-H. Yang, Y. Sun, Y. Zhang, B. Yan, S.S.P. Parkin, Giant facet-dependent spin-orbit torque and spin Hall conductivity in the triangular antiferromagnet IrMn3, Sci. Adv. 2 (2016) e1600759, https://doi.org/10.1126/sciadv.1600759.
[105] W. Zhang, M.B. Jungfleisch, F. Freimuth, W. Jiang, J. Sklenar, J.E. Pearson, J.B. Ketterson, Y. Mokrousov, A. Hoffmann, All-electrical manipulation of magnetization dynamics in a ferromagnet by antiferromagnets with anisotropic spin Hall effects, Phys. Rev. B 92 (2015) 144405.
[106] A.R. Mellnik, J.S. Lee, A. Richardella, J.L. Grab, P.J. Mintun, M.H. Fischer, A. Vaezi, A. Manchon, E.A. Kim, N. Samarth, D.C. Ralph, Spin-transfer torque generated by a topological insulator, Nature 511 (2014) 449,
[107] M.B. Jungfleisch, W. Zhang, J. Sklenar, J. Ding, W. Jiang, H. Chang, F.Y. Fradin, J.E. Pearson, J.B. Ketterson, V. Novosad, M. Wu, A. Hoffmann, Large spin-wave bullet in a ferrimagnetic insulator driven by the spin Hall effect, Phy. Rev. Lett. 116 (2016) 57601.
[108] T. Chiba, G.E.W. Bauer, S. Takahashi, Current-induced spin-torque resonance of magnetic insulators, Phys. Rev. Appl. 2 (2014) 34003, https://doi.org/10.1103/PhysRevApplied.2.034003.
[109] T. Chiba, M. Schreier, G.E.W. Bauer, S. Takahashi, Current-induced spin torque resonance of magnetic insulators affected by field-like spin-orbit torques and out-of-plane magnetizations, J. Appl. Phys. 117 (2015) 17C715, https://doi.org/10.1063/1.4913632.
[110] M. Schreier, T. Chiba, A. Niedermayr, J. Lotze, H. Huebl, S. Geprägs, S. Takahashi, G.E.W. Bauer, R. Gross, S.T.B. Goennenwein, Current-induced spin torque resonance of a magnetic insulator, Phys. Rev. B 92 (2015) 144411, https://doi.org/10.1103/PhysRevB.92.144411.
[111] Y. Wang, R. Ramaswamy, Y. Hyunsoo, FMR-related phenomena in spintronic devices, J. Phys. D Appl. Phys. 51 (2018) 273002. Iopscience.Iop.Org. (n.d.), https://iopscience.iop.org/article/10.1088/1361-6463/aac7b5/pdf.
[112] W. Rippard, M. Pufall, S. Kaka, S. Russek, T. Silva, Direct-Current Induced Dynamics in Co$_{90}$Fe$_{10}$ / Ni$_{80}$Fe$_{20}$ Point Contacts, Phys. Rev. Lett. 92 (2004) 027201, https://doi.org/10.1103/PhysRevLett.92.027201.
[113] V.E. Demidov, S. Urazhdin, A. Zholud, A.V. Sadovnikov, S.O. Demokritov, Nanoconstriction-based spin-Hall nano-oscillator, Appl. Phys. Lett. 105 (2014) 172410, https://doi.org/10.1063/1.4901027.
[114] S.M. Mohseni, S.R. Sani, J. Persson, T.N.A. Nguyen, S. Chung, Y. Pogoryelov, P.K. Muduli, E. Iacocca, A. Eklund, R.K. Dumas, S. Bonetti, A. Deac, M.A. Hoefer, J. Akerman, Spin torque-generated magnetic droplet solitons, Science 339 (2013) 1295–1298, https://doi.org/10.1126/science.1230155.

[115] V.E. Demidov, S. Urazhdin, H. Ulrichs, V. Tiberkevich, A. Slavin, D. Baither, G. Schmitz, S.O. Demokritov, Magnetic nano-oscillator driven by pure spin current, Nat, Mater. 11 (2012) 1028.

[116] M. Haidar, A.A. Awad, M. Dvornik, R. Khymyn, A. Houshang, J. Åkerman, A single layer spin-orbit torque nano-oscillator, Nat. Commun. 10 (2019) 1–6, https://doi.org/10.1038/s41467-019-10120-4.

[117] A.A. Awad, P. Dürrenfeld, A. Houshang, M. Dvornik, E. Iacocca, R.K. Dumas, J. Åkerman, Long-range mutual synchronization of spin Hall nano-oscillators, Nat. Phys. 13 (2017) 292–299, https://doi.org/10.1038/nphys3927.

[118] N. Sato, K. Schultheiss, L. Körber, N. Puwenberg, T. Mühl, A.A. Awad, S.S.P.K. Arekapudi, O. Hellwig, J. Fassbender, H. Schultheiss, Domain wall based spin-Hall nano-oscillators, Phys. Rev. Lett. 123 (2019) 57204, https://doi.org/10.1103/PhysRevLett.123.057204.

[119] S. Langenfeld, V. Tshitoyan, Z. Fang, A. Wells, Exchange magnon induced resistance asymmetry in permalloy spin-Hall oscillators, Appl. Phys. Lett. 108 (2016) 192402, https://doi.org/10.1063/1.4948921.

[120] T. Sebastian, K. Schultheiss, B. Obry, B. Hillebrands, H. Schultheiss, Micro-focused Brillouin light scattering: imaging spin waves at the nanoscale, Front, Phys. 3 (2015) 1589.

[121] J. Fassbender, Magnetization dynamics investigated by Time-Resolved Kerr effect magnetometry, in: Spin Dynamics in Confined Magnetic Structures II, Springer, Berlin, Heidelberg, 2003, pp. 59–92, https://doi.org/10.1007/3-540-46097-7_3.

[122] J. Jersch, V.E. Demidov, H. Fuchs, K. Rott, P. Krzysteczko, J. Münchenberger, G. Reiss, S.O. Demokritov, Mapping of localized spin-wave excitations by near-field Brillouin light scattering, Appl. Phys. Lett. 97 (2010) 152502, https://doi.org/10.1063/1.3502599.

[123] C.M. Schneider, A. Krasyuk, S.A. Nepijko, A. Oelsner, G. Schönhense, Accessing fast magnetization dynamics by XPEEM: status and perspectives, J. Magn. Magn. Mater. 304 (2006) 6–9, https://doi.org/10.1016/J.JMMM.2006.02.013.

[124] G. Schönhense, Surface magnetism studied by photoelectron spectromicroscopy with high spatial and time resolution, J. Electron Spectrosc. Relat. Phenom. 137–140 (2004) 769–783, https://doi.org/10.1016/J.ELSPEC.2004.02.161.

[125] P. Fischer, Viewing spin structures with soft X-ray microscopy, Mater. Today 13 (2010) 14–22, https://doi.org/10.1016/S1369-7021(10)70161-9.

[126] B.P. Tonner, D. Dunham, T. Droubay, J. Kikuma, J. Denlinger, E. Rotenberg, A. Warwick, The development of electron spectromicroscopy, J. Electron Spectrosc. Relat. Phenom. 75 (1995) 309–332, https://doi.org/10.1016/0368-2048(95)02523-5.

[127] S. Anders, H.A. Padmore, R.M. Duarte, T. Renner, T. Stammler, A. Scholl, M.R. Scheinfein, J. Stöhr, L. Séve, B. Sinkovic, Photoemission electron microscope for the study of magnetic materials, Rev. Sci. Instrum. 70 (1999) 3973–3981, https://doi.org/10.1063/1.1150023.

[128] S. Imada, A. Sekiyama, S. Suga, Soft x-ray spectroscopy and microspectroscopy of correlated materials: photoemission and magnetic circular dichroism, J. Phys. Condens. Matter 19 (2007) 125204, https://doi.org/10.1088/0953-8984/19/12/125204.

[129] A. Locatelli, E. Bauer, Recent advances in chemical and magnetic imaging of surfaces and interfaces by XPEEM, J. Phys. Condens. Matter 20 (2008) 093002, https://doi.org/10.1088/0953-8984/20/9/093002.

[130] X.M. Cheng, D.J. Keavney, Studies of nanomagnetism using synchrotron-based x-ray photoemission electron microscopy (X-PEEM), Reports Prog. Phys. 75 (2012) 026501, https://doi.org/10.1088/0034-4885/75/2/026501.

[131] K. Zeissler, S.K. Walton, S. Ladak, D.E. Read, T. Tyliszczak, L.F. Cohen, W.R. Branford, The non-random walk of chiral magnetic charge carriers in artificial spin ice, Sci. Rep. 3 (2013) 1252, https://doi.org/10.1038/srep01252.
[132] M. Weigand, B. Van Waeyenberge, A. Vansteenkiste, M. Curcic, V. Sackmann, H. Stoll, T. Tyliszczak, K. Kaznatcheev, D. Bertwistle, G. Woltersdorf, C.H. Back, G. Schütz, Vortex core switching by coherent excitation with single in-plane magnetic field pulses, Phys. Rev. Lett. 102 (2009) 077201, https://doi.org/10.1103/PhysRevLett.102.077201.
[133] D. Backes, F. Macià, S. Bonetti, R. Kukreja, H. Ohldag, A.D. Kent, Direct observation of a localized magnetic soliton in a spin-transfer nanocontact, Phys. Rev. Lett. 115 (2015) 127205, https://doi.org/10.1103/PhysRevLett.115.127205.
[134] C. Behncke, C.F. Adolff, S. Wintz, M. Hänze, B. Schulte, M. Weigand, S. Finizio, J. Raabe, G. Meier, Tunable geometrical frustration in magnonic vortex crystals, Sci. Rep. 8 (2018) 186, https://doi.org/10.1038/s41598-017-17480-1.
[135] V. Sluka, T. Schneider, R.A. Gallardo, A. Kákay, M. Weigand, T. Warnatz, R. Mattheis, A. Roldán-Molina, P. Landeros, V. Tiberkevich, A. Slavin, G. Schütz, A. Erbe, A. Deac, J. Lindner, J. Raabe, J. Fassbender, S. Wintz, Emission and propagation of 1D and 2D spin waves with nanoscale wavelengths in anisotropic spin textures, Nat. Nanotechnol. 14 (2019) 328–333, https://doi.org/10.1038/s41565-019-0383-4.
[136] H. Stoll, M. Noske, M. Weigand, K. Richter, B. Krüger, R.M. Reeve, M. Hänze, C.F. Adolff, F.-U. Stein, G. Meier, M. Kläui, G. Schütz, Imaging spin dynamics on the nanoscale using X-Ray microscopy, Front. Phys. 3 (2015) 26, https://doi.org/10.3389/fphy.2015.00026.
[137] S. Bonetti, X-ray imaging of spin currents and magnetisation dynamics at the nanoscale, J. Phys. Condens. Matter 29 (2017) 133004, https://doi.org/10.1088/1361-648X/aa5a13.
[138] J. Drisko, T. Marsh, J. Cumings, Topological frustration of artificial spin ice, Nat. Commun. 8 (2017) 14009, https://doi.org/10.1038/ncomms14009.
[139] Y.-L. Wang, Z.-L. Xiao, A. Snezhko, J. Xu, L.E. Ocola, R. Divan, J.E. Pearson, G.W. Crabtree, W.-K. Kwok, Rewritable artificial magnetic charge ice, Science 352 (2016) 962–966, https://doi.org/10.1126/science.aad8037.
[140] J.C. Gartside, D.M. Arroo, D.M. Burn, V.L. Bemmer, A. Moskalenko, L.F. Cohen, W.R. Branford, Realization of ground state in artificial kagome spin ice via topological defect-driven magnetic writing, Nat. Nanotechnol. 13 (2018) 53–58, https://doi.org/10.1038/s41565-017-0002-1.
[141] Y.-L. Wang, X. Ma, J. Xu, Z.-L. Xiao, A. Snezhko, R. Divan, L.E. Ocola, J.E. Pearson, B. Janko, W.-K. Kwok, Switchable geometric frustration in an artificial-spin-ice–superconductor heterosystem, Nat. Nanotechnol. 13 (2018) 560–565, https://doi.org/10.1038/s41565-018-0162-7.
[142] A. Haldar, D. Kumar, A.O. Adeyeye, A reconfigurable waveguide for energy-efficient transmission and local manipulation of information in a nanomagnetic device, Nat. Nanotechnol. 11 (2016) 437–443, https://doi.org/10.1038/nnano.2015.332.
[143] R. Macêdo, G.M. Macauley, F.S. Nascimento, R.L. Stamps, Apparent ferromagnetism in the pinwheel artificial spin ice, Phys. Rev. B 98 (2018) 14437.
[144] Y. Li, G.W. Paterson, G.M. Macauley, F.S. Nascimento, C. Ferguson, S.A. Morley, M.C. Rosamond, E.H. Linfield, D.A. MacLaren, R. Macêdo, others, Superferromagnetism and domain-wall topologies in Artificial "Pinwheel" spin ice, ACS Nano 13 (2018) 2213–2222,
[145] S. Zhang, J. Li, I. Gilbert, J. Bartell, M.J. Erickson, Y. Pan, P.E. Lammert, C. Nisoli, K.K. Kohli, R. Misra, others, Perpendicular magnetization and generic realization of the Ising model in artificial spin ice, Phys. Rev. Lett. 109 (2012) 87201.

[146] R.D. Fraleigh, S. Kempinger, P.E. Lammert, S. Zhang, V.H. Crespi, P. Schiffer, N. Samarth, Characterization of switching field distributions in Ising-like magnetic arrays, Phys. Rev. B 95 (2017) 144416.

[147] S. Kempinger, R.D. Fraleigh, P. Lammert, S. Zhang, V.H. Crespi, P. Schiffer, N. Samarth, Imaging the stochastic microstructure and dynamic development of correlations in perpendicular artificial spin ice, ArXiv Prepr, (2019). https://arxiv.org/abs/1904.06625.

[148] D. Louis, D. Lacour, M. Hehn, V. Lomakin, T. Hauet, F. Montaigne, A tunable magnetic metamaterial based on the dipolar four-state Potts model, Nat. Mater. 17 (2018) 1076–1080, https://doi.org/10.1038/s41563-018-0199-x.

[149] J. Sklenar, Y. Lao, A. Albrecht, J.D. Watts, C. Nisoli, G.-W. Chern, P. Schiffer, Field-induced phase coexistence in an artificial spin ice, Nat. Phys. 15 (2019) 191–195, https://doi.org/10.1038/s41567-018-0348-9.

[150] M. Blume, V.J. Emery, R.B. Griffiths, Ising model for the λ transition and phase separation in He 3-He 4 mixtures, Phys. Rev. A 4 (1971) 1071.

[151] A. Farhan, M. Saccone, C.F. Petersen, S. Dhuey, R.V. Chopdekar, Y.-L. Huang, N. Kent, Z. Chen, M.J. Alava, T. Lippert, A. Scholl, S. van Dijken, Emergent magnetic monopole dynamics in macroscopically degenerate artificial spin ice, Sci. Adv. 5 (2019) eaav6380https://doi.org/10.1126/sciadv.aav6380.

[152] E. Östman, H. Stopfel, I.-A. Chioar, U.B. Arnalds, A. Stein, V. Kapaklis, B. Hjörvarsson, Interaction modifiers in artificial spin ices, Nat, Phys. 14 (2018) 375.

[153] A. Farhan, C.F. Petersen, S. Dhuey, L. Anghinolfi, Q.H. Qin, M. Saccone, S. Velten, C. Wuth, S. Gliga, P. Mellado, others, Nanoscale control of competing interactions and geometrical frustration in a dipolar trident lattice, Nat, Commun. 8 (2017) 995.

[154] Y. Lao, F. Caravelli, M. Sheikh, J. Sklenar, D. Gardeazabal, J.D. Watts, A.M. Albrecht, A. Scholl, K. Dahmen, C. Nisoli, others, Classical topological order in the kinetics of artificial spin ice, Nat. Phys. 14 (2018) 723.

[155] H. Stopfel, E. Östman, I.-A. Chioar, D. Greving, U.B. Arnalds, T.P.A. Hase, A. Stein, B. Hjörvarsson, V. Kapaklis, Magnetic order and energy-scale hierarchy in artificial spin-ice structures, Phys. Rev. B 98 (2018) 14435.

[156] I. Gilbert, Y. Lao, I. Carrasquillo, L. O'Brien, J.D. Watts, M. Manno, C. Leighton, A. Scholl, C. Nisoli, P. Schiffer, Emergent reduced dimensionality by vertex frustration in artificial spin ice. Nat. Phys. 12 (2016) 162–165, https://doi.org/10.1038/nphys3520.

[157] A. Farhan, A. Scholl, C.F. Petersen, L. Anghinolfi, C. Wuth, S. Dhuey, R.V. Chopdekar, P. Mellado, M.J. Alava, S. Van Dijken, Thermodynamics of emergent magnetic charge screening in artificial spin ice, Nat, Commun. 7 (2016) 12635.

[158] S. Gliga, G. Hrkac, C. Donnelly, J. Büchi, A. Kleibert, J. Cui, A. Farhan, E. Kirk, R.V. Chopdekar, Y. Masaki, N.S. Bingham, A. Scholl, R.L. Stamps, L.J. Heyderman, Emergent dynamic chirality in a thermally driven artificial spin ratchet, Nat. Mater. 16 (2017) 1106–1111, https://doi.org/10.1038/nmat5007.

[159] R. Streubel, N. Kent, S. Dhuey, A. Scholl, S. Kevan, P. Fischer, Spatial and temporal correlations of XY macro spins, Nano Lett. 18 (2018) 7428–7434.

[160] H. Arava, N. Leo, D. Schildknecht, J. Cui, J. Vijayakumar, P.M. Derlet, A. Kleibert, L.J. Heyderman, Engineering relaxation pathways in building blocks of artificial spin ice for computation, Phys. Rev. Appl. 11 (2019) 054086, https://doi.org/10.1103/PhysRevApplied.11.054086.

[161] H. Arava, P.M. Derlet, J. Vijayakumar, J. Cui, N.S. Bingham, A. Kleibert, L.J. Heyderman, Computational logic with square rings of nanomagnets, Nanotechnology 29 (2018) 265205.

[162] P. Gypens, J. Leliaert, B. Van Waeyenberge, Balanced magnetic logic gates in a Kagome spin ice, Phys. Rev. Appl. 9 (2018) 34004.

[163] W.R. Branford, S. Ladak, D.E. Read, K. Zeissler, L.F. Cohen, Emerging chirality in artificial spin ice, Science 335 (2012) 1597–1600, https://doi.org/10.1126/science.1211379.
[164] B.L. Le, D.W. Rench, R. Misra, L. O'Brien, Effects of exchange bias on magnetotransport in permalloy kagome artificial spin ice, New J. Phys. 17 (2015) 23047.
[165] G.-W. Chern, Magnetotransport in artificial kagome spin ice, Phys. Rev. Appl. 8 (2017) 064006, https://doi.org/10.1103/PhysRevApplied.8.064006.
[166] J. Park, B.L. Le, J. Sklenar, G.-W. Chern, J.D. Watts, P. Schiffer, Magnetic response of brickwork artificial spin ice, Phys. Rev. B 96 (2017) 024436, https://doi.org/10.1103/PhysRevB.96.024436.
[167] Ansys, (n.d.) https://www.ansys.com. https://www.ansys.com.
[168] M.B. Jungfleisch, J. Sklenar, J. Ding, J. Park, J.E. Pearson, V. Novosad, P. Schiffer, A. Hoffmann, High-frequency dynamics modulated by collective magnetization reversal in artificial spin ice, Phys. Rev. Appl. 8 (2017) 064026, https://doi.org/10.1103/PhysRevApplied.8.064026.
[169] J. Sklenar, V.S. Bhat, L.E. DeLong, J.B. Ketterson, Broadband ferromagnetic resonance studies on an artificial square spin-ice island array, J. Appl. Phys. 113 (2013) 17B530, https://doi.org/10.1063/1.4800740.
[170] X. Zhou, G.-L. Chua, N. Singh, A.O. Adeyeye, Large area artificial spin ice and anti-spin ice Ni $_{80}$ Fe $_{20}$ structures: static and dynamic behavior, Adv. Funct. Mater. 26 (2016) 1437–1444, https://doi.org/10.1002/adfm.201505165.
[171] V.S. Bhat, F. Heimbach, I. Stasinopoulos, D. Grundler, Magnetization dynamics of topological defects and the spin solid in a kagome artificial spin ice, Phys. Rev. B 93 (2016) 140401, https://doi.org/10.1103/PhysRevB.93.140401.
[172] W. Bang, F. Montoncello, M.B. Jungfleisch, A. Hoffmann, L. Giovannini, J.B. Ketterson, Angular-dependent spin dynamics of a triad of permalloy macrospins, Phys. Rev. B 99 (2019) 14415, https://doi.org/10.1103/PhysRevB.99.014415.
[173] W. Bang, M.B. Jungfleisch, F. Montoncello, B.W. Farmer, P.N. Lapa, A. Hoffmann, L. Giovannini, L.E. De Long, J.B. Ketterson, Coupled macrospins: mode dynamics in symmetric and asymmetric vertices, AIP Adv. 8 (2017) 56020.
[174] F. Montoncello, L. Giovannini, W. Bang, J.B. Ketterson, M.B. Jungfleisch, A. Hoffmann, B.W. Farmer, L.E. De Long, Mutual influence between macrospin reversal order and spin-wave dynamics in isolated artificial spin-ice vertices. Phys. Rev. B 97 (2018) 014421, https://doi.org/10.1103/PhysRevB.97.014421.
[175] V.S. Bhat, D. Grundler, Angle-dependent magnetization dynamics with mirror-symmetric excitations in artificial quasicrystalline nanomagnet lattices, Phys. Rev. B 98 (2018) 174408, https://doi.org/10.1103/PhysRevB.98.174408.
[176] C.I.L. de Araujo, R.C. Silva, I.R.B. Ribeiro, F.S. Nascimento, J.F. Felix, S.O. Ferreira, L.A.S. Mól, W.A. Moura-Melo, A.R. Pereira, Magnetic vortex crystal formation in the antidot complement of square artificial spin ice, Appl. Phys. Lett. 104 (2014) 92402, https://doi.org/10.1063/1.4867530.
[177] I.R.B. Ribeiro, J.F. Felix, L.C. Figueiredo, P.C. Morais, S.O. Ferreira, W.A. Moura-Melo, A.R. Pereira, A. Quindeau, C.I.L. de Araujo, Investigation of ferromagnetic resonance and magnetoresistance in anti-spin ice structures, J. Phys. Condens. Matter 28 (2016) 456002, https://doi.org/10.1088/0953-8984/28/45/456002.
[178] Y. Li, G. Gubbiotti, F. Casoli, F.J.T. Gonçalves, S.A. Morley, M.C. Rosamond, E.H. Linfield, C.H. Marrows, S. McVitie, R.L. Stamps, Brillouin light scattering study of magnetic-element normal modes in a square artificial spin ice geometry, J. Phys. D Appl. Phys. 50 (2017) 015003, https://doi.org/10.1088/1361-6463/50/1/015003.
[179] G. Gubbiotti, X. Zhou, Z. Haghshenasfard, M.G. Cottam, A.O. Adeyeye, Reprogrammable magnonic band structure of layered permalloy/Cu/permalloy nanowires, Phys. Rev. B 97 (2018) 134428.

[180] S. Tacchi, M. Madami, G. Gubbiotti, G. Carlotti, A.O. Adeyeye, S. Neusser, B. Botters, D. Grundler, Magnetic normal modes in squared antidot array with circular holes: a combined brillouin light scattering and broadband ferromagnetic resonance study, IEEE Trans, Magn. 46 (2010) 172–178.
[181] S. Neusser, B. Botters, M. Becherer, D. Schmitt-Landsiedel, D. Grundler, Spin-wave localization between nearest and next-nearest neighboring holes in an antidot lattice, Appl, Phys. Lett. 93 (2008) 122501.
[182] D. Grundler, No Title, Priv. Commun. (n.d.).
[183] A. Haldar, A.O. Adeyeye, Vortex chirality control in circular disks using dipole-coupled nanomagnets, Appl. Phys. Lett. 106 (2015) 32404, https://doi.org/10.1063/1.4906142.
[184] K. Vogt, H. Schultheiss, S. Jain, J.E. Pearson, A. Hoffmann, S.D. Bader, B. Hillebrands, Spin waves turning a corner, Appl. Phys. Lett. 101 (2012) 42410, https://doi.org/10.1063/1.4738887.
[185] K. Vogt, F.Y. Fradin, J.E. Pearson, T. Sebastian, S.D. Bader, B. Hillebrands, A. Hoffmann, H. Schultheiss, Realization of a spin-wave multiplexer, Nat. Commun. 5 (2014) 3727, https://doi.org/10.1038/ncomms4727.
[186] M.B. Jungfleisch, W. Zhang, J. Ding, W. Jiang, J. Sklenar, J.E. Pearson, J.B. Ketterson, A. Hoffmann, All-electrical detection of spin dynamics in magnetic antidot lattices by the inverse spin Hall effect, Appl. Phys. Lett. 108 (2016) 052403, https://doi.org/10.1063/1.4941392.
[187] A. Hamadeh, O. d'Allivy Kelly, C. Hahn, H. Meley, R. Bernard, A.H. Molpeceres, V.V. Naletov, M. Viret, A. Anane, V. Cros, S.O. Demokritov, J.L. Prieto, M. Muñoz, G. de Loubens, O. Klein, Full control of the spin-wave damping in a magnetic insulator using spin-orbit torque, Phys. Rev. Lett. 113 (2014) 197203, https://doi.org/10.1103/PhysRevLett.113.197203.
[188] V.E. Demidov, H. Ulrichs, S.V. Gurevich, S.O. Demokritov, V.S. Tiberkevich, A.N. Slavin, A. Zholud, S. Urazhdin, Synchronization of spin Hall nano-oscillators to external microwave signals, Nat. Commun. 5 (2014) 3179, https://doi.org/10.1038/ncomms4179.
[189] M. Collet, X. de Milly, O. d'Allivy Kelly, V.V. Naletov, R. Bernard, P. Bortolotti, J.B. Youssef, V.E. Demidov, S.O. Demokritov, J.L. Prieto, M. Muñoz, V. Cros, A. Anane, G. de Loubens, O. Klein, Generation of coherent spin-wave modes in yttrium iron garnet microdiscs by spin-orbit torque, Nat. Commun. 7 (2016) 10377, https://doi.org/10.1038/ncomms10377.
[190] R.H. Liu, W.L. Lim, S. Urazhdin, Spectral characteristics of the microwave emission by the spin Hall nano-oscillator, Phys. Rev. Lett. 110 (2013) 147601, https://doi.org/10.1103/PhysRevLett.110.147601.
[191] Z. Luo, T.P. Dao, A. Hrabec, J. Vijayakumar, A. Kleibert, M. Baumgartner, E. Kirk, J. Cui, T. Savchenko, G. Krishnaswamy, L.J. Heyderman, P. Gambardella, Chirally coupled nanomagnets, Science 363 (2019) 1435–1439, https://doi.org/10.1126/science.aau7913.
[192] C.O. Avci, K. Garello, A. Ghosh, M. Gabureac, S.F. Alvarado, P. Gambardella, Unidirectional spin Hall magnetoresistance in ferromagnet/normal metal bilayers, Nat. Phys. 11 (2015) 570–575, https://doi.org/10.1038/nphys3356.
[193] H. Nakayama, Y. Kanno, H. An, T. Tashiro, S. Haku, A. Nomura, K. Ando, Rashba-Edelstein magnetoresistance in metallic heterostructures, Phys. Rev. Lett. 117 (2016) 116602.
[194] M.B. Jungfleisch, W. Zhang, J. Sklenar, W. Jiang, J.E. Pearson, J.B. Ketterson, A. Hoffmann, Interface-driven spin-torque ferromagnetic resonance by Rashba coupling at the interface between non-magnetic materials, Phys. Rev. B 93 (2016) 224419.

[195] J. Grollier, D. Querlioz, M.D. Stiles, Spintronic nanodevices for bioinspired computing, Proc. IEEE 104 (2016) 2024–2039.
[196] G. Bourianoff, D. Pinna, M. Sitte, K. Everschor-Sitte, Potential implementation of reservoir computing models based on magnetic skyrmions, AIP Adv. 8 (2018) 055602, https://doi.org/10.1063/1.5006918.
[197] D. Prychynenko, M. Sitte, K. Litzius, B. Krüger, G. Bourianoff, M. Kläui, J. Sinova, K. Everschor-Sitte, Magnetic skyrmion as a nonlinear resistive element: a potential building block for reservoir computing, Phys. Rev. Appl. 9 (2018) 014034, https://doi.org/10.1103/PhysRevApplied.9.014034.
[198] D. Pinna, G. Bourianoff, K. Everschor-Sitte, Reservoir Computing with Random Skyrmion Textures, http://arxiv.org/abs/1811.12623, 2018. [(Accessed 3 September 2019)].
[199] V.S. Bhat, F. Heimbach, I. Stasinopoulos, D. Grundler, Angular-dependent magnetization dynamics of kagome artificial spin ice incorporating topological defects, Phys. Rev. B 96 (2017) 014426, https://doi.org/10.1103/PhysRevB.96.014426.

CHAPTER SIX

Effective flexoelectric and flexomagnetic response of ferroics

Eugene A. Eliseev[a], Anna N. Morozovska[b,c,*], Victoria V. Khist[d,e], Victor Polinger[f,*]

[a]Institute for Problems of Materials Science, National Academy of Sciences of Ukraine, Kyiv, Ukraine
[b]Institute of Physics, National Academy of Sciences of Ukraine, Kyiv, Ukraine
[c]Bogolyubov Institute for Theoretical Physics, National Academy of Sciences of Ukraine, Kyiv, Ukraine
[d]National Technical University of Ukraine "Igor Sikorsky Kyiv Polytechnic Institute", Kyiv, Ukraine
[e]Institute of Magnetism, National Academy of Sciences of Ukraine and Ministry of Education and Science of Ukraine, Kyiv, Ukraine
[f]Department of Chemistry, University of Washington, Seattle, WA, United States
*Corresponding authors: e-mail address: anna.n.morozovska@gmail.com; vpolinger@msn.com

Contents

Solid State Physics, Volume 70
ISSN 0081-1947
https://doi.org/10.1016/bs.ssp.2019.09.002

1. Introduction

1.1 Definition and symmetry of the flexoelectric coupling

Flexoelectricity in solids, its fundamental theoretical aspects, existing and potential applications, attracts permanent attention of researchers [1]. The static flexoelectric effect is the appearance of elastic strain u_{ij} in response to electric polarization gradient $\partial P_k/\partial x_l$ (direct effect), and, vice versa, the polarization P_i appears as a response to the strain gradient $\partial u_{ij}/\partial x_l$ (the inverse effect) [2–5]. For a ferroic, in the continuum-medium approximation, its free energy includes static flexoelectric coupling (for brevity "**flexocoupling**") in the form of Lifshitz invariant [6–9],

$$F_{FL} = \int_V d^3 r \frac{f_{ijkl}}{2}\left(P_k \frac{\partial u_{ij}}{\partial x_l} - u_{ij}\frac{\partial P_k}{\partial x_l}\right) \tag{1}$$

Here the strain tensor components $u_{ij} = (\partial U_i/\partial x_j + \partial U_j/\partial x_i)/2$ are symmetrized derivatives of the displacement vector components U_i. Notably, the antisymmetric of the displacement gradient, $(\partial U_i/\partial x_j - \partial U_j/\partial x_i)$, could not affect physical properties in general, and the flexoelectric response in particular of the bulk material [9], and so the form (1) is the comprehensive one. Due to the evident index-permutation symmetry of the strain tensor, $u_{ij} \equiv u_{ji}$, all components of the static flexocoupling tensor f_{ijkl} are symmetrical (i.e., index-permutative) with respect to the permutation of the first pair of indices,

$$f_{ijkl} \equiv f_{jikl}. \tag{2a}$$

The static flexoelectric effect is allowed by symmetry in all 32 crystalline point groups [1,5]. For instance, Shu et al. [10], Quang and He [11], consider possible symmetries of the flexoelectric tensor and derived the number of its independent components for each symmetry. Markedly, Shu et al. [10], and Quang and He [11] used the permutation invariance (2a) of the flexocoupling tensor f_{ijkl} with respect to the first pair of indices.

Under the continuous medium approximation, one can apply integration by parts to Eq. (1) and neglect surface integrals an infinite bulk ferroic (see Appendix A). For infinite ferroic this operation reveals more, previously "hidden" index-permutation symmetry properties [12], namely:

$$f_{ijkl} = f_{ilkj} \tag{2b}$$

Note that Eq. (2b) can be valid for an infinite material only, and violates for confined and nanosized systems. The full analysis of the symmetry (2b) on flexoelectric coupling is presented in Ref. [13].

Using the lattice dynamics theory, Kvasov and Tagantsev [14] predicted the existence of dynamic flexoelectric effect on example of a cubic lattice. They argued that dynamic equations of state for a condensed media can be derived from the minimization of Lagrange function, $L=F-T$, where the Landau-Gingburg-Devonshire (**LGD**) free energy F is given by expression

$$F=\int_V d^3r\left(\alpha P_iP_i+\alpha_{ijkl}P_iP_jP_kP_l+\frac{g_{ijkl}}{2}\left(\frac{\partial P_i}{\partial x_j}\frac{\partial P_k}{\partial x_l}\right)\right.$$
$$\left.-q_{ijkl}u_{ij}P_kP_l+\frac{c_{ijkl}}{2}u_{ij}u_{kl}-P_iE_i-N_{ij}u_{ij}\right)+F_{FL} \tag{3a}$$

Hereinafter the summation is over all repeating indexes; the coefficient α is temperature dependent for ferroics, q_{mnij} is electrostriction tensor, the higher-order coefficients α_{ijkl} are assumed to be temperature independent, g_{ijkl} are gradient coefficients, c_{ijkl} are elastic compliances. N_{ij} is the anisotropic external load, E_i is electric field that obeys electrostatic equation, $\varepsilon_b\varepsilon_0\frac{\partial E_i}{\partial x_i}=-\frac{\partial P_i}{\partial x_i}$ (ε_b is background permittivity [15] and $\varepsilon_0=8.85\times10^{-12}$ F/m is the vacuum dielectric constant). The kinetic energy T is given by

$$T=\frac{\mu}{2}\left(\frac{\partial P_i}{\partial t}\right)^2+M_{ij}\frac{\partial P_i}{\partial t}\frac{\partial U_j}{\partial t}+\frac{\rho}{2}\left(\frac{\partial U_i}{\partial t}\right)^2, \tag{3b}$$

which includes the dynamic flexoelectric coupling that is described by the second-rank tensor M_{ij} [14]. U_i is elastic displacement, μ is a dynamic coefficient and ρ is the density of a ferroelectric.

The flexoelectric coupling plays an important at the nanoscale [1], especially in ferroelectric thin films [16–20], nanoparticles [7,21,22], and curved domain structures, such as polar vortices [23] and labyrinthine domains [24,25].

1.2 Definition and symmetry of the flexomagnetic coupling

Inhomogeneous strains coupled to magnetic fields, which should exist in systems with inhomogeneously distributed magnetic moments (e.g., in the vicinity of the domain walls, surfaces and interfaces), can give rise to the flexomagnetic coupling [26,27].

The flexomagnetic effect can be defined similarly to the static flexoelectric effect. Using the similarity, one can define the flexomagnetic effect as the appearance of elastic strain u_{ij} in response to electric magnetization gradient

$\partial M_k/\partial x_l$ (direct effect), and, vice versa, the magnetization M_i appears as a response to the strain gradient $\partial u_{ij}/\partial x_l$ (the inverse effect). For a ferroic, in the continuum-medium approximation, its LGD free energy can include the flexomagnetic coupling in the form of Lifshitz invariant:

$$F_{FL} = \int_V d^3r \frac{m_{ijkl}}{2}\left(M_k \frac{\partial u_{ij}}{\partial x_l} - u_{ij}\frac{\partial M_k}{\partial x_l}\right). \tag{4a}$$

Due to the evident index-permutation symmetry of the strain tensor, $u_{ij} \equiv u_{ji}$, all components of the flexomagnetic tensor m_{ijkl} are symmetrical (i.e., index-permutative) with respect to the permutation of the first pair of indices,

$$m_{ijkl} \equiv m_{jikl}. \tag{4b}$$

Under the continuous medium approximation, one can apply integration by parts to Eq. (4a) and neglect surface integrals an infinite bulk ferroic. For infinite ferroic this operation reveals more index-permutation symmetry properties, similar to Eq. (2b):

$$m_{ijkl} = m_{ilkj} \tag{4c}$$

Similarly to Eq. (2b), Eq. (4c) can be valid for an infinite material only, and violates for confined and nanosized systems.

Similarly to the flexoelectric coupling, one can expect that the role of the flexomagnetic effect can be especially important at micro- and nanoscale, in ferromagnetic or antiferromagnetic thin films, nanoparticles, magnetic vortices, skyrmions, and other curved domain structures, their kinetics and thermodynamics. Actually, Chen et al. [28] discussed the role of flexomagnetic effect in the appearance of magnetization at the ferroelectric domain walls. Naimov and Bratkovsky [29] used a conception of flexomagnetic coupling when considering the influence of inhomogeneous strain on the band structure of graphene. Spurgeon et al. [30] experimentally studied heterostructure consisting of magnetic and ferroelectric layers, and demonstrated the presence of inhomogeneous strain in their system, attributed to the influence of flexomagnetic coupling. Sidhardh and Ray [31] modeled a magnetic response of bended nano-cantilever taking into account the flexomagnetic coupling. Ghobadi et al. [32] included the flexomagnetic coupling into the phenomenological consideration of nanoplate under magnetic flux.

However, the numerical value of the flexomagnetic coupling constants is much less studied in comparison with the flexoelectric one, and only a few pioneer papers exist [27,28]. Specifically, Lukashev et al. calculated a flexomagnetic coefficient of about 1.95 μ_B Å for the antiperovskite Mn_3GaN, and defined it as the coupling between the strain gradient and magnetic dipole moment per Mn atom.

It is worth to underline that the time and/or space inversion operations alone in a symmetry group exclude the flexomagnetic effect existence. For a flexomagnetic effect to exist these operations should be coupled with each other in the material symmetry group. The structure of the flexomagnetic effect tensor has been established by Eliseev et al. [33] using the symmetry theory combined with continuum media LGD approach. Previously the combined approach allowed to establish the structure and some properties of the piezomagnetic [34,35], magnetoelectric [36–38], inhomogeneous magnetoelectric [39,40] and flexomagnetoelectric [41] effects.

1.3 The chapter motivation and structure

The symmetry analysis allows one to establish the impact of the material symmetry on the effective flexoelectric or flexomagnetic response (for brevity "**flexoresponse**") of different ferroics, performed in the work. Remarkably, effective flexocoupling constants are the only physical quantities that can be directly determined from bending experiments. All other experimental methods, such as Raman and inelastic neutron scattering, are indirect.

To the best of our knowledge, for arbitrary orientation of the sample, calculations of the effective flexoresponse and its symmetry analysis are missing. In this chapter we calculate and visualize the effective flexoresponse of the bended plate for arbitrary sample orientation, analyze its anisotropy, and angular dependences. The calculations are of practical use, because, if the angular dependence of the flexoresponse is known, there is no need to measure it by continuously changing the orientation of the crystal, but measuring it along several crystallographic directions would be enough.

Using the direct method, described in Ref. [34], here we study the structure of flexocoupling tensors. These results are presented in Section 2. In Section 3 we calculate and visualize the effective flexoresponse of a cubic and tetragonal ferroic plate measured in bending experiment. Further we discuss the problem of determining flexocoupling constants from inelastic neutron and Raman scattering experiments [42–49] in Section 4. Section 5 is a brief summary.

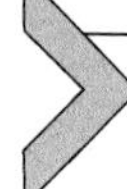

2. Flexocoupling in ferroics

2.1 Flexoelectric coupling in ferroics with cubic and lower symmetry

Direct application of point group symmetry operations to the flexoelectric tensor f_{ijkl} (shortly, the "**direct method**" [35]) along with the index-permutation symmetry (2a), leads to the system of linear algebraic equations:

$$\begin{cases} f_{ijkl} = C_{ii'} C_{jj'} C_{kk'} C_{ll'} f_{i'j'k'l'}, \\ f_{ijkl} = f_{jikl}. \end{cases} \tag{5a}$$

The first line in Eq. (5a) includes point group symmetry, where matrix elements $C_{ii'}$ are concrete point group symmetry operations in the matrix form, and the second line reflects the index-permutation symmetry (2a). Solution of Eq. (5a) gives the detailed structure of f_{ijkl} for all 32 point symmetry groups, including the number of nonzero and independent elements, (see Table B1 in Appendix B, and compare it with the results of Shu et al. [10], and Quang and He [11]).

More complex the system of linear algebraic equations,

$$\begin{cases} f_{ijkl} = C_{ii'} C_{jj'} C_{kk'} C_{ll'} f_{i'j'k'l'}, \\ f_{ijkl} = f_{jikl}, \\ f_{ijkl} = f_{ilkj} \end{cases} \tag{5b}$$

includes the point group symmetry (the first line), evident index-permutation symmetry (2a) (the second line) and recently established hidden index-permutation symmetry (2b) (the third line). Solution of Eq. (5b) gives the detailed structure of f_{ijkl} for all 32 point symmetry groups (see Table B2 in Appendix B).

The meaning of the results [13] follows from the comparison of the Tables 1 and 2, presented below for several point symmetries. Namely, for a cubic point symmetry (corresponding to the paraelectric phase of most perovskites) the index-permutation symmetry (2b) reduces the number of f_{ijkl} independent elements from 3 to 2, and for the tetragonal point symmetry (corresponding to the polar long-range ordered phase of many ferroelectric perovskites), it reduces from 8 to 5 independent elements. For the orthorhombic point symmetry (the often case) the index-permutation symmetry (2b) reduces the number of independent elements f_{ijkl} from 15 to 9.

Table 1 Structure of the static flexoelectric effect tensor calculated from Eq. (5a) for several point symmetry groups.

Point symmetry	Number of nonzero elements	Number of independent elements	Nonzero elements and relations between them
Cubic	21	3 (without "hidden" index-permutation symmetry)	$f_{1111}=f_{2222}=f_{3333}$,
$m3m$, 432, $4'3m$			$f_{1122}=f_{2211}=f_{1133}=f_{3311}=f_{2233}=f_{3322}$,
			$f_{1221}=f_{2112}=f_{1331}=f_{3113}=f_{2332}=f_{3223}=$ $f_{1212}=f_{2121}=f_{1313}=f_{3131}=f_{2323}=f_{3232}$
Tetragonal	21	8 (without "hidden" index-permutation symmetry)	$f_{3333}, f_{1111}=f_{2222}, f_{1122}=f_{2211}$,
$4'2m$, 422, $4mm$, $4/mmm$			$f_{1133}=f_{2233}, f_{3311}=f_{3322}$,
			$f_{1212}=f_{2112}=f_{1221}=f_{2121}$,
			$f_{1313}=f_{3113}=f_{2323}=f_{3223}$,
			$f_{1331}=f_{3131}=f_{2332}=f_{3232}$
Orthorhombic	21	15 (without "hidden" index-permutation symmetry)	$f_{1111}, f_{1122}, f_{1133}, f_{2211}$,
222, $mm2$, mmm			$f_{2222}, f_{2233}, f_{3311}, f_{3322}, f_{3333}$,
			$f_{1212}=f_{2112}, f_{1221}=f_{2121}, f_{1313}=f_{3113}$,
			$f_{1331}=f_{3131}, f_{2323}=f_{3223}, f_{2332}=f_{3232}$

Table 2 Structure of the static flexoelectric effect tensor calculated from Eq. (5b) for several point symmetry groups.

Point symmetry	Number of nonzero elements	Number of independent elements	Nonzero elements and relations between them
Cubic[a]	21	2 ("hidden" index-permutation symmetry is included)	$f_{1111}=f_{2222}=f_{3333}$,
$m3m$, 432, $4'3m$			$f_{1122}=f_{2211}=f_{1133}=f_{3311}=f_{2233}=f_{3322}=$ $f_{1221}=f_{2112}=f_{1331}=f_{3113}=f_{2332}=f_{3223}=$ $f_{1212}=f_{2121}=f_{1313}=f_{3131}=f_{2323}=f_{3232}$
Tetragonal	21	5 ("hidden" index-permutation symmetry is included)	$f_{3333}, f_{1111}=f_{2222}$,
$4'2m$, 422, $4mm$, $4/mmm$			$f_{1122}=f_{1221}=f_{2121}=f_{2211}=f_{2112}=f_{1212}$,
			$f_{1133}=f_{1331}=f_{3131}=f_{2233}=f_{3232}=f_{2332}$,
			$f_{1313}=f_{3311}=f_{3113}=f_{2323}=f_{3322}=f_{3223}$
orthorhombic	21	9 ("hidden" index-permutation symmetry is included)	$f_{1111}, f_{2222}, f_{3333}$
222, $mm2$, mmm			$f_{1122}=f_{1221}=f_{2121}, f_{2211}=f_{2112}=f_{1212}$,
			$f_{2233}=f_{2332}=f_{3232}, f_{3311}=f_{1313}=f_{3113}$,
			$f_{3322}=f_{3223}=f_{2323}, f_{1133}=f_{1331}=f_{3131}$

[a]There are other cubic and tetragonal point groups, not listed in Tables 1 and 2, but listed in Appendix B.

The number of nonzero elements is 21 for all considered examples, but it can be different for other groups (see comparative Table B3 in Appendix B).

2.2 Flexomagnetic coupling in ferroics with cubic symmetry

Direct application of point group symmetry operations to the flexomagnetic tensor m_{ijkl} along with the index-permutation symmetry (4a), (4b), (4c), leads to the system of linear algebraic equations:

$$\begin{cases} m_{ijkl} = (-1)^R \det\left(\hat{C}\right) C_{ii'} C_{jj'} C_{kk'} C_{ll'} f_{i'j'k'l'}, \\ m_{ijkl} = m_{jikl}, \\ \left\{ m_{ijkl} = m_{ilkj} \right\} \end{cases} \tag{5c}$$

Here we introduced determinant of the transformation matrix, $\det\left(\hat{C}\right) = \pm 1$ $\det\left(\hat{A}\right) = \pm 1$; the factor R denotes either the presence ($R = 1$) or the absence ($R = 0$) of the time-reversal operation coupled to the space transformation C_{ij}. The first line in Eq. (5c) includes point group symmetry, where matrix elements $C_{ii'}$ are concrete point group symmetry operations in the matrix form, and the second and third lines reflect the index-permutation symmetries $m_{ijkl} = m_{jikl}$ and $m_{ijkl} = m_{ilkj}$, respectively. Solution of Eq. (5c) gives the detailed structure of m_{ijkl} for all 90 magnetic symmetry groups, including the number of nonzero and independent elements.

The meaning of the results follows from the analyses of Table 3, presented below for two cubic point symmetries, $m'3m'$ and $m'3m$. For these groups the index-permutation symmetry $m_{ijkl} = m_{ilkj}$ reduces the number of m_{ijkl} independent elements from 3 to 2, and from 2 to 1, respectively. The total number of nonzero elements is 21 and 18 for the considered cubic groups.

The results of the calculations of flexomagnetic tensor structure are given in Appendix D for all the magnetic symmetry groups.

3. Effective flexoresponse

3.1 Effective flexoelectric response

The example of experimental determination of f_{ijkl} is an incipient paraelectric $SrTiO_3$ having $m3m$ symmetry (see Refs. [50,51]). Another case is a ferroelectric family $Ba_xSr_{1-x}TiO_3$ with $x \geq 0.5$ with $m3m$ parent symmetry and $4mm$ symmetry in a polar tetragonal phase, for which the flexoelectric coefficients have been calculated from the first principles in

Table 3 Structure of the static flexomagnetic effect tensor calculated from Eq. (5c) for several point symmetry groups.

Point symmetry	Number of nonzero elements	Number of independent elements	Nonzero elements and relations between them
Cubic $m'3m'$	21	3 (without "hidden" index-permutation symmetry)	$m_{1111}=m_{2222}=m_{3333}$, $m_{1122}=m_{2211}=m_{1133}=m_{3311}=m_{2233}=m_{3322}$, $m_{1221}=m_{2112}=m_{1331}=m_{3113}=m_{2332}=m_{3223}=$ $m_{1212}=m_{2121}=m_{1313}=m_{3131}=m_{2323}=m_{3232}$
Cubic $m'3m$	18	2 (without "hidden" index-permutation symmetry)	$m_{1122}=m_{2233}=m_{3311}=-m_{1133}=-m_{2211}=-m_{3322}$, $m_{1212}=m_{2112}=m_{2323}=m_{3223}=m_{3131}=m_{1331}=$ $-m_{2121}=-m_{1221}=-m_{3232}=-m_{2332}=-m_{3113}=-m_{1313}$
Cubic[a] $m'3m'$	21	2 ("hidden" index-permutation symmetry $m_{ijkl}=m_{ilkj}$ is included)	$m_{1111}=m_{2222}=m_{3333}$, $m_{1122}=m_{2211}=m_{1133}=m_{3311}=m_{2233}=m_{3322}=$ $m_{1221}=m_{2112}=m_{1331}=m_{3113}=m_{2332}=m_{3223}=$ $m_{1212}=m_{2121}=m_{1313}=m_{3131}=m_{2323}=m_{3232}$
Cubic $m'3m$	18	1 ("hidden" index-permutation symmetry $m_{ijkl}=m_{ilkj}$ is included)	$m_{1122}=m_{2233}=m_{3311}=$ $=m_{2121}=m_{1221}=m_{3232}=m_{2332}=m_{3113}=m_{1313}$ $m_{1212}=m_{2112}=m_{2323}=m_{3223}=m_{3131}=m_{1331}=$ $=m_{1133}=m_{2211}=m_{3322}=-m_{1122}$

[a]There are other cubic point groups, not listed in Table 3.

Refs. [52,53]. In what follows below, we visualize the effective flexoelectric response f_{ij}^{eff} for $Ba_xSr_{1-x}TiO_3$ with $x=0$, 0.5, and 1, using experimental results [52], ab initio calculations [53,54], and elastic constants c_{ij} determined experimentally in Ref. [54,55], respectively.

Typical for measurements of effective flexoresponse [51,52], the three-knife load experimental setup and the corresponding orientation of the sample is shown in Fig. 1.

For a plate with its surface inclined at an angle α with respect to crystallographic axes, the effective flexoresponse (that is a measurable value) has the following two components:

$$f_{13}^{eff}=\tilde{f}_{1133}-\frac{\tilde{c}_{1133}}{\tilde{c}_{1313}\tilde{c}_{3333}-\tilde{c}_{1333}^{2}}\left(\tilde{c}_{1313}\tilde{f}_{3333}-\tilde{c}_{1333}\tilde{f}_{1333}\right) \tag{6a}$$

$$f_{23}^{eff}=\tilde{f}_{2233}-\frac{\tilde{c}_{2233}}{\tilde{c}_{1313}\tilde{c}_{3333}-\tilde{c}_{1333}^{2}}\left(\tilde{c}_{1313}\tilde{f}_{3333}-\tilde{c}_{1333}\tilde{f}_{1333}\right) \tag{6b}$$

Here we introduced the tensor components in the rotated coordinate frame, linked to the sample as shown in Fig. 1A. For the derivation of

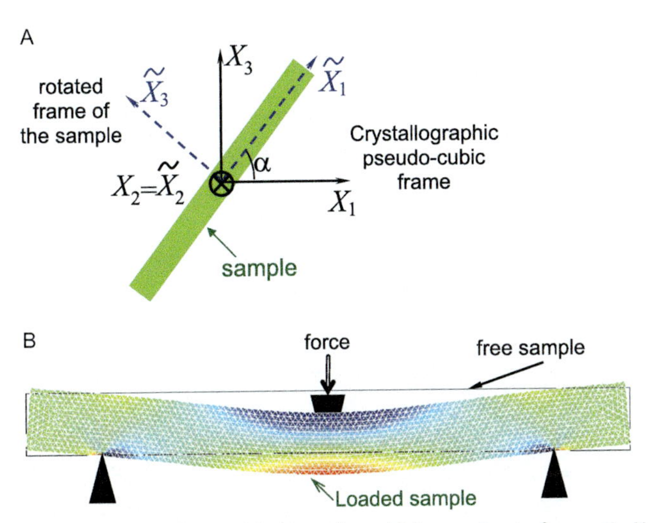

Fig. 1 (A) Rotation of crystallographic (pseudo-cubic) coordinate frame X_1, X_2, X_3 about the axis $\tilde{X}_2$ X_2 through an angle α with respect to the sample-related frame $\tilde{X}_1$,$\tilde{X}_2$,$\tilde{X}_3$ (The coordinate axis $X_2 \equiv \tilde{X}_2$ is perpendicular to the page. The turn is in the clock-wise direction, indicated by the crossed circle "⊗" as in the right-hand rule). (B) Typical three-knife experiment of the plate bending. Black solid lines represent cross-section of the initial plate position, rainbow-colored shape represents the distorted plate and strain distributions in the loaded plate.

Eqs. (6a), (6b) see Appendix C. Hereinafter $\widetilde{f}_{ijkl}$ and $\widetilde{c}_{ijkl}$ are the "rotated" tensors of flexoelectric coefficients and elastic stiffness, respectively.

For **cubic groups** $m3m$, 432, and $\bar{4}3m$, the "rotated" tensors $\widetilde{f}_{ijkl}$ and $\widetilde{c}_{ijkl}$ have the following angular dependence:

$$\widetilde{f}_{1133}=f_{1122}+\frac{\sin^2(2\alpha)}{2}\Delta f_{11},\ \widetilde{f}_{1333}=\frac{\sin(4\alpha)}{2}\Delta f_{11}, \tag{7a}$$

$$\widetilde{f}_{3333}=f_{1111}-\frac{\sin^2(2\alpha)}{2}\Delta f_{11},\ \widetilde{f}_{2233}=f_{1122}, \tag{7b}$$

$$\widetilde{c}_{1133}=c_{1133}+\frac{\sin^2(2\alpha)}{2}\Delta c_{11},\widetilde{c}_{1313}=c_{1313}+\frac{\sin^2(2\alpha)}{2}\Delta c_{11}, \tag{8a}$$

$$\widetilde{c}_{1333}=\frac{\sin(4\alpha)}{2}\Delta c_{11},\widetilde{c}_{2233}=c_{1122},\widetilde{c}_{3333}=c_{1111}-\frac{\sin^2(2\alpha)}{2}\Delta c_{11}. \tag{8b}$$

Here f_{ijkl} and c_{ijkl} are tensors components of flexoelectric coupling and elastic stiffness in the crystallographic frame. Also, in Eqs. (7a), (7b), (8a), (8b) for flexoelectric coupling and elastic stiffness we introduce anisotropy factors

$$\Delta f_{11}=f_{1111}-f_{3311}-f_{1313}-f_{1331}, \tag{9a}$$

$$\Delta c_{11}=c_{1111}-c_{1133}-2c_{1313}. \tag{9b}$$

Under the condition of hidden symmetry (2b), the expressions (9a) simplify, $\Delta f_{11}=f_{1111}-3f_{1133}$, while Eqs. (7a), (7b), (8a), (8b) remain unaffected.

Calculated for several **cubic** symmetries ($m3m$, 432, $\bar{4}3m$), angular dependence of the effective flexoresponse is shown in Fig. 2. Listed in Table 4 for the case of a thin plate made of $SrTiO_3$, the corresponding components are $f_{1122}=2.64\,V$ and $f_{1212}=2.185\,V$. Distinguished from our predictions listed in Tables 1 and 2, these values are not equal, $f_{1122}\neq f_{1212}$, so $f_{1122}-f_{1212}=0.455\,V$. This difference is due to the data collected for a thin plate, finite in all three dimensions, where the conditions for the hidden index-permutation symmetry do not apply. Fig. 2 shows angular dependences of the flexoresponse of the cubic perovskite $SrTiO_3$ with the flexoelectric constants $f_{ij}^{eff}(\alpha)$ according to Eqs. (6a), (6b)–(9a), (9b), and (5a). In Fig. 2A, we use experimental values of flexoelectric constants measured by Zubko [26] and listed in Table 4, namely, $f_{1111}=0.1\,V$, $f_{1122}=2.64\,V$, and $f_{1212}=2.185\,V$. The inequality $f_{1122}\neq f_{1212}$, means no index-permutation symmetry condition (2b) applies. As the effective

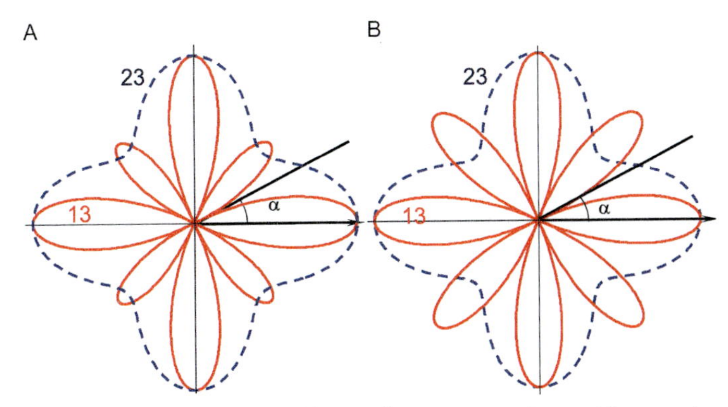

Fig. 2 Angular dependence of the effective flexoresponse in a cubic crystal, symmetry $m3m$, for (A) a thin plate of $SrTiO_3$ and (B) in bulk of an infinite crystal $SrTiO_3$. The angle α represents rotation of the coordinate frame with respect to the plate-related crystallographic pseudo-cubic axis $\widetilde{X}_2 \equiv X_2$ (see Fig. 1). Solid and dashed curves represent the flexoresponse in two perpendicular directions, f_{13}^{eff} and f_{23}^{eff}, respectively. The corresponding values of f_{ijkl} and c_{ijkl} are from the Table 4. The plot (A) is calculated from Eq. (5a). For the plot (B), calculated from Eq. (5b), we use symmetrized off-diagonal flexoelectric coefficients $f_{1122} = f_{1212} = 2.413\,\text{V}$.

Table 4 Tensor components of flexoelectric coupling (f_{ijkl}) and elastic stiffness (c_{ijkl}) for $SrTiO_3$.

Tensor	Numerical values			Reference and notation
f_{ijkl} (V)	$f_{1111} = -3.39$	$f_{1122} = 1.51$	$f_{1212} = 1.13$	[a]Zubko et al. [51], experiment at RT
f_{ijkl} (V)	$f_{1111} = 0.1$	$f_{1122} = 2.64$	$f_{1212} = 2.185$	[b]Zubko et al. [52], experiment at RT
c_{ijkl} (10^{11} Pa)	$c_{1111} = 3.19$	$c_{1122} = 1.03$	$c_{1212} = 1.24$	Bell and Rupprecht [55], experiment at RT

[a]Initial values, obtained from misinterpretation of experimental results.
[b]Corrected experimental values.
RT means room temperature.

flexoresponse was measured for a thin plate, this inequality can be interpreted as manifestation of the plate confinement in the perpendicular direction $\widetilde{X}_3$.

Fig. 2B shows the flexoresponse as it is expected to be in the bulk of an infinite $SrTiO_3$. The corresponding angular dependences $f_{ij}^{eff}(\alpha)$ result from the same Eqs. (6a), (6b)–(9a), (9b) and (5b) as above, but this time under assumption of the hidden symmetry when the index-permutation rule

(2b) applies [11]. The latter requires $f_{1122}=f_{1212}$, and here we use the average off-diagonal values $f_{1122}=f_{1212}=2.413\,\text{V}$ with the same diagonal value of $f_{1111}=0.1\,\text{V}$. Comparing solid curves in Fig. 2A and B, the flexoelectric component f_{13}^{eff} is found to be significantly affected by the hidden symmetry (in bulk, all of its eight lobes are of almost the same length), while there is no any significant change in the flexoelectric component f_{23}^{eff} shown by the dashed curve. To summarize, the distinctive difference of Fig. 2A for a plate from Fig. 2B in bulk is a manifestation of the confinement effect in the thin plate in the perpendicular direction $\widetilde{X}_3$.

For a **tetragonal** crystal, symmetry groups $4'2m$, 422, $4mm$, and $4/mmm$, the "rotated" tensors $\widetilde{f}_{ijkl}$ and $\widetilde{c}_{ijkl}$ have the following angular dependence:

$$\widetilde{f}_{1133}=\cos^2(\alpha)f_{1133}+\sin^2(\alpha)f_{3311}+\frac{\sin^2(2\alpha)}{4}(\Delta f_{11}+\Delta f_{33}), \tag{10a}$$

$$\widetilde{f}_{1333}=\frac{\sin(2\alpha)}{2}\left(\cos^2(\alpha)\Delta f_{33}-\sin^2(\alpha)\Delta f_{11}\right), \tag{10b}$$

$$\widetilde{f}_{3333}=\cos^2(\alpha)f_{3333}+\sin^2(\alpha)f_{1111}-\frac{\sin^2(2\alpha)}{4}(\Delta f_{11}+\Delta f_{33}), \tag{10c}$$

$$\widetilde{f}_{2233}=\sin^2(\alpha)f_{1122}+\cos^2(\alpha)f_{1133}, \tag{10d}$$

$$\widetilde{c}_{1133}=c_{1133}+\frac{\sin^2(2\alpha)}{4}(\Delta c_{11}+\Delta c_{33}), \tag{11a}$$

$$\widetilde{c}_{1313}=c_{1313}+\frac{\sin^2(2\alpha)}{4}(\Delta c_{11}+\Delta c_{33}), \tag{11b}$$

$$\widetilde{c}_{1333}=\frac{\sin(2\alpha)}{2}\left(\cos^2(\alpha)\Delta c_{33}-\sin^2(\alpha)\Delta c_{11}\right), \tag{11c}$$

$$\widetilde{c}_{2233}=\sin^2(\alpha)c_{1122}+\cos^2(\alpha)c_{1133}, \tag{11d}$$

$$\widetilde{c}_{3333}=\cos^2(\alpha)c_{3333}+\sin^2(\alpha)c_{1111}-\frac{\sin^2(2\alpha)}{4}(\Delta c_{11}+\Delta c_{33}). \tag{11e}$$

Similar to Eqs. (7a), (7b) and (8a), (8b), in Eqs. (10a)–(10d) we introduce anisotropy factors

$$\Delta f_{11}=f_{1111}-f_{3311}-f_{1313}-f_{1331},\quad \Delta f_{33}=f_{3333}-f_{1133}-f_{1313}-f_{1331}, \tag{12a}$$

$$\Delta c_{11}=c_{1111}-c_{1133}-2c_{1313},\quad \Delta c_{33}=c_{3333}-c_{1133}-2c_{1313}. \tag{12b}$$

Under condition of hidden symmetry (2b), we have $f_{1313}=f_{3311}=f_{3113}=f_{2323}=f_{3322}=f_{3223}$, and, therefore, expressions (12a) simplify to $\Delta f_{11}=f_{1111}-2f_{3311}-f_{1313}$, $\Delta f_{33}=f_{3333}-2f_{1133}-f_{1313}$.

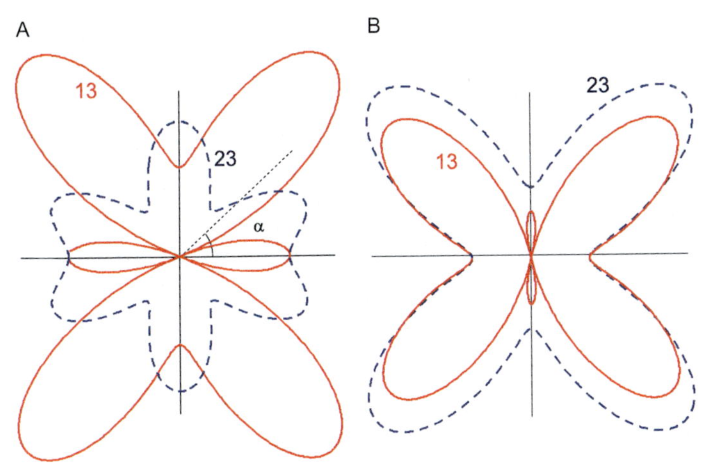

Fig. 3 Angular dependence of the effective flexoresponse for $Ba_{0.5}Sr_{0.5}TiO_3$. The case of a finite plate (A) and of an infinite crystal (B) of 4*mm* symmetry. The angle α represents rotation of the coordinate frame with respect to the plate-related crystallographic pseudo-cubic axis $\widetilde{X}_2 \equiv X_2$ (see Fig. 1). Solid and dashed curves represent the flexoresponse in two perpendicular directions, f_{13}^{eff} and f_{23}^{eff}, respectively. Values of f_{ijkl} and c_{ijkl} are given in Table 5. The plot (A) follows from Eq. (5a). For the plot (B), resulting from Eq. (5b), we use symmetrized off-diagonal flexoelectric coefficients $f_{1122} = f_{1212} = 1.585\,V$.

For this case (symmetry groups $4'2m$, 422, 4*mm*, and 4/*mmm*), the angular dependence of the effective constants f_{13}^{eff} and f_{23}^{eff} is shown in Figs. 3 and 4 for the particular examples of $BaTiO_3$ and $Ba_{0.5}Sr_{0.5}TiO_3$ plates, respectively.

As above in Fig. 2A, angular dependence $f_{ij}^{eff}(\alpha)$ in Fig. 3A and Fig. 4A, follows Eqs. (6a), (6b), (10a), (10b), (10c), (10d)–(12a), (12b) and (5a), without hidden index-permutation symmetry condition (2b) involved. Used in our calculations coefficients f_{ijkl} and c_{ijkl} are from Table 5. According to Ponomareva et al. [53], for $Ba_{0.5}Sr_{0.5}TiO_3$ we have $f_{1111} \sim f_{1122} \sim 4\,V$, while for $BaTiO_3$, according to Maranganti and Sharma [54], $f_{1111} \ll |f_{1122}| \sim 4\,V$. Moreover, in these materials f_{ijkl} have opposite signs (compare the first and the second lines in Table 5). Therefore, for plates of $BaTiO_3$ and $Ba_{0.5}Sr_{0.5}TiO_3$ we expect different effective flexoelectric responses $f_{ij}^{eff}(\alpha)$. From the table, obtained from ab initio calculations at 0 K in cubic approximation [53,54], the difference $f_{1122} - f_{1212}$ is nonzero and rather high, of the order of $(1 \div 3)$ V. Notably, even under periodic boundary conditions, DFT calculations do not reproduce the hidden index-permutation symmetry of the flexoelectric tensor.

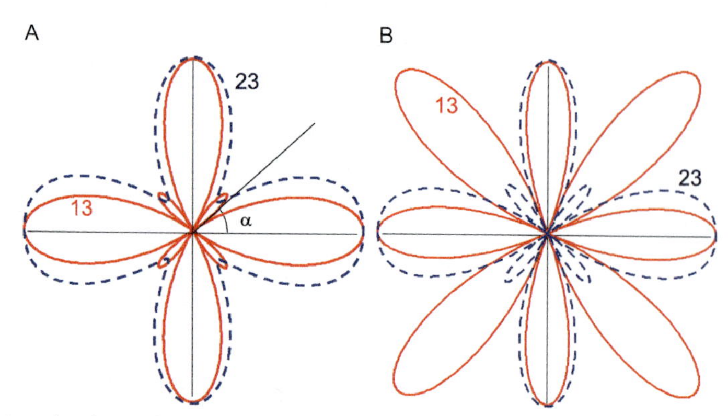

Fig. 4 Angular dependence of the effective flexoresponse for $BaTiO_3$ in tetragonal phase of crystal symmetry 4*mm*. The case of a finite plate (A) and of an infinite crystal (B). The angle α represents rotation of the coordinate frame with respect to the plate-related crystallographic pseudo-cubic axis $\widetilde{X}_2 \equiv X_2$ (see Fig. 1). Solid and dashed curves represent the flexoresponse in two perpendicular directions, f_{13}^{eff} and f_{23}^{eff}, respectively. Values of f_{ijkl} and c_{ijkl} are from Table 5. The plot (A) follows Eq. (5a). For the plot (B) resulting from Eq. (5b), we use symmetrized off-diagonal flexoelectric coefficients $f_{1122}=f_{1212}=1.585\,V$.

Shown in Fig. 3B and Fig. 4B, angular dependences $f_{ij}^{eff}(\alpha)$ were calculated from Eqs. (6a), (6b), (10a)–(10d), (11a)–(11e), (12a), (12b) and (5b), for an infinite crystal where the condition (2b) of hidden index-permutation symmetry applies. For Fig. 3B we used symmetrized off-diagonal coefficients, $f_{1122}=f_{1212}=1.585$ V, and the diagonal value $f_{1111}=5.12$ V calculated for $Ba_{0.5}Sr_{0.5}TiO_3$ by Ponomareva et al. [53]. For Fig. 4B the off-diagonal coefficients, $f_{1122}=f_{1212}=-0.935$ V, and the diagonal one is $f_{1111}=0.04$ V, calculated for $BaTiO_3$ by Maranganti and Sharma [54].

Comparing Fig. 3A with Fig. 3B we can see that $f_{ij}^{eff}(\alpha)$ are very different for solid and dashed curves. This means that both f_{13}^{eff} and f_{23}^{eff} are strongly affected by the index-permutation hidden symmetry (2b) of f_{ijkl}. Namely, the hidden symmetry decreases the number of lobes and decreases the differences between f_{13}^{eff} and f_{23}^{eff}.

Similarly, in Fig. 4A and B we see that angular dependences $f_{ij}^{eff}(\alpha)$ are very different for solid (f_{13}^{eff}) and dashed (f_{23}^{eff}) curves. Correspondingly, both flexoelectric components, f_{13}^{eff} and f_{23}^{eff}, are affected by the index-permutation symmetry (2b). It makes the angular dependences of f_{13}^{eff} and f_{23}^{eff} more complex. With the hidden symmetry small additional lobes are bigger for both components. At the same time, differences between solid and dashed curves in Fig. 4 [as well as between Fig. 4A and B], are much smaller than the differences between solid and dashed curves in Fig. 3.

Table 5 Tensor components for flexoelectric coupling, f_{ijkl}, and elastic stiffness, c_{ijkl}, for $Ba_xSr_{1-x}TiO_3$ with $x = 0.5$ and 1.

Tensor	Values of components						Reference and notation
f_{ijkl} (V)	$f_{1111} = 5.12$	$f_{1122} = 3.32$	n/a	n/a	$f_{1212} = 0.05$	n/a	Ponomareva et al. [53], ab initio calculations at 0 K in cubic approximation for $Ba_{0.5}Sr_{0.5}TiO_3$
f_{ijkl} (V)	$f_{1111} = 0.04$	$f_{1122} = 1.39$	n/a	n/a	$f_{1212} = 0.48$	n/a	Maranganti and Sharma [54] ab initio calculations at 0 K in cubic approximation for $BaTiO_3$
c_{ijkl} (10^{11} Pa)	$c_{1111} = 2.43$	$c_{1122} = 1.28$	$c_{1133} = 1.23$	$c_{1212} = 1.20$	$c_{1313} = 0.55$	$c_{3333} = 1.48$	Shaefer et al. [56], experiment at room temperature for $BaTiO_3$

3.2 Effective flexomagnetic response

The future measurements of effective flexoresponse can be performed using the three-knife load experimental setup [51,52], and the corresponding orientation of the sample is shown in Fig. 1. For a plate with its surface inclined at an angle α with respect to crystallographic axes, the effective flexomagnetic response (that is a measurable value) has the following two components:

$$m_{13}^{eff} = \widetilde{m}_{1133} - \frac{\widetilde{c}_{1133}}{\widetilde{c}_{1313}\widetilde{c}_{3333} - \widetilde{c}_{1333}^{2}}(\widetilde{c}_{1313}\widetilde{m}_{3333} - \widetilde{c}_{1333}\widetilde{m}_{1333}) \tag{13a}$$

$$m_{23}^{eff} = \widetilde{m}_{2233} - \frac{\widetilde{c}_{2233}}{\widetilde{c}_{1313}\widetilde{c}_{3333} - \widetilde{c}_{1333}^{2}}(\widetilde{c}_{1313}\widetilde{m}_{3333} - \widetilde{c}_{1333}\widetilde{m}_{1333}) \tag{13b}$$

Eqs. (13a), (13b) follow from Eqs. (6a), (6b) after substitution $\widetilde{f}_{ijkl} \rightarrow \widetilde{m}_{ijkl}$ in the rotated coordinate frame, linked to the sample as shown in Fig. 1A. The angle α represents rotation of the coordinate frame with respect to the plate-related crystallographic pseudo-cubic axis $\widetilde{X}_2 \equiv X_2$.

For **cubic groups** $m'3m'$, the "rotated" tensors $\widetilde{m}_{ijkl}$ and $\widetilde{c}_{ijkl}$ have the following angular dependence:

$$\widetilde{m}_{1133} = m_{1122} + \frac{\sin^2(2\alpha)}{2}\Delta m_{11}, \widetilde{m}_{1333} = \frac{\sin(4\alpha)}{2}\Delta m_{11}, \tag{14a}$$

$$\widetilde{m}_{3333} = m_{1111} - \frac{\sin^2(2\alpha)}{2}\Delta m_{11}, \widetilde{m}_{2233} = m_{1122}, \tag{14b}$$

$$\widetilde{c}_{1133} = c_{1133} + \frac{\sin^2(2\alpha)}{2}\Delta c_{11}, \widetilde{c}_{1313} = c_{1313} + \frac{\sin^2(2\alpha)}{2}\Delta c_{11}, \tag{15a}$$

$$\widetilde{c}_{1333} = \frac{\sin(4\alpha)}{2}\Delta c_{11}, \widetilde{c}_{2233} = c_{1122}, \widetilde{c}_{3333} = c_{1111} - \frac{\sin^2(2\alpha)}{2}\Delta c_{11}. \tag{15b}$$

Here m_{ijkl} and c_{ijkl} are tensors components of flexomagnetic coupling and elastic stiffness in the crystallographic frame. Also, in Eqs. (14a), (14b), (15a), (15b) for flexomagnetic coupling and elastic stiffness we introduce anisotropy factors

$$\Delta m_{11} = m_{1111} - m_{3311} - m_{1313} - m_{1331}, \tag{16a}$$

$$\Delta c_{11} = c_{1111} - c_{1133} - 2c_{1313}. \tag{16b}$$

Under the condition of hidden symmetry (2b), the expression (16a) simplifies, $\Delta m_{11} = m_{1111} - 3m_{1133}$.

For **cubic groups** $m'3m$, the "rotated" tensors $\widetilde{m}_{ijkl}$ and $\widetilde{c}_{ijkl}$ have the following angular dependence:

$$\widetilde{m}_{1133} = -\cos(2\alpha)m_{1122}, \widetilde{m}_{2233} = \cos(2\alpha)m_{1122}, \widetilde{m}_{1333} = \frac{\sin(2\alpha)}{2}m_{1212}, \widetilde{m}_{3333} = 0. \tag{17}$$

Eqs. (14a), (14b), (15a), (15b), (16a), (16b), (17) follow from Eqs. (7a), (7b)–(12a), (12b) after substitution $f_{ijkl} \to m_{ijkl}$ and $\widetilde{f}_{ijkl} \to \widetilde{m}_{ijkl}$.

In Appendix D we derived the following expression for the effective flexomagnetic coefficient:

$$m_{ijml} = \chi^{(P)}_{kn} \eta_{nm} f_{ijkl}, \tag{18}$$

where $\chi^{(P)}_{mk}$ is the tensor of dielectric susceptibility and η_{kl} is the components of linear magnetoelectric coupling tensor.

Angular dependence of the effective flexomagnetic response coefficients m^{eff}_{13} and m^{eff}_{23} for a thin plate of $EuTiO_3$ are shown in Fig. 5A and B,

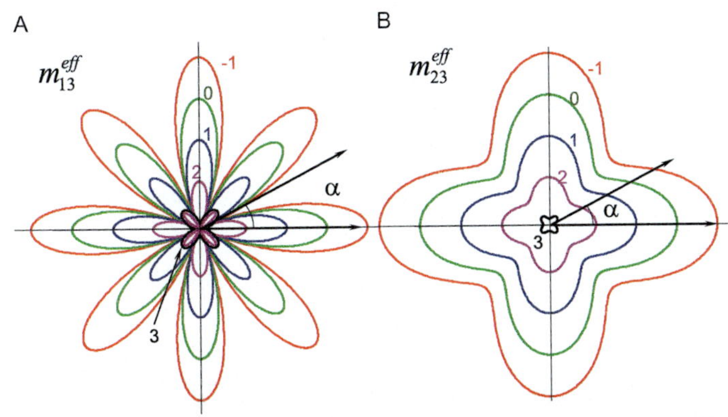

Fig. 5 Angular dependence of the effective flexoresponse coefficients (A) f^{eff}_{13} and (B) f^{eff}_{23} for a thin plate of $EuTiO_3$. The angle α represents rotation of the coordinate frame with respect to the plate-related crystallographic pseudo-cubic axis $\widetilde{X}_2 \equiv X_2$ (see Fig. 1). Different curves correspond to different values of ratio m_{1111}/m_{1212} (shown near the curves), and we supposed $m_{1122} = m_{1212}$.

respectively. From comparison of different curves one can see that the eight lobes become longer with the ratio m_{1111}/m_{1212} increase.

The dependence of the effective flexoresponse coefficients m_{13}^{eff} and m_{23}^{eff} on the ratio m_{1111}/m_{1212} for a thin plate of $EuTiO_3$ at the two different crystallographic direction, $\alpha=0$ ([100]) and $\alpha=\pi/4$ ([110]) are shown in Fig. 6A and B, respectively. Solid and dashed curves correspond to $m_{1122}=m_{1212}$ and $m_{1122}=2m_{1212}$, respectively. The distinctive difference between the solid and dashed curves can be a manifestation of the confinement effect in the thin plate in the perpendicular direction $\widetilde{X}_3$.

To conclude this section, the symmetry of flexoelectric and flexomagnetic tensors is shown to strongly affect angular dependences of the effective flexoresponse $f_{ij}^{eff}(\alpha)$ and $m_{ij}^{eff}(\alpha)$. This result can be used in bending experiments for the determination of flexocoupling constants f_{ijkl} and m_{ijkl}. However, likely valid in bulk, the index-permutation symmetry relations, $f_{ijkl}=f_{ilkj}$ (and possibly $m_{ijkl}=m_{ilkj}$) holds neither in experiments [51,52] conducted for finite thin plates, nor in *ab initio* calculations for a 3D-confined cell [53,54]. Hence, to verify the obtained theoretical results, it seems important to conduct new experimental studies of the effective flexoresponse of various crystalline plates with different orientations of crystallographic axes. Other options are discussed in the next section.

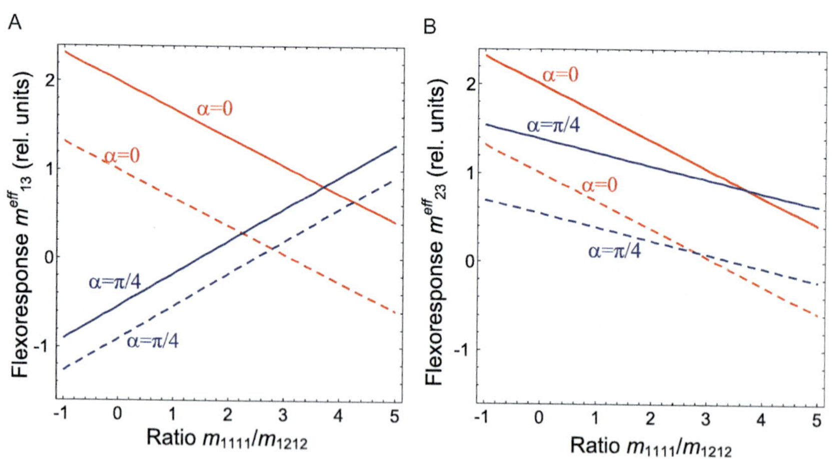

Fig. 6 The dependence of the effective flexoresponse coefficients (A) m_{13}^{eff} and (B) m_{23}^{eff} on the ratio m_{1111}/m_{1212} for a thin plate of $EuTiO_3$ at the two different crystallographic direction, $\alpha=0$ ([100]) and $\alpha=\pi/4$ ([110]). Solid and dashed curves correspond to $m_{1122}=m_{1212}$ and $m_{1122}=2m_{1212}$, respectively.

4. About determination of flexocoupling constants

Following from the general theory, symmetry properties of the flexoelectric tensor [12] are in challenging contradiction with available experimental data [51,52] and ab initio calculations [53,54]. This is just a part of the bigger problem. Measured values of flexocoupling constants are orders of magnitude higher than theoretically predicted [56]. Established by Yudin et al. [57] upper limits for f_{ijkl}, as well as the values for bulk ferroics calculated from the first principles [58–62], are several orders of magnitude smaller than those measured experimentally in ferroelectric ceramics [63–65] and thin films [66], ferroelectric relaxor polymers [67] and electrets [68], incipient ferroelectrics [52] and biological membranes [69,70]. Stengel [71], Abdollahi et al. [72] and Rahmati et al. [73] argued that the giant flexoelectric effect can result from special electric boundary conditions. According to Bersuker [74], in perovskite ceramics the anomalously high flexoelectric coupling is due to the pseudo Jahn-Teller effect. Steming from vibronic coupling [75], the pseudo Jahn-Teller effect softens polar modes promoting both static and dynamic flexoeffect in ferroics. The available information about numerical values of M_{ij} is controversial: there are microscopic theories in which the dynamic effect is absent [72]. On the other hand, its determination from the soft phonon spectra leads to nonzero M_{ij} [76,77]. Besides, the impact of dynamic flexoelectric effect appears to be comparable to that of the static one.

Notably, that measurements of phonon dispersion of soft modes by inelastic neutron and Raman scattering, combined with theoretical calculations allow one to extract the valuable information about the flexoelectric coefficients [78–80]. At the same time, scattering spectra can be readily obtained for a plate or a beam, or for a bulk sample to exclude any size effect. To illustrate the statement, consider analytical expressions of Ref. [77] for a bulk material. Dynamic susceptibility $\widetilde{\chi}_{ij}(\mathbf{k},\omega)$ is the constant of proportionality relating external electric field $\delta\widetilde{E}_j^{ext}(\mathbf{k},\omega)$ and elastic force, $\delta\widetilde{F}_j^{ext}(\mathbf{k},\omega)$, to the induced elastic displacement, $\delta\widetilde{U}_i(\mathbf{k},\omega)$, and polarization fluctuations, $\delta\widetilde{P}_i(\mathbf{k},\omega)$,

$$\begin{pmatrix} \delta\widetilde{U}_i(\mathbf{k},\omega) \\ \delta\widetilde{P}_i(\mathbf{k},\omega) \end{pmatrix} = \widetilde{\chi}_{ij}(\mathbf{k},\omega) \begin{pmatrix} \delta\widetilde{F}_j^{ext}(\mathbf{k},\omega) \\ \delta\widetilde{E}_j^{ext}(\mathbf{k},\omega) \end{pmatrix}, \tag{19}$$

where the indexes are $i, j=1, 2, 3$. The expression for the matrix 6×6 of inverse susceptibility, $\widetilde{\chi}_{ij}^{-1}(\mathbf{k},\omega)$, valid for a dielectric or paraelectric solid, can be presented in the block form:

$$\widetilde{\chi}_{ij}^{-1}(\mathbf{k},\omega)=\begin{pmatrix} c_{imjl}k_l k_m-\rho\omega^2\delta_{ij} & f_{iljk}k_l k_m-M_{ij}\omega^2 \\ f_{jnli}k_l k_n-M_{ij}\omega^2 & (2\alpha-\mu\omega^2)\delta_{ij}+\dfrac{k_i k_j}{\varepsilon_b\varepsilon_0 k^2}+g_{imjl}k_m k_l \end{pmatrix} \tag{20}$$

Each of the four elements in Eq. (20) is a matrix 3×3, and therefore the comprehensive analytical form of direct matrix $\widetilde{\chi}_{ij}(\mathbf{k},\omega)$ cannot be derived in the general case of arbitrary point symmetry and arbitrary $\mathbf{k}$-orientation. Analytical expressions for $\widetilde{\chi}_{ij}(\mathbf{k},\omega)$ are available only in several particular cases [77].

Noteworthy, experimental determination of $\widetilde{\chi}_{ij}(\mathbf{k},\omega)$ can open the way for indirect measuring poorly known values of static and dynamic flexoelectric coupling constants, f_{ijkl} and M_{ij} (see e.g., [77–80]). As an example, for a cubic crystal, three constants of the static flexoeffect, f_{1111}, f_{1212} and f_{1122}, and one constant of dynamic flexoeffect, $M_{11}=M_{22}=M_{33}$, can be determined from the fitting of three acoustic and one optic soft phonon modes at a given direction of the wave vector $\mathbf{k}$. If the hidden symmetry (2b) exists, for any direction of $\mathbf{k}$, the best fitting corresponds to $f_{1212}=f_{1122}$.

Hence, it is important to know the contribution of the flexocoupling to $\widetilde{\chi}(\mathbf{k},\omega)$. The index-permutation hidden symmetry (2b) can be of unique importance for bulk samples, since it reduces the number of independent components of f_{ijkl}, significantly simplifying fitting of scattering spectra and facilitating unambiguous determination of f_{ijkl} from scattering experiments. The analysis of magnon spectra can give us information about m_{ijkl}.

5. Summary

Using a direct matrix method we establish the structure, including the number of nonzero independent elements, of the static flexoelectric coupling tensor for all 32 point groups, and the poorly studied flexomagnetic coupling tensor for all 90 magnetic classes. For several point symmetries most important for applications we visualize the effective flexoelectric/flexomagnetic response of the bended plate, analyze its anisotropy, and angular dependences for arbitrary orientation of the sample. With known angular dependence, there is no need to measure effective flexoresponse by continuously changing orientation of the crystal; it is enough to measure it along several crystallographic directions. Flexocoupling constants can be determined (as fitting parameters) from the fitting of inelastic neutron and Raman scattering spectra [77,78]. Obtained results open the way for unambiguous determination of flexoelectric/flexomagnetic tensors from bending and scattering experiments, opening the way for its novel applications.

Authors' contribution

E.A.E. jointly with V.V.K. performed calculations of the flexoelectric tensor symmetry using direct matrix method. Also E.A.E. performed all analytical calculations of effective flexoresponse. V.P. performed calculations of the flexoelectric tensor symmetry and flexocoupling energy using group-theoretical approach. A.N.M. generated research idea, performed calculations of susceptibility, analyze obtained results, and wrote the manuscript. A.N.M. and V.P. contributed equally to the results discussion and manuscript improvement.

Appendix A. Derivation of relation (2b)

Let us consider the contribution of flexoelectric coupling to the free energy

$$\int\limits_{V\to\infty} d^3r \frac{f_{ijkl}}{2}\left(u_{ij}\frac{\partial P_k}{\partial x_l} - P_k\frac{\partial u_{ij}}{\partial x_l}\right) = \left|\begin{array}{l}\text{integration in parts}\\ \text{in the second integral}\end{array}\right|$$
$$= \int\limits_{V\to\infty} d^3r f_{ijkl} u_{ij}\frac{\partial P_k}{\partial x_l} \quad (A.1)$$

When integrating in parts in Eq. (A.1) we neglected the surface integrals like $-\int_S d^2r\frac{f_{ijkl}}{2}u_{ij}P_k n_l$ hereinafter, since it can be done for a bulk material in a continuum media approximation. Since $u_{ij} = (\partial U_i/\partial x_j + \partial U_j/\partial x_i)/2$, the transformations of Eq. (A.1) yields

$$
\begin{aligned}
&\int\limits_{V\to\infty} d^3r\,\frac{f_{ijkl}}{2}\left(\frac{\partial U_i}{\partial x_j}+\frac{\partial U_j}{\partial x_i}\right)\frac{\partial P_k}{\partial x_l}=\left|\begin{array}{l}\text{renaming indices}\\ "i"\to"\tilde{j}"\text{and}"j"\to"\tilde{i}"\\ \text{in the second term}\end{array}\right|=\int\limits_{V\to\infty} d^3r\left(\frac{f_{ijkl}}{2}\frac{\partial U_i}{\partial x_j}+\frac{f_{\tilde{j}\tilde{i}kl}}{2}\frac{\partial U_{\tilde{i}}}{\partial x_{\tilde{j}}}\right)\frac{\partial P_k}{\partial x_l}=\\
&\left|\begin{array}{l}\text{renaming indices}\\ "\tilde{j}"\to"j"\text{and}"\tilde{i}"\to"i"\\ \text{in the second term}\end{array}\right|=\int\limits_{V\to\infty} d^3r\left(\frac{f_{ijkl}+f_{jikl}}{2}\right)\frac{\partial U_i}{\partial x_j}\frac{\partial P_k}{\partial x_l}=\\
&=\left|\begin{array}{l}\text{integration in parts}\\ \text{for both multipliers }\dfrac{\partial U_i}{\partial x_j}\text{ and }\dfrac{\partial P_k}{\partial x_l}\end{array}\right|=\int\limits_{V\to\infty} d^3r\left(\frac{f_{ijkl}+f_{jikl}}{2}\right)\frac{\partial U_i}{\partial x_l}\frac{\partial P_k}{\partial x_j}=\\
&=\left|\begin{array}{l}\text{renaming indices}\\ "l"\to"\tilde{j}"\text{and}"j"\to"\tilde{l}"\\ \text{in both terms}\end{array}\right|=\int\limits_{V\to\infty} d^3r\left(\frac{f_{i\tilde{l}k\tilde{j}}+f_{\tilde{l}ik\tilde{j}}}{2}\right)\frac{\partial U_i}{\partial x_{\tilde{j}}}\frac{\partial P_k}{\partial x_{\tilde{l}}}=\left|\begin{array}{l}\text{omit}\\ \text{tilda}\end{array}\right|=\int\limits_{V\to\infty} d^3r\left(\frac{f_{ilkj}+f_{likj}}{2}\right)\frac{\partial U_i}{\partial x_j}\frac{\partial P_k}{\partial x_l}\\
&=\frac{1}{2}\int\limits_{V\to\infty} d^3r\left(f_{ilkj}\frac{\partial U_i}{\partial x_j}+f_{ljki}\frac{\partial U_j}{\partial x_i}\right)\frac{\partial P_k}{\partial x_l}
\end{aligned}
\qquad (A.2)
$$

Comparing underlined by "blue" and "red" steps of the transformations (A.2) for an arbitrary coordinate-dependent function $\frac{\partial U_i}{\partial x_j}\frac{\partial P_k}{\partial x_l}$, one leads to the relation

$$f_{ijkl} + f_{jikl} = f_{ilkj} + f_{likj}, \tag{A.3}$$

since $f_{ijkl} \equiv f_{jikl}$ and $f_{ilkj} \equiv f_{likj}$, due to the symmetry of the strain tensor, $u_{ij} \equiv u_{ji}$, Eq. (A.3) elementary leads to Eqs. (3a), (3b):

$$f_{ijkl} = f_{jikl} = f_{ilkj} = f_{likj} \tag{A.4}$$

Hence we proved that besides the relation $f_{ijkl} \equiv f_{jikl}$ there is one more relation $f_{ijkl} = f_{ilkj}$, Eqs. (3a), (3b), that imposes additional constrains on the structure of the flexoelectric tensor.

Appendix B. Structure of flexoelectric tensor allowing for point symmetry, "evident" index-permutation symmetry and "hidden" index-permutation symmetry

B1 Structure of flexoelectric tensor (FET) allowing for point symmetry (PS) and "evident" index-permutation symmetry (EIPS)

Table B1 Number of elements of EIPS FET for different point groups.

Point group	Nonzero	Independent	Nonzero element and relations between them
m3m, 432 $\overline{4}$3m	21	3	$\overline{f}_{1111}=\overline{f}_{2222}=\overline{f}_{3333}$; $\overline{f}_{1122}=\overline{f}_{2211}=\overline{f}_{1133}=\overline{f}_{3311}=\overline{f}_{2233}=\overline{f}_{3322}$; $\overline{f}_{1212}=\overline{f}_{2112}=\overline{f}_{1221}=\overline{f}_{2121}=$ $\overline{f}_{1313}=\overline{f}_{3113}=\overline{f}_{1331}=\overline{f}_{3131}$ $\overline{f}_{2323}=\overline{f}_{3223}=\overline{f}_{2332}=\overline{f}_{3232}$;
23, m3	21	5	$\overline{f}_{1111}=\overline{f}_{2222}=\overline{f}_{3333}$; $\overline{f}_{1122}=\overline{f}_{2233}=\overline{f}_{3311}$; $\overline{f}_{1212}=\overline{f}_{2112}=\overline{f}_{1331}=\overline{f}_{3131}=\overline{f}_{2323}=\overline{f}_{3223}$; $\overline{f}_{1133}=\overline{f}_{2211}=\overline{f}_{3322}$; $\overline{f}_{1221}=\overline{f}_{2121}=\overline{f}_{1313}=\overline{f}_{3113}=\overline{f}_{2332}=\overline{f}_{3232}$
$\overline{6}m2$, 622, 6mm, 6/mmm	21	7	$\overline{f}_{1111}=\overline{f}_{2222}$; $\overline{f}_{1122}=\overline{f}_{2211}$; $\overline{f}_{1133}=\overline{f}_{2233}$; $\overline{f}_{1212}=\overline{f}_{2112}=\overline{f}_{1221}=\overline{f}_{2121}=\frac{\overline{f}_{1111}}{2}-\frac{\overline{f}_{1122}}{2}$ $\overline{f}_{1313}=\overline{f}_{3113}=\overline{f}_{2323}=\overline{f}_{2323}$; $\overline{f}_{1331}=\overline{f}_{3131}=\overline{f}_{2332}=\overline{f}_{3232}$; $\overline{f}_{3311}=\overline{f}_{3322}$; $\overline{f}_{3333}$

6, $\overline{6}$, 6/m	39	12	$\bar{f}_{1111}=\bar{f}_{2222};\bar{f}_{1112}=-\bar{f}_{2221};$ $\bar{f}_{1121}=-\bar{f}_{2212};\bar{f}_{1122}=\bar{f}_{2211};\bar{f}_{1133}=\bar{f}_{2233};$ $\bar{f}_{1211}=\bar{f}_{2111}=-\bar{f}_{1222}=-\bar{f}_{2122}=-\frac{\bar{f}_{1112}}{2}-\frac{\bar{f}_{1121}}{2};$ $\bar{f}_{1212}=\bar{f}_{2112}=\bar{f}_{1221}=\bar{f}_{2121}=\frac{\bar{f}_{1111}}{2}-\frac{\bar{f}_{1122}}{2};$ $\bar{f}_{1313}=\bar{f}_{3113}=\bar{f}_{2323}=\bar{f}_{2323};$ $\bar{f}_{1323}=\bar{f}_{3123}=-\bar{f}_{2313}=-\bar{f}_{3213};$ $\bar{f}_{1331}=\bar{f}_{3131}=\bar{f}_{2332}=\bar{f}_{3232};$ $\bar{f}_{1332}=\bar{f}_{3132}=-\bar{f}_{2331}=-\bar{f}_{3231};$ $\bar{f}_{3312}=-\bar{f}_{3321};\bar{f}_{3311}=\bar{f}_{3322};\bar{f}_{3333}$
3, $\overline{3}$	71	18	$\bar{f}_{1111}=\bar{f}_{2222};\bar{f}_{1112}=-\bar{f}_{2221};\bar{f}_{1113}=-\bar{f}_{1223}=-\bar{f}_{2123}=-\bar{f}_{2213};$ $\bar{f}_{1121}=-\bar{f}_{2212};\bar{f}_{1122}=\bar{f}_{2211};\bar{f}_{1123}=\bar{f}_{1213}=\bar{f}_{2113}=-\bar{f}_{2223};$ $\bar{f}_{1131}=-\bar{f}_{1232}=-\bar{f}_{2132}=-\bar{f}_{2231};$ $\bar{f}_{1132}=\bar{f}_{1231}=\bar{f}_{2131}=-\bar{f}_{2232};\bar{f}_{1133}=\bar{f}_{2233};$ $\bar{f}_{1211}=\bar{f}_{2111}=-\bar{f}_{1222}=-\bar{f}_{2122}=-\frac{\bar{f}_{1112}}{2}-\frac{\bar{f}_{1121}}{2};$ $\bar{f}_{1212}=\bar{f}_{2112}=\bar{f}_{1221}=\bar{f}_{2121}=\frac{\bar{f}_{1111}}{2}-\frac{\bar{f}_{1122}}{2};$ $\bar{f}_{1311}=\bar{f}_{3111}=-\bar{f}_{1322}=-\bar{f}_{3122}=-\bar{f}_{2321}=-\bar{f}_{3221}=-\bar{f}_{2312}=-\bar{f}_{3212};$ $\bar{f}_{1312}=\bar{f}_{1321}=\bar{f}_{2311}=\bar{f}_{3112}=\bar{f}_{3121}=\bar{f}_{3211}=-\bar{f}_{2322}=-\bar{f}_{3222};$ $\bar{f}_{1313}=\bar{f}_{3113}=\bar{f}_{2323}=\bar{f}_{2323};$ $\bar{f}_{1323}=\bar{f}_{3123}=-\bar{f}_{2313}=-\bar{f}_{3213};$ $\bar{f}_{1331}=\bar{f}_{3131}=\bar{f}_{2332}=\bar{f}_{3232};$ $\bar{f}_{1332}=\bar{f}_{3132}=-\bar{f}_{2331}=-\bar{f}_{3231};$ $\bar{f}_{3312}=-\bar{f}_{3321};\bar{f}_{3311}=\bar{f}_{3322};\bar{f}_{3333}$

Continued

Table B1 Number of elements of EIPS FET for different point groups.—cont'd

Point group	Nonzero	Independent	Nonzero element and relations between them
32, 3m, $\overline{3}m$	37	10	$\overline{f}_{1111}=\overline{f}_{2222};\overline{f}_{1113}=-\overline{f}_{1223}=-\overline{f}_{2123}=-\overline{f}_{2213};$ $\overline{f}_{1122}=\overline{f}_{2211};\overline{f}_{1131}=-\overline{f}_{1232}=-\overline{f}_{2132}=-\overline{f}_{2231};\overline{f}_{1133}=\overline{f}_{2233};$ $\overline{f}_{1212}=\overline{f}_{2112}=\overline{f}_{1221}=\overline{f}_{2121}=\frac{\overline{f}_{1111}}{2}-\frac{\overline{f}_{1122}}{2};$ $\overline{f}_{1311}=\overline{f}_{3111}=-\overline{f}_{1322}=-\overline{f}_{3122}=-\overline{f}_{2321}=-\overline{f}_{3221}=-\overline{f}_{2312}=-\overline{f}_{3212};$ $\overline{f}_{1313}=\overline{f}_{3113}=\overline{f}_{2323}=\overline{f}_{2323};$ $\overline{f}_{1331}=\overline{f}_{3131}=\overline{f}_{2332}=\overline{f}_{3232};$ $\overline{f}_{3311}=\overline{f}_{3322};\overline{f}_{3333}$
$\overline{4}\,2m$, 422, 4mm, 4/mmm	21	8	$\overline{f}_{1111}=\overline{f}_{2222};\overline{f}_{3333};$ $\overline{f}_{1122}=\overline{f}_{2211};\overline{f}_{1133}=\overline{f}_{2233};\overline{f}_{3311}=\overline{f}_{3322};$ $\overline{f}_{1212}=\overline{f}_{2112}=\overline{f}_{1221}=\overline{f}_{2121};$ $\overline{f}_{1313}=\overline{f}_{3113}=\overline{f}_{2323}=\overline{f}_{3223};$ $\overline{f}_{1331}=\overline{f}_{3131}=\overline{f}_{2332}=\overline{f}_{3232}$
4,$\overline{4}$, 4/m	39	14	$\overline{f}_{1111}=\overline{f}_{2222};\overline{f}_{1122}=\overline{f}_{2211};\overline{f}_{1133}=\overline{f}_{2233};\overline{f}_{3311}=\overline{f}_{3322};$ $\overline{f}_{1212}=\overline{f}_{2112}=\overline{f}_{1221}=\overline{f}_{2121};$ $\overline{f}_{1313}=\overline{f}_{3113}=\overline{f}_{2323}=\overline{f}_{3223};$ $\overline{f}_{1331}=\overline{f}_{3131}=\overline{f}_{2332}=\overline{f}_{3232};$ $\overline{f}_{1211}=\overline{f}_{2111}=-\overline{f}_{1222}=-\overline{f}_{2122};$ $\overline{f}_{1323}=\overline{f}_{3123}=-\overline{f}_{2313}=-\overline{f}_{3213}$ $\overline{f}_{1332}=\overline{f}_{3132}=-\overline{f}_{2331}=-\overline{f}_{3231};$ $\overline{f}_{1112}=-\overline{f}_{2221};\overline{f}_{1121}=-\overline{f}_{2212};$ $\overline{f}_{3312}=-\overline{f}_{3321};\overline{f}_{3333}$

222, mm2, mmm	21	15	$\overline{f}_{1111};\overline{f}_{1122};\overline{f}_{1133};$ $\overline{f}_{2211};\overline{f}_{2222};\overline{f}_{2233};$ $\overline{f}_{3311};\overline{f}_{3322};\overline{f}_{3333};$ $\overline{f}_{1212}=\overline{f}_{2112};\overline{f}_{1221}=\overline{f}_{2121};$ $\overline{f}_{1313}=\overline{f}_{3113};\overline{f}_{1331}=\overline{f}_{3131};$ $\overline{f}_{2323}=\overline{f}_{3223};\overline{f}_{2332}=\overline{f}_{3232}$
2, m, 2/m	41	28	$\overline{f}_{1111};\overline{f}_{1112};\overline{f}_{1121};\overline{f}_{1122};\overline{f}_{1133};$ $\overline{f}_{2211};\overline{f}_{2212};\overline{f}_{2221};\overline{f}_{2222};\overline{f}_{2233};$ $\overline{f}_{3311};f_{3312};\overline{f}_{3321};\overline{f}_{3322};\overline{f}_{3333};$ $\overline{f}_{1212}=\overline{f}_{2112};\overline{f}_{1221}=\overline{f}_{2121};\overline{f}_{1211}=\overline{f}_{2111};\overline{f}_{1222}=\overline{f}_{2122};\overline{f}_{1233}=\overline{f}_{2133};$ $\overline{f}_{1313}=\overline{f}_{3113};\overline{f}_{1331}=\overline{f}_{3131};\overline{f}_{1323}=\overline{f}_{3123};\overline{f}_{1332}=\overline{f}_{3132};$ $\overline{f}_{2313}=\overline{f}_{3213};\overline{f}_{2323}=\overline{f}_{3223};\overline{f}_{2331}=\overline{f}_{3231};\overline{f}_{2332}=\overline{f}_{3232}$
1, $\overline{1}$	81	54	$\overline{f}_{1111},\overline{f}_{1112},\overline{f}_{1113},\overline{f}_{1121},\overline{f}_{1122},\overline{f}_{1123},\overline{f}_{1131},\overline{f}_{1132},\overline{f}_{1133};$ $\overline{f}_{1211}=\overline{f}_{2111},\overline{f}_{1212}=\overline{f}_{2112},\overline{f}_{1213}=\overline{f}_{2113},$ $\overline{f}_{1221}=\overline{f}_{2121},\overline{f}_{1222}=\overline{f}_{2122},\overline{f}_{1223}=\overline{f}_{2123},$ $\overline{f}_{1231}=\overline{f}_{2131},\overline{f}_{1232}=\overline{f}_{2132},\overline{f}_{1233}=\overline{f}_{2133},$ $\overline{f}_{1311}=\overline{f}_{3111},\overline{f}_{1312}=\overline{f}_{3112},\overline{f}_{1313}=\overline{f}_{3113},$ $\overline{f}_{1321}=\overline{f}_{3121},\overline{f}_{1322}=\overline{f}_{3122},\overline{f}_{1323}=\overline{f}_{3123},$ $\overline{f}_{1331}=\overline{f}_{3131},\overline{f}_{1332}=\overline{f}_{3132},\overline{f}_{1333}=\overline{f}_{3133},$ $\overline{f}_{2211},\overline{f}_{2212},\overline{f}_{2213},\overline{f}_{2221},\overline{f}_{2222},\overline{f}_{2223},\overline{f}_{2231},\overline{f}_{2232},\overline{f}_{2233};$ $\overline{f}_{2311}=\overline{f}_{3211},\overline{f}_{2312}=\overline{f}_{3212},\overline{f}_{2313}=\overline{f}_{3213},$ $\overline{f}_{2321}=\overline{f}_{3221},\overline{f}_{2322}=\overline{f}_{3222},\overline{f}_{2323}=\overline{f}_{3223},$ $\overline{f}_{2331}=\overline{f}_{3231},\overline{f}_{2332}=\overline{f}_{3232},\overline{f}_{2333}=\overline{f}_{3233},$ $\overline{f}_{3311},\overline{f}_{3312},\overline{f}_{3313},\overline{f}_{3321},\overline{f}_{3322},\overline{f}_{3323},\overline{f}_{3331},\overline{f}_{3332},\overline{f}_{3333}$

B2 Structure of flexoelectric tensor allowing for point symmetry (PS) and "hidden" index-permutation symmetry (HIPS)

Table B2 Number of elements of HIPS FET for different point groups.

Point group	Nonzero	Independent	Nonzero element and relations between them
m3m, 432, $\overline{4}3m$	21	2	$f_{1111}=f_{2222}=f_{3333}$; $f_{1122}=f_{2211}=f_{1133}=f_{3311}=f_{2233}=f_{3322}=$ $f_{1221}=f_{2112}=f_{1331}=f_{3113}=f_{2332}=f_{3223}=$ $f_{1212}=f_{2121}=f_{1313}=f_{3131}=f_{2323}=f_{3232}$
23, m3	21	3	$f_{1111}=f_{2222}=f_{3333}$; $f_{1122}=f_{1221}=f_{2121}=f_{2233}=f_{2332}=f_{3232}=f_{3311}=f_{3113}=f_{1313}$; $f_{2211}=f_{2112}=f_{1212}=f_{3322}=f_{3223}=f_{2323}=f_{1133}=f_{1331}=f_{3131}$
$\overline{6}m2$, 622, 6mm, 6/mmm	21	4	$f_{1111}=f_{2222}$; f_{3333}; $f_{1122}=f_{1221}=f_{2121}=f_{2211}=f_{2112}=f_{1212}=\frac{f_{1111}}{3}$; $f_{1133}=f_{1331}=f_{3131}=f_{2233}=f_{2332}=f_{3232}$; $f_{1313}=f_{3113}=f_{3311}=f_{2323}=f_{3223}=f_{3322}$
6, $\overline{6}$, 6/m	35	6	$f_{1111}=f_{2222}$; f_{3333}; $f_{1121}=-f_{2212}, f_{1222}=f_{2122}=f_{2221}=\frac{f_{1121}}{3}=-f_{1112}=-f_{1211}=-f_{2111}$; $f_{1122}=f_{1221}=f_{2121}=f_{2211}=f_{2112}=f_{1212}=\frac{f_{1111}}{3}$; $f_{1133}=f_{1331}=f_{3131}=f_{2233}=f_{2332}=f_{3232}$; $f_{1313}=f_{3311}=f_{3113}=f_{3322}=f_{3223}=f_{2323}$; $f_{2313}=f_{3213}=f_{3312}, f_{1323}=f_{3123}=f_{3321}=-f_{2313}$;

$3, \overline{3}$	67	10	$f_{1111}=f_{2222}; f_{3333};$ $f_{1121}=-f_{2212},$ $f_{1112}=f_{1211}=f_{2111}=-\frac{f_{1121}}{3}=-f_{1222}=-f_{2122}=-f_{2221};$ $f_{1122}=f_{1221}=f_{2121}=f_{2211}=f_{2112}=f_{1212}=\frac{f_{1111}}{3}$ $f_{1113}=f_{1311}=f_{3111};$ $f_{1223}=f_{1322}=f_{2123}=f_{2321}=f_{3122}=f_{3212}=f_{2213}=f_{2312}=f_{3221}=-f_{1113};$ $f_{1123}=f_{1321}=f_{3121}=f_{1213}=f_{1312}=f_{2113}=f_{2311}=f_{3112}=f_{3211};$ $f_{2223}=f_{2322}=f_{3222}=-f_{1123};$ $f_{1131}=-f_{1232}=-f_{2132}=-f_{2231};$ $f_{1132}=f_{1231}=f_{2131}=-f_{2232};$ $f_{1133}=f_{2233}=f_{1331}=f_{2332}=f_{3131}=f_{3232};$ $f_{1313}=f_{3113}=f_{3311}=f_{2323}=f_{3223}=f_{3322};$ $f_{1323}=f_{3123}=f_{3321}=-f_{2313}=-f_{3213}=-f_{3312};$
$32, 3m, \overline{3}m$	37	6	$f_{1111}=f_{2222}; f_{3333};$ $f_{1122}=f_{1221}=f_{2121}=f_{2211}=f_{2112}=f_{1212}=\frac{f_{1111}}{3};$ $f_{1113}=f_{1311}=f_{3111}=-f_{1223}=-f_{1322}=-f_{2123}=-f_{2321}=-f_{3122}=-f_{3221}=-f_{2213}=-f_{2312}=-f_{3212};$ $f_{1131}, f_{1232}=f_{2132}=f_{2231}=-f_{1131};$ $f_{1133}=f_{2233}=f_{1331}=f_{2332}=f_{3131}=f_{3232};$ $f_{1313}=f_{3113}=f_{3311}=f_{2323}=f_{3223}=f_{3322};$
$\overline{4}2m, 422,$ 4mm, 4/mmm	21	5	$f_{1111}=f_{2222}; f_{3333};$ $f_{1122}=f_{1221}=f_{2121}=f_{2211}=f_{2112}=f_{1212};$ $f_{1133}=f_{1331}=f_{3131}=f_{2233}=f_{3232}=f_{2332};$ $f_{1313}=f_{3311}=f_{3113}=f_{2323}=f_{3322}=f_{3223};$

Continued

Table B2 Number of elements of HIPS FET for different point groups.—cont'd

Point group	Nonzero	Independent	Nonzero element and relations between them
4,$\overline{4}$, 4/m	35	8	$f_{1111}=f_{2222}; f_{3333};$ $f_{1122}=f_{1221}=f_{2121}=f_{2211}=f_{2112}=f_{1212};$ $f_{1133}=f_{1331}=f_{3131}=f_{2233}=f_{2332}=f_{3232};$ $f_{3311}=f_{3113}=f_{1313}=f_{3322}=f_{3223}=f_{2323};$ $f_{1121}=-f_{2212};$ $f_{1112}=f_{1211}=f_{2111}=-f_{1222}=-f_{2122}=-f_{2221};$ $f_{2313}=f_{3213}=f_{3312}=-f_{1323}=-f_{3123}=-f_{3321};$
222, mm2, mmm	21	9	$f_{1111}, f_{2222}, f_{3333};$ $f_{1122}=f_{1221}=f_{2121}; f_{2233}=f_{2332}=f_{3232}; f_{3311}=f_{3113}=f_{1313};$ $f_{2211}=f_{2112}=f_{1212}; f_{3322}=f_{3223}=f_{2323}; f_{1133}=f_{1331}=f_{3131};$
2, m, 2/m	41	16	$f_{1111}, f_{2222}, f_{3333};$ $f_{1122}=f_{1221}=f_{2121}; f_{2211}=f_{2112}=f_{1212};$ $f_{1112}=f_{1211}=f_{2111}; f_{1222}=f_{2122}=f_{2221};$ $f_{1133}=f_{1331}=f_{3131}; f_{1313}=f_{3113}=f_{3311};$ $f_{1233}=f_{2133}=f_{2331}=f_{1332}=f_{3132}=f_{3231};$ $f_{1323}=f_{3123}=f_{3321}; f_{2313}=f_{3213}=f_{3312};$ $f_{2233}=f_{2332}=f_{3232}; f_{2323}=f_{3223}=f_{3322};$ $f_{1121}; f_{2212};$

1, $\bar{1}$	81	30	$f_{1111}, f_{1112}=f_{1211}=f_{2111}, f_{1113}=f_{1311}=f_{3111};$ $f_{1121}, f_{1122}=f_{1221}=f_{2121}, f_{1123}=f_{1321}=f_{3121};$ $f_{1131}, f_{1132}=f_{1231}=f_{2131}, f_{1133}=f_{1331}=f_{3131};$ $f_{1212}=f_{2112}=f_{2211}, f_{1213}=f_{1312}=f_{3112}=f_{3211}=f_{2311}=f_{2113};$ $f_{1222}=f_{2122}=f_{2221}, f_{1223}=f_{1322}=f_{3122}=f_{3221}=f_{2321}=f_{2123};$ $f_{1232}=f_{2132}=f_{2231}, f_{1233}=f_{1332}=f_{3132}=f_{3231}=f_{2331}=f_{2133};$ $f_{1313}=f_{3113}=f_{3311}, f_{1323}=f_{3123}=f_{3321}, f_{1333}=f_{3133}=f_{3331};$ $f_{2212}, f_{2213}=f_{2312}=f_{3212};$ $f_{2222}, f_{2223}=f_{2322}=f_{3222};$ $f_{2232}, f_{2233}=f_{2332}=f_{3232};$ $f_{2313}=f_{3213}=f_{3312}, f_{2323}=f_{3223}=f_{3322}, f_{2333}=f_{3233}=f_{3332};$ $f_{3313}, f_{3323}, f_{3333}.$

B3 Comparison of EIPS FET $\bar{f}_{ijkl}$ and HIPS FET f_{ijkl} for different point groups

Table B3 The comparison of EIPS FET $\bar{f}_{ijkl}$ and HIPS FET f_{ijkl} for different point groups.

	EIPS FET		HIPS FET		
Point group	**Nonzero**	**Independent**	**Nonzero**	**Independent**	**"Difference" between two tensors**
m3m, 432 $\bar{4}$3m	21	3	21	2	$\bar{f}_{1122} \neq \bar{f}_{1212}; f_{1122} = f_{1212}$
23, m3	21	5	21	3	$\bar{f}_{1122} \neq \bar{f}_{1221}, \bar{f}_{2211} \neq \bar{f}_{1212}; f_{1122} = f_{1221}, f_{2211} = f_{1212}$
$\bar{6}m2$, 622, 6mm, 6/mmm	21	7	21	4	$\bar{f}_{1212} = \bar{f}_{2112} = \bar{f}_{1221} = \bar{f}_{2121} = \frac{\bar{f}_{1111}}{2} - \frac{\bar{f}_{1122}}{2} \neq \bar{f}_{1122}$ $\bar{f}_{1133} \neq \bar{f}_{1331}, \bar{f}_{2233} \neq \bar{f}_{2332},$ $f_{1221} = f_{2121} = f_{2211} = f_{2112} = f_{1212} = \frac{f_{1111}}{3} = f_{1122}$ $f_{1133} = f_{1331}, f_{2233} = f_{2332},$
6, $\bar{6}$, 6/m	39	12	35	6	$\bar{f}_{1122} \neq \bar{f}_{1221} \neq \frac{\bar{f}_{1111}}{3}; \bar{f}_{1133} \neq \bar{f}_{1331}; \bar{f}_{3311} \neq \bar{f}_{1313}; \bar{f}_{3312} \neq \bar{f}_{3213}$ $\bar{f}_{1332} = \bar{f}_{3132} = -\bar{f}_{2331} = -\bar{f}_{3231} \neq 0$ $f_{1122} = f_{1221} = \frac{f_{1111}}{3}; f_{1133} = f_{1331}; f_{1313} = f_{3311}; f_{3312} = f_{3213}$ $f_{1332} = f_{3132} = f_{2331} = f_{3231} = 0$
3, $\bar{3}$	71	18	67	10	$\bar{f}_{1112} \neq \bar{f}_{1211} = -\frac{\bar{f}_{1112}}{2} - \frac{\bar{f}_{1121}}{2}; \bar{f}_{1113} \neq \bar{f}_{1311};$ $\bar{f}_{1122} \neq \bar{f}_{1221} = \frac{\bar{f}_{1111}}{2} - \frac{\bar{f}_{1122}}{2}; \bar{f}_{1123} \neq \bar{f}_{1321};$ $\bar{f}_{1133} \neq \bar{f}_{1331}; \bar{f}_{3311} \neq \bar{f}_{3113}; \bar{f}_{3123} \neq \bar{f}_{3321}$ $\bar{f}_{1332} = \bar{f}_{3132} = -\bar{f}_{2331} = -\bar{f}_{3231} \neq 0;$ $f_{1112} = f_{1211} = -\frac{f_{1121}}{3}; f_{1113} = f_{1311};$ $f_{1122} = f_{1221} = \frac{f_{1111}}{3}; f_{1123} = f_{1321}$ $f_{1133} = f_{1331}; f_{3311} = f_{3113}; f_{3123} = f_{3321}$ $f_{1332} = f_{3132} = f_{2331} = f_{3231} = 0$

32, 3m, $\overline{3}m$	37	10	37	6	$\overline{f}_{1122} \neq \overline{f}_{1221} = \frac{\overline{f}_{1111}}{2} - \frac{\overline{f}_{1122}}{2}; \overline{f}_{1133} \neq \overline{f}_{1331};$ $\overline{f}_{1113} \neq \overline{f}_{1311}; \overline{f}_{1313} \neq \overline{f}_{3311};$ $f_{1122} = f_{1221} = \frac{f_{1111}}{3}; f_{1133} = f_{1331};$ $f_{1113} = f_{1311}; f_{1313} = f_{3311}$
$\overline{4}2m$, 422, 4mm, 4/mmm	21	8	21	5	$\overline{f}_{1122} \neq \overline{f}_{1221}, \overline{f}_{1133} \neq \overline{f}_{1331}, \overline{f}_{2233} \neq \overline{f}_{2332},$ $f_{1122} = f_{1221}, f_{1133} = f_{1331}, f_{2233} = f_{2332}$
4,$\overline{4}$, 4/m	39	14	35	8	$\overline{f}_{1122} \neq \overline{f}_{1221}; \overline{f}_{1133} \neq \overline{f}_{1331}; \overline{f}_{3311} \neq \overline{f}_{1313};$ $\overline{f}_{1211} \neq \overline{f}_{1112}; \overline{f}_{3213} \neq \overline{f}_{3312}$ $\overline{f}_{1332} = \overline{f}_{3132} = -\overline{f}_{2331} = -\overline{f}_{3231} \neq 0;$ $f_{1122} = f_{1221}; f_{1133} = f_{1331}; f_{3311} = f_{1313};$ $f_{1112} = f_{1211}; f_{3213} = f_{3312}$ $f_{1332} = f_{3132} = f_{2331} = f_{3231} = 0$
222, mm2, mmm	21	15	21	9	$\overline{f}_{1122} \neq \overline{f}_{1221}, \overline{f}_{1133} \neq \overline{f}_{1331}, \overline{f}_{2233} \neq \overline{f}_{2332},$ $\overline{f}_{2211} \neq \overline{f}_{2112}, \overline{f}_{3311} \neq \overline{f}_{3113}, \overline{f}_{3322} \neq \overline{f}_{3223};$ $f_{1122} = f_{1221}, f_{1133} = f_{1331}, f_{2233} = f_{2332},$ $f_{2211} = f_{2112}, f_{3311} = f_{3113}, f_{3322} = f_{3223}$
2, m, 2/m	41	28	41	16	$\overline{f}_{1112} \neq \overline{f}_{1211}; \overline{f}_{1122} \neq \overline{f}_{1221}; \overline{f}_{1133} \neq \overline{f}_{1331}; \overline{f}_{1222} \neq \overline{f}_{2221};$ $\overline{f}_{1233} \neq \overline{f}_{1332} \neq \overline{f}_{2331} \neq \overline{f}_{1233}; \overline{f}_{1313} \neq \overline{f}_{3311}; \overline{f}_{1323} \neq \overline{f}_{3321};$ $\overline{f}_{2211} \neq \overline{f}_{2112}; \overline{f}_{2233} \neq \overline{f}_{2332}; \overline{f}_{2313} \neq \overline{f}_{3312}; \overline{f}_{2323} \neq \overline{f}_{3322};$ $f_{1112} = f_{1211}; f_{1122} = f_{1221}; f_{1133} = f_{1331}; f_{1222} = f_{2122};$ $f_{1233} = f_{2331} = f_{1332}; f_{1313} = f_{3311}; f_{1323} = f_{3321};$ $f_{2211} = f_{2112}; f_{2233} = f_{2332}; f_{2313} = f_{3312}; f_{2323} = f_{3322};$

Continued

Table B3 The comparison of EIPS FET $\bar{f}_{ijkl}$ and HIPS FET f_{ijkl} for different point groups.—cont'd

Point group	EIPS FET		HIPS FET		"Difference" between two tensors
	Nonzero	Independent	Nonzero	Independent	
$1, \bar{1}$	81	54	81	30	$\bar{f}_{1112} \neq \bar{f}_{1211}, \bar{f}_{1113} \neq \bar{f}_{1311}, \bar{f}_{1122} \neq \bar{f}_{1221}, \bar{f}_{1123} \neq \bar{f}_{1321},$ $\bar{f}_{1132} \neq \bar{f}_{1231}, \bar{f}_{1133} \neq \bar{f}_{1331}, \bar{f}_{1212} \neq \bar{f}_{2211},$ $\bar{f}_{1213} \neq \bar{f}_{1312}, \bar{f}_{1213} \neq \bar{f}_{2311}, \bar{f}_{1222} \neq \bar{f}_{2221}, \bar{f}_{1223} \neq \bar{f}_{1322},$ $\bar{f}_{1223} \neq \bar{f}_{2321}, \bar{f}_{1232} \neq \bar{f}_{2231},$ $\bar{f}_{1233} \neq \bar{f}_{1332}, \bar{f}_{1233} \neq \bar{f}_{2331},$ $\bar{f}_{1312} \neq \bar{f}_{2311}, \bar{f}_{1313} \neq \bar{f}_{3311}, \bar{f}_{1323} \neq \bar{f}_{3321}, \bar{f}_{1322} \neq \bar{f}_{2321}$ $\bar{f}_{1332} \neq \bar{f}_{2331}, \bar{f}_{1333} \neq \bar{f}_{3331},$ $\bar{f}_{2213} \neq \bar{f}_{2312}, \bar{f}_{2223} \neq \bar{f}_{2322}, \bar{f}_{2233} \neq \bar{f}_{2332};$ $\bar{f}_{2313} \neq \bar{f}_{3312}, \bar{f}_{2323} \neq \bar{f}_{3322}, \bar{f}_{2333} \neq \bar{f}_{3332}$ $f_{1112} = f_{1211}, f_{1113} = f_{1311}, f_{1122} = f_{1221}, f_{1123} = f_{1321};$ $f_{1132} = f_{1231}, f_{1133} = f_{1331}; f_{1212} = f_{2211}, f_{1222} = f_{2221},$ $f_{1213} = f_{1312} = f_{3112} = f_{3211} = f_{2311} = f_{2113}, f_{1232} = f_{2231},$ $f_{1223} = f_{1322} = f_{3122} = f_{3221} = f_{2321} = f_{2123}, f_{1323} = f_{3321},$ $f_{1233} = f_{1332} = f_{3132} = f_{3231} = f_{2331} = f_{2133}, f_{1333} = f_{3331},$ $f_{1313} = f_{3311}, f_{2213} = f_{2312}, f_{2223} = f_{2322}, f_{2233} = f_{2332},$ $f_{2313} = f_{3312}, f_{2323} = f_{3322}, f_{2333} = f_{3332}$

Appendix C. Effective flexoelectric response of the plate

The coordinate frame (see Fig. 1A) transformation is

$$\begin{aligned}\widetilde{X}_1 &= \cos(\alpha)X_1 + \sin(\alpha)X_3\\ \widetilde{X}_3 &= -\sin(\alpha)X_1 + \cos(\alpha)X_3\end{aligned} \tag{C.1}$$

For 4/mmm point symmetry group transformation yields

$$\widetilde{f}_{1133} = f_{1133} + \sin^2(\alpha)\begin{pmatrix} f_{3311} - f_{1133} + \\ \cos^2(\alpha)(f_{1111} - f_{1133} + f_{3333} - f_{3311} - 2(f_{1313} + f_{1331})) \end{pmatrix} \tag{C.2a}$$

$$\widetilde{f}_{2233} = f_{1122} + \cos^2(\alpha)(f_{1133} - f_{1122}) \tag{C.2b}$$

$$\widetilde{f}_{1333} = \cos(\alpha)\sin(\alpha)\begin{pmatrix} +\cos^2(\alpha)(f_{3333} - f_{1133} - f_{1313} - f_{1331}) \\ -\sin^2(\alpha)(f_{1111} - f_{3311} - f_{1313} - f_{1331}) \end{pmatrix} \tag{C.2c}$$

$$\widetilde{f}_{3333} = f_{3333} + \sin^2(\alpha)\begin{pmatrix} f_{1111} - f_{3333} - \\ \cos^2(\alpha)(f_{1111} - f_{1133} + f_{3333} - f_{3311} - 2(f_{1313} + f_{1331})) \end{pmatrix} \tag{C.2d}$$

For m3m point symmetry group transformation yields

$$\widetilde{f}_{1133} = f_{1122} + 2\sin^2(\alpha)\cos^2(\alpha)(f_{1111} - f_{1122} - (f_{1212} + f_{1221})) \tag{C.3a}$$

$$\begin{aligned}\widetilde{f}_{1313} &= f_{1212}\\ &\quad - \sin^2(\alpha)(f_{1212} - f_{1221}) + 2\sin^2(\alpha)\cos^2(\alpha)(f_{1111} - f_{1122} - (f_{1212} + f_{1221}))\end{aligned} \tag{C.3b}$$

$$\begin{aligned}\widetilde{f}_{1331} &= f_{1221}\\ &\quad - \sin^2(\alpha)(f_{1221} - f_{1212}) + 2\sin^2(\alpha)\cos^2(\alpha)(f_{1111} - f_{1122} - (f_{1212} + f_{1221}))\end{aligned} \tag{C.3c}$$

$$\widetilde{f}_{2233} = f_{1122} \tag{C.3d}$$

$$\widetilde{f}_{1333} = \cos(\alpha)\sin(\alpha)\left(\cos^2(\alpha) - \sin^2(\alpha)\right)(f_{1111} - f_{1122} - f_{1212} - f_{1221}) \tag{C.3e}$$

$$\widetilde{f}_{3333} = f_{1111} - 2\sin^2(\alpha)\cos^2(\alpha)(f_{1111} - f_{1122} - (f_{1212} + f_{1221})) \tag{C.3f}$$

Below we consider the bending problem for a plate (see Fig. 1B), having developed surface rotated on angle α from principal (pseudo-) cubic direction (see Fig. 1A). Below we omit the tilde sings for clarity and consider only normal component of polarization, P_3. Equations of state could be obtained after the variation of the corresponding thermodynamic potential in the following form

$$\sigma_{ij} = c_{ijkl} u_{kl} + f_{ijkl} P_{k,l} \tag{C.4a}$$

$$\alpha_3 P_3 - g_{3333} P_{3,33} - f_{ij3l} u_{ij,l} = E_3 \tag{C.4b}$$

Using compatibility conditions and considering the case of strain field, depending only on X_3, one could get the following restrictions on the strain tensor components:

$$u_{11,33} = u_{12,33} = u_{22,33} = 0 \tag{C.5}$$

while other components could have an arbitrary dependence on X_3. The relations (C.5a) means that three components of strain tensor have linear dependences

$$u_{11} = u_{11}^{(0)} + \frac{X_3}{R_1} \tag{C.6a}$$

$$u_{12} = u_{12}^{(0)} \tag{C.6b}$$

$$u_{22} = u_{22}^{(0)} + \frac{X_3}{R_2} \tag{C.6c}$$

Here constants, $u_{11}^{(0)}$, $u_{11}^{(0)}$, $u_{11}^{(0)}$, R_1 and R_2 should be determined from boundary conditions at side faces of plate (usually in Saint-Venant approximation). However, as far as we are interested in the strain gradient estimation in order to get expression for polarization from (C.4b), "an exact" expressions for constants from Eqs. (C.6a), (C.6b), (C.6c) are not necessary, since radii of plate curvature R_1 and R_2 could be estimated from the shape of the stressed plate. Next we recall boundary conditions for stresses at the developed surface of the plate, namely $\sigma_{ij} n_j |_S = 0$, and condition of mechanical equilibrium, $\sigma_{ij,j} = 0$. For the case of 1D dependences these two relations give the following

$$\sigma_{13} = \sigma_{23} = \sigma_{33} = 0 \tag{C.7}$$

Finally, taking into account Eqs. (C.2a)–(C.2d), (C.3a)–(C.3f), and (C.6a)–(C.6c), one could get the evident form of equations of state, $\sigma_{13} = 2c_{1313} u_{13} + c_{1333} u_{33} + f_{1333} P_{3,3}$ and $\sigma_{33} = c_{1133} u_{11} + 2c_{1333} u_{13} + c_{2233} u_{22} + c_{3333} u_{33} + f_{3333} P_{3,3}$. So that relations (C.7) give the system of equations for the unknown strains u_{13} and u_{33}:

$$2c_{1313} u_{13} + c_{1333} u_{33} = -f_{1333} P_{3,3} \tag{C.8a}$$

$$2c_{1333} u_{13} + c_{3333} u_{33} = -c_{1133} u_{11} - c_{2233} u_{22} - f_{3333} P_{3,3} \tag{C.8b}$$

The solution of (C.8a), (C.8b) is

$$2u_{13} = \frac{+c_{1333}(c_{1133}u_{11} + c_{2233}u_{22} + f_{3333}P_{3,3}) - c_{3333}f_{1333}P_{3,3}}{c_{1313}c_{3333} - c_{1333}^2} \quad \text{(C.9a)}$$

$$u_{33} = \frac{-c_{1313}(c_{1133}u_{11} + c_{2233}u_{22} + f_{3333}P_{3,3}) + c_{1333}f_{1333}P_{3,3}}{c_{1313}c_{3333} - c_{1333}^2} \quad \text{(C.9b)}$$

Next, using evident form of equation of state (C.4b), $\alpha_3 P_3 - g_{3333}P_{3,33} - f_{1133}u_{11,3} - f_{2233}u_{22,3} - f_{3333}u_{33,3} - 2f_{1333}u_{13,3} = E_3$, one could get the following equation for polarization inside the slab:

$$\begin{aligned}\alpha_3 P_3 - \left(g_{3333} - \frac{c_{1313}f_{3333}^2 + c_{3333}f_{1333}^2 - 2c_{1333}f_{1333}f_{3333}}{c_{1313}c_{3333} - c_{1333}^2}\right)P_{3,33} = \\ E_3 + \left(f_{1133} - \frac{(f_{3333}c_{1313} - f_{1333}c_{1333})c_{1133}}{c_{1313}c_{3333} - c_{1333}^2}\right)u_{11,3} \\ + \left(f_{2233} - \frac{(f_{3333}c_{1313} - f_{1333}c_{1333})c_{2233}}{c_{1313}c_{3333} - c_{1333}^2}\right)u_{22,3}\end{aligned} \quad \text{(C.10)}$$

One could see that according to Eqs. (C.6a)–(C.6c) there is a uniform gradient of strain leading to a sort of constant "flexoelectric field" in the right-hand side of Eq. (C.10)

$$\begin{aligned}E_3^{(flexo)} = \left(f_{1133} - \frac{(f_{3333}c_{1313} - f_{1333}c_{1333})c_{1133}}{c_{1313}c_{3333} - c_{1333}^2}\right)\frac{1}{R_1} \\ + \left(f_{2233} - \frac{(f_{3333}c_{1313} - f_{1333}c_{1333})c_{2233}}{c_{1313}c_{3333} - c_{1333}^2}\right)\frac{1}{R_2}\end{aligned} \quad \text{(C.11)}$$

Two terms from (C.11) represent the flexoelectric response to plate bending into two perpendicular directions, leading to plate transforming to "cup" like shape with two main values of curvature, $1/R_1$ and $1/R_2$. Hence we could introduce coefficients of flexoelectric response in the following form:

$$f_{13}^{eff} = f_{1133} - \frac{c_{1133}}{c_{1313}c_{3333} - c_{1333}^2}(f_{3333}c_{1313} - f_{1333}c_{1333}) \quad \text{(C.12a)}$$

$$f_{23}^{eff} = f_{2233} - \frac{c_{2233}}{c_{1313}c_{3333} - c_{1333}^2}(f_{3333}c_{1313} - f_{1333}c_{1333}) \quad \text{(C.12b)}$$

Appendix D. Components of the effective flexoresponse

Let us consider linearized model (corresponding to paraelectric and paramagnetic phase) with the following set of equations of state for polarization vector P_i, magnetization vector M_i and strain tensor u_{ij}:

$$F = \eta_{kl} M_l P_k + \alpha_{lk}^{(M)} M_k M_l - M_k H_k \ldots + \alpha_{km}^{(P)} P_k P_m - P_m E_m + \frac{f_{ijkl}}{2}\left(P_k \frac{\partial u_{ij}}{\partial x_l} - u_{ij}\frac{\partial P_k}{\partial x_l}\right) \ldots + \frac{c_{ijkl}}{2} u_{ij} u_{kl} \tag{D.1}$$

$$\alpha_{km}^{(P)} P_m + f_{ijkl}\frac{\partial u_{ij}}{\partial x_l} = E_k - \eta_{kl} M_l \tag{D.2a}$$

$$\alpha_{lk}^{(M)} M_k = H_l - \eta_{kl} P_k \tag{D.2b}$$

$$\sigma_{ij} = c_{ijkl} u_{kl} - f_{ijkl}\frac{\partial P_k}{\partial x_l} \tag{D.2c}$$

Here η_{kl} are the components of linear magnetoelectric coupling tensor. One could get from (D.2a).

$$P_m = -\chi_{mk}^{(P)} f_{ijkl}\frac{\partial u_{ij}}{\partial x_l} + \chi_{mk}^{(P)} E_k - \chi_{mk}^{(P)} \eta_{kl} M_l \tag{D.3}$$

Here $\chi_{mk}^{(P)}$ is the tensor of dielectric susceptibility, inverse to dielectric stiffness tensor $\alpha_{km}^{(P)}$. Substitution of (D.3) into (D.2b) and (D.2c) gives the following equations:

$$\left(\alpha_{pk}^{(M)} - \eta_{mp}\chi_{mn}^{(P)}\eta_{nk}\right) M_k = H_p + \eta_{mp}\chi_{mn}^{(P)} f_{ijnl}\frac{\partial u_{ij}}{\partial x_l} - \eta_{mp}\chi_{mn}^{(P)} E_n \tag{D.4a}$$

Substitution of (D.3) into (D.2c) yields:

$$\sigma_{ij} = c_{ijkl} u_{kl} + f_{ijkl}\chi_{kn}^{(P)} f_{pqnm}\frac{\partial^2 u_{pq}}{\partial x_l \partial x_m} - \chi_{kn}^{(P)} f_{ijkl}\frac{\partial E_n}{\partial x_l} + \chi_{kn}^{(P)} \eta_{nm} f_{ijkl}\frac{\partial M_m}{\partial x_l} \tag{D.4b}$$

Comparing Eqs. (D.4a) and (D.4b), one could get the following expression for the effective flexomagnetic coefficient:

$$m_{ijml}^{(FM)} = \chi_{kn}^{(P)} \eta_{nm} f_{ijkl}, \tag{D.5}$$

where $\chi_{mk}^{(P)}$ is the tensor of dielectric susceptibility and η_{kl} are the components of linear magnetoelectric coupling tensor.

Appendix E. The evident form of flexomagnetic tensor (FMT) "*m*"

To find out the nonzero components FMT we will use the system of linear equations obtained from the transformation laws for the axial forth rank tensor m_{ijkl} describing flexomagnetic effects (see Eq. (5c)). For the case when the matrices $\hat{C}$ represent all the elements of the material point symmetry group (considered hereinafter) the identity (5c) should be valid for nonzero components of the flexo-tensors. To minimize the calculation work, it is enough to use only the generating elements of the material point symmetry group, in order to obtain the evident form of the flexo-tensors for a given group. General relations (internal symmetry) are listed below. Namely, FMT is invariant to permutation of the first and the second indices (due to the symmetry of strain tensor):

$$\overline{\mathrm{m}}_{1211}=\overline{\mathrm{m}}_{2111},\overline{\mathrm{m}}_{1212}=\overline{\mathrm{m}}_{2112},\overline{\mathrm{m}}_{1213}=\overline{\mathrm{m}}_{2113}, \quad \text{(E.1a)}$$

$$\overline{\mathrm{m}}_{1221}=\overline{\mathrm{m}}_{2121},\overline{\mathrm{m}}_{1222}=\overline{\mathrm{m}}_{2122},\overline{\mathrm{m}}_{1223}=\overline{\mathrm{m}}_{2123}, \quad \text{(E.1b)}$$

$$\overline{\mathrm{m}}_{1231}=\overline{\mathrm{m}}_{2131},\overline{\mathrm{m}}_{1232}=\overline{\mathrm{m}}_{2132},\overline{\mathrm{m}}_{1233}=\overline{\mathrm{m}}_{2133}, \quad \text{(E.1c)}$$

$$\overline{\mathrm{m}}_{1311}=\overline{\mathrm{m}}_{3111},\overline{\mathrm{m}}_{1312}=\overline{\mathrm{m}}_{3112},\overline{\mathrm{m}}_{1313}=\overline{\mathrm{m}}_{3113}, \quad \text{(E.1d)}$$

$$\overline{\mathrm{m}}_{1321}=\overline{\mathrm{m}}_{3121},\overline{\mathrm{m}}_{1322}=\overline{\mathrm{m}}_{3122},\overline{\mathrm{m}}_{1323}=\overline{\mathrm{m}}_{3123}, \quad \text{(E.1e)}$$

$$\overline{\mathrm{m}}_{1331}=\overline{\mathrm{m}}_{3131},\overline{\mathrm{m}}_{1332}=\overline{\mathrm{m}}_{3132},\overline{\mathrm{m}}_{1333}=\overline{\mathrm{m}}_{3133}, \quad \text{(E.1f)}$$

$$\overline{\mathrm{m}}_{2311}=\overline{\mathrm{m}}_{3211},\overline{\mathrm{m}}_{2312}=\overline{\mathrm{m}}_{3212},\overline{\mathrm{m}}_{2313}=\overline{\mathrm{m}}_{3213} \quad \text{(E.1g)}$$

$$\overline{\mathrm{m}}_{2321}=\overline{\mathrm{m}}_{3221},\overline{\mathrm{m}}_{2322}=\overline{\mathrm{m}}_{3222},\overline{\mathrm{m}}_{2323}=\overline{\mathrm{m}}_{3223}, \quad \text{(E.1h)}$$

$$\overline{\mathrm{m}}_{2331}=\overline{\mathrm{m}}_{3231},\overline{\mathrm{m}}_{2332}=\overline{\mathrm{m}}_{3232},\overline{\mathrm{m}}_{2333}=\overline{\mathrm{m}}_{3233}. \quad \text{(E.1i)}$$

Symmetry of the media imposes additional restrictions of the evident form of tensor (see Table E1).

Table E1 Number of elements of FMT for different syngonies.

Syngony	Nonzero	Independent	Nonzero element and relations between them
$1, \overline{1}'$	81	54	$\overline{m}_{1111}; \overline{m}_{1112}; \overline{m}_{1113}; \overline{m}_{1121}; \overline{m}_{1122}; \overline{m}_{1123}; \overline{m}_{1131}; \overline{m}_{1132}; \overline{m}_{1133};$ $\overline{m}_{1211} = \overline{m}_{2111}; \overline{m}_{1212} = \overline{m}_{2112}; \overline{m}_{1213} = \overline{m}_{2113};$ $\overline{m}_{1221} = \overline{m}_{2121}; \overline{m}_{1222} = \overline{m}_{2122}; \overline{m}_{1223} = \overline{m}_{2123};$ $\overline{m}_{1231} = \overline{m}_{2131}; \overline{m}_{1232} = \overline{m}_{2132}; \overline{m}_{1233} = \overline{m}_{2133};$ $\overline{m}_{1311} = \overline{m}_{3111}; \overline{m}_{1312} = \overline{m}_{3112}; \overline{m}_{1313} = \overline{m}_{3113};$ $\overline{m}_{1321} = \overline{m}_{3121}; \overline{m}_{1322} = \overline{m}_{3122}; \overline{m}_{1323} = \overline{m}_{3123};$ $\overline{m}_{1331} = \overline{m}_{3131}; \overline{m}_{1332} = \overline{m}_{3132}; \overline{m}_{1333} = \overline{m}_{3133};$ $\overline{m}_{2211}; \overline{m}_{2212}; \overline{m}_{2213}; \overline{m}_{2221}; \overline{m}_{2222}; \overline{m}_{2223}; \overline{m}_{2231}; \overline{m}_{2232}; \overline{m}_{2233};$ $\overline{m}_{2311} = \overline{m}_{3211}; \overline{m}_{2312} = \overline{m}_{3212}; \overline{m}_{2313} = \overline{m}_{3213};$ $\overline{m}_{2321} = \overline{m}_{3221}; \overline{m}_{2322} = \overline{m}_{3222}; \overline{m}_{2323} = \overline{m}_{3223};$ $\overline{m}_{2331} = \overline{m}_{3231}; \overline{m}_{2332} = \overline{m}_{3232}; \overline{m}_{2333} = \overline{m}_{3233}.$ $\overline{m}_{3311}; \overline{m}_{3312}; \overline{m}_{3313}; \overline{m}_{3321}; \overline{m}_{3322}; \overline{m}_{3323}; \overline{m}_{3331}; \overline{m}_{3332}; \overline{m}_{3333}.$
$2/m, 2'/m'$	0	–	No effect, $\overline{m}_{ijkl} \equiv 0;$
$2, 2/m', m'$	41	28	$\overline{m}_{1111}; \overline{m}_{1112}; \overline{m}_{1121}; \overline{m}_{1122}; \overline{m}_{1133};$ $\overline{m}_{1211} = \overline{m}_{2111}; \overline{m}_{1212} = \overline{m}_{2112};$ $\overline{m}_{1221} = \overline{m}_{2121}; \overline{m}_{1222} = \overline{m}_{2122};$ $\overline{m}_{1233} = \overline{m}_{2133}; \overline{m}_{1313} = \overline{m}_{3113}; \overline{m}_{1323} = \overline{m}_{3123};$ $\overline{m}_{1331} = \overline{m}_{3131}; \overline{m}_{1332} = \overline{m}_{3132};$ $\overline{m}_{2211}; \overline{m}_{2212}; \overline{m}_{2221}; \overline{m}_{2222}; \overline{m}_{2233};$ $\overline{m}_{2313} = \overline{m}_{3213}; \overline{m}_{2323} = \overline{m}_{3223};$ $\overline{m}_{2331} = \overline{m}_{3231}; \overline{m}_{2332} = \overline{m}_{3232};$ $\overline{m}_{3311}; \overline{m}_{3312}; \overline{m}_{3321}; \overline{m}_{3322}; \overline{m}_{3333}.$

$2', 2'/m, m$	40	26	$\overline{m}_{1113}$; $\overline{m}_{1123}$; $\overline{m}_{1131}$; $\overline{m}_{1132}$; $\overline{m}_{1213} = \overline{m}_{2113}$; $\overline{m}_{1223} = \overline{m}_{2123}$; $\overline{m}_{1231} = \overline{m}_{2131}$; $\overline{m}_{1232} = \overline{m}_{2132}$; $\overline{m}_{1311} = \overline{m}_{3111}$; $\overline{m}_{1312} = \overline{m}_{3112}$; $\overline{m}_{1321} = \overline{m}_{3121}$; $\overline{m}_{1322} = \overline{m}_{3122}$; $\overline{m}_{1333} = \overline{m}_{3133}$; $\overline{m}_{2213}$; $\overline{m}_{2223}$; $\overline{m}_{2231}$; $\overline{m}_{2232}$; $\overline{m}_{2311} = \overline{m}_{3211}$; $\overline{m}_{2312} = \overline{m}_{3212}$; $\overline{m}_{2321} = \overline{m}_{3221}$; $\overline{m}_{2322} = \overline{m}_{3222}$; $\overline{m}_{2333} = \overline{m}_{3233}$; $\overline{m}_{3313}$; $\overline{m}_{3323}$; $\overline{m}_{3331}$; $\overline{m}_{3332}$.
$222, m'm'2, m'm'm'$	21	15	$\overline{m}_{1111}$; $\overline{m}_{1122}$; $\overline{m}_{1133}$; $\overline{m}_{1212} = \overline{m}_{2112}$; $\overline{m}_{1221} = \overline{m}_{2121}$; $\overline{m}_{1313} = \overline{m}_{3113}$; $\overline{m}_{1331} = \overline{m}_{3131}$; $\overline{m}_{2211}$; $\overline{m}_{2222}$; $\overline{m}_{2233}$; $\overline{m}_{2323} = \overline{m}_{3223}$; $\overline{m}_{2332} = \overline{m}_{3232}$; $\overline{m}_{3311}$; $\overline{m}_{3322}$; $\overline{m}_{3333}$.
$22'2', mm2, m2'm', mmm'$	20	13	$\overline{m}_{1112}$; $\overline{m}_{1121}$; $\overline{m}_{1211} = \overline{m}_{2111}$; $\overline{m}_{1222} = \overline{m}_{2122}$; $\overline{m}_{1233} = \overline{m}_{2133}$; $\overline{m}_{1323} = \overline{m}_{3123}$; $\overline{m}_{1332} = \overline{m}_{3132}$; $\overline{m}_{2212}$; $\overline{m}_{2221}$; $\overline{m}_{2313} = \overline{m}_{3213}$; $\overline{m}_{2331} = \overline{m}_{3231}$; $\overline{m}_{3312}$; $\overline{m}_{3321}$.

Continued

Table E1 Number of elements of FMT for different syngonies.—cont'd

Syngony	Nonzero	Independent	Nonzero element and relations between them
$mm'2'$(alternative choice of coordinate system)	20	13	$\overline{m}_{1113}$; $\overline{m}_{1131}$; $\overline{m}_{1223}=\overline{m}_{2123}$;$\overline{m}_{1232}=\overline{m}_{2132}$; $\overline{m}_{1311}=\overline{m}_{3111}$; $\overline{m}_{1322}=\overline{m}_{3122}$; $\overline{m}_{1333}=\overline{m}_{3133}$; $\overline{m}_{2213}$; $\overline{m}_{2231}$; $\overline{m}_{2312}=\overline{m}_{3212}$; $\overline{m}_{2321}=\overline{m}_{3221}$; $\overline{m}_{3313}$; $\overline{m}_{3331}$
mmm, $m'm'm$	0	–	No effect, $\overline{m}_{ijkl}\equiv 0$;
$\overline{4}2m$, $\overline{4}m'2'$, $4'2'2$, $4'mm'$, $4'/m'mm'$	20	7	$\overline{m}_{1112}=\overline{m}_{2221}$; $\overline{m}_{1121}=\overline{m}_{2212}$; $\overline{m}_{1211}=\overline{m}_{2111}=\overline{m}_{1222}=\overline{m}_{2122}$; $\overline{m}_{1233}=\overline{m}_{2133}$ $\overline{m}_{1323}=\overline{m}_{3123}=\overline{m}_{2313}=\overline{m}_{3213}$; $\overline{m}_{1332}=\overline{m}_{3132}=\overline{m}_{2331}=\overline{m}_{3231}$; $\overline{m}_{3312}=\overline{m}_{3321}$.
$\overline{4}m2$, $\overline{4}2'm'$, $4'22'$, $4'm'm$, $4'/m'm'm$	20	7	Variant of the previous case (difference is only in axes designations)
$\overline{4}'2m'$, 422, $4m'm'$, $4/m'm'm'$	21	8	$\overline{m}_{1111}=\overline{m}_{2222}$; $\overline{m}_{1122}=\overline{m}_{2211}$; $\overline{m}_{1133}=\overline{m}_{2233}$; $\overline{m}_{1212}=\overline{m}_{1221}=\overline{m}_{2112}=\overline{m}_{2121}$; $\overline{m}_{1313}=\overline{m}_{2323}=\overline{m}_{3113}=\overline{m}_{3223}$; $\overline{m}_{1331}=\overline{m}_{2332}=\overline{m}_{3131}=\overline{m}_{3232}$; $\overline{m}_{3311}=\overline{m}_{3322}$; $\overline{m}_{3333}$.

$\overline{4}'2'm, 42'2'$, $4mm$, $4/m'mm$	18	6	$\overline{m}_{1112}$; $\overline{m}_{1121}$; $\overline{m}_{1211} = \overline{m}_{2111}$; $\overline{m}_{1222} = \overline{m}_{2122} = -\overline{m}_{1211}$; $\overline{m}_{1323} = \overline{m}_{3123}$; $\overline{m}_{1332} = \overline{m}_{3132}$; $\overline{m}_{2212} = -\overline{m}_{1121}$; $\overline{m}_{2221} = -\overline{m}_{1112}$; $\overline{m}_{2313} = \overline{m}_{3213} = -\overline{m}_{1323}$; $\overline{m}_{2331} = \overline{m}_{3231} = -\overline{m}_{1332}$; $\overline{m}_{3312}$; $\overline{m}_{3321} = -\overline{m}_{3312}$.
$4/m$, $4'/m$, $4/mmm$, $4/mm'm'$, $4'/mmm'$	0	–	No effect, $\overline{m}_{ijkl} \equiv 0$;
$4, \overline{4}', 4/m'$	39	14	$\overline{m}_{1111} = \overline{m}_{2222}$; $\overline{m}_{1112}$; $\overline{m}_{1121}$; $\overline{m}_{1122} = \overline{m}_{2211}$; $\overline{m}_{1133} = \overline{m}_{2233}$; $\overline{m}_{1211} = \overline{m}_{2111}$; $\overline{m}_{1212} = \overline{m}_{2112} = \overline{m}_{1221} = \overline{m}_{2121}$; $\overline{m}_{1222} = \overline{m}_{2122} = -\overline{m}_{1211}$; $\overline{m}_{1313} = \overline{m}_{3113} = \overline{m}_{2323} = \overline{m}_{3223}$; $\overline{m}_{1323} = \overline{m}_{3123}$; $\overline{m}_{1331} = \overline{m}_{3131} = \overline{m}_{2332} = \overline{m}_{3232}$; $\overline{m}_{1332} = \overline{m}_{3132}$; $\overline{m}_{2212} = -\overline{m}_{1121}$; $\overline{m}_{2221} = -\overline{m}_{1112}$; $\overline{m}_{2313} = \overline{m}_{3213} = -\overline{m}_{1323}$; $\overline{m}_{2331} = \overline{m}_{3231} = -\overline{m}_{1332}$; $\overline{m}_{3311} = \overline{m}_{3322}$; $\overline{m}_{3312}$; $\overline{m}_{3321} = -\overline{m}_{3312}$; $\overline{m}_{3333}$.

Continued

Table E1 Number of elements of FMT for different syngonies.—cont'd

Syngony	Nonzero	Independent	Nonzero element and relations between them
$4', \overline{4}, 4'/m'$	40	14	$\overline{m}_{1111};\ \overline{m}_{1122};\ \overline{m}_{1133};$ $\overline{m}_{1112}=\overline{m}_{2221};\ \overline{m}_{1121}=\overline{m}_{2212};$ $\overline{m}_{1211}=\overline{m}_{2111}=\overline{m}_{1222}=\overline{m}_{2122};\ \overline{m}_{1212}=\overline{m}_{2112};$ $\overline{m}_{1221}=\overline{m}_{2121}=-\overline{m}_{1212};$ $\overline{m}_{1233}=\overline{m}_{2133};\ \overline{m}_{1313}=\overline{m}_{3113};$ $\overline{m}_{1323}=\overline{m}_{3123}=\overline{m}_{2313}=\overline{m}_{3213};$ $\overline{m}_{1331}=\overline{m}_{3131};\ \overline{m}_{1332}=\overline{m}_{3132}=\overline{m}_{2331}=\overline{m}_{3231};$ $\overline{m}_{2211}=-\overline{m}_{1122};\ \overline{m}_{2222}=-\overline{m}_{1111};$ $\overline{m}_{2233}=-\overline{m}_{1133};$ $\overline{m}_{2323}=\overline{m}_{3223}=-\overline{m}_{1313};$ $\overline{m}_{2332}=\overline{m}_{3232}=-\overline{m}_{1331};$ $\overline{m}_{3311};\ \overline{m}_{3312}=\overline{m}_{3321};\ \overline{m}_{3322}=-\overline{m}_{3311}.$
$3, \overline{3}'$	71	18	$\overline{m}_{1111}=\overline{m}_{2222};$ $\overline{m}_{1112};\ \overline{m}_{1113};\ \overline{m}_{1121};\ \overline{m}_{1131};$ $\overline{m}_{1122}=\overline{m}_{2211};\ \overline{m}_{1133}=\overline{m}_{2233};$ $\overline{m}_{1123}=\overline{m}_{1213}=\overline{m}_{2113};\ \overline{m}_{1132}=\overline{m}_{1231}=\overline{m}_{2131};$ $\overline{m}_{1211}=\overline{m}_{2111}=\frac{-\overline{m}_{1112}-\overline{m}_{1121}}{2};$ $\overline{m}_{1212}=\overline{m}_{1221}=\overline{m}_{2112}=\overline{m}_{2121}=\frac{\overline{m}_{1111}-\overline{m}_{1122}}{2};$ $\overline{m}_{1222}=\overline{m}_{2122}=\frac{\overline{m}_{1112}+\overline{m}_{1121}}{2};\ \overline{m}_{1223}=\overline{m}_{2123}=\overline{m}_{2213}=-\overline{m}_{1113};$ $\overline{m}_{1232}=\overline{m}_{2132}=\overline{m}_{2231}=-\overline{m}_{1131};$ $\overline{m}_{1311}=\overline{m}_{3111};$ $\overline{m}_{1312}=\overline{m}_{1321}=\overline{m}_{2311}=\overline{m}_{3112}=\overline{m}_{3121}=\overline{m}_{3211};$ $\overline{m}_{1313}=\overline{m}_{3113}=\overline{m}_{2323}=\overline{m}_{3223};$ $\overline{m}_{1322}=\overline{m}_{2312}=\overline{m}_{3122}=\overline{m}_{3212}=-\overline{m}_{1311}$ $\overline{m}_{1323}=\overline{m}_{3123};\ \overline{m}_{1331}=\overline{m}_{3131}=\overline{m}_{2332}=\overline{m}_{3232};\ \overline{m}_{1332}=\overline{m}_{3132};$ $\overline{m}_{2212}=-\overline{m}_{1121};$ $\overline{m}_{2221}=-\overline{m}_{1112};\ \overline{m}_{2223}=-\overline{m}_{1123};\ \overline{m}_{2232}=-\overline{m}_{1132};$ $\overline{m}_{2322}=\overline{m}_{3222}=-\overline{m}_{1312};\ \overline{m}_{2331}=\overline{m}_{3231}=-\overline{m}_{1332};$ $\overline{m}_{3311}=\overline{m}_{3322};\ \overline{m}_{3312};\ \overline{m}_{3321}=-\overline{m}_{3312};\ \overline{m}_{3333}.$

$32, 3m', \overline{3}'m'$	37	10	$\overline{m}_{1111} = \overline{m}_{2222}$; $\overline{m}_{1113}$; $\overline{m}_{1131}$; $\overline{m}_{1122} = \overline{m}_{2211}$; $\overline{m}_{1133} = \overline{m}_{2233}$; $\overline{m}_{1212} = \overline{m}_{1221} = \overline{m}_{2112} = \overline{m}_{2121} = \frac{\overline{m}_{1111} - \overline{m}_{1122}}{2}$; $\overline{m}_{1223} = \overline{m}_{2123} = \overline{m}_{2213} = -\overline{m}_{1113}$; $\overline{m}_{1232} = \overline{m}_{2132} = \overline{m}_{2231} = -\overline{m}_{1131}$; $\overline{m}_{1311} = \overline{m}_{3111}$; $\overline{m}_{1313} = \overline{m}_{3113} = \overline{m}_{2323} = \overline{m}_{3223}$; $\overline{m}_{1322} = \overline{m}_{2312} = \overline{m}_{2321} = \overline{m}_{3122} = \overline{m}_{3212} = \overline{m}_{3221} = -\overline{m}_{1311}$; $\overline{m}_{1331} = \overline{m}_{2332} = \overline{m}_{3131} = \overline{m}_{3232}$; $\overline{m}_{3311} = \overline{m}_{3322}$; $\overline{m}_{3333}$.
$32', 3m, \overline{3}'m$	34	8	$\overline{m}_{1112}$; $\overline{m}_{1121}$; $\overline{m}_{1123} = \overline{m}_{1213} = \overline{m}_{2113}$; $\overline{m}_{1132} = \overline{m}_{1231} = \overline{m}_{2131}$; $\overline{m}_{1211} = \overline{m}_{2111} = \frac{-\overline{m}_{1112} - \overline{m}_{1121}}{2}$; $\overline{m}_{1222} = \overline{m}_{2122} = \frac{\overline{m}_{1112} + \overline{m}_{1121}}{2}$; $\overline{m}_{1312} = \overline{m}_{3112} = \overline{m}_{1321} = \overline{m}_{3121} = \overline{m}_{2311} = \overline{m}_{3211}$; $\overline{m}_{1323} = \overline{m}_{3123}$; $\overline{m}_{1332} = \overline{m}_{3132}$; $\overline{m}_{2212} = -\overline{m}_{1121}$; $\overline{m}_{2221} = -\overline{m}_{1112}$; $\overline{m}_{2223} = -\overline{m}_{1123}$; $\overline{m}_{2232} = -\overline{m}_{1132}$; $\overline{m}_{2313} = \overline{m}_{3213} = -\overline{m}_{1323}$; $\overline{m}_{2322} = \overline{m}_{3222} = -\overline{m}_{1312}$; $\overline{m}_{2331} = \overline{m}_{3231} = -\overline{m}_{1332}$. $\overline{m}_{3312}$; $\overline{m}_{3321} = -\overline{m}_{3312}$.
$\overline{6}, 6', 6'/m$	32	6	$\overline{m}_{1113}$; $\overline{m}_{1131}$; $\overline{m}_{1123} = \overline{m}_{1213} = \overline{m}_{2113}$; $\overline{m}_{1132} = \overline{m}_{1231} = \overline{m}_{2131}$; $\overline{m}_{1223} = \overline{m}_{2123} = \overline{m}_{2213} = -\overline{m}_{1113}$; $\overline{m}_{1232} = \overline{m}_{2132} = -\overline{m}_{1131}$; $\overline{m}_{1311} = \overline{m}_{3111}$; $\overline{m}_{1312} = \overline{m}_{1321} = \overline{m}_{2311} = \overline{m}_{3112} = \overline{m}_{3121} = \overline{m}_{3211}$; $\overline{m}_{1322} = \overline{m}_{2312} = \overline{m}_{2321} = \overline{m}_{3122} = \overline{m}_{3212} = \overline{m}_{3221} = -\overline{m}_{1311}$; $\overline{m}_{2322} = \overline{m}_{3222} = -\overline{m}_{1312}$; $\overline{m}_{2223} = -\overline{m}_{1123}$; $\overline{m}_{2231} = -\overline{m}_{1131}$; $\overline{m}_{2232} = -\overline{m}_{1132}$.

Continued

Table E1 Number of elements of FMT for different syngonies.—cont'd

Syngony	Nonzero	Independent	Nonzero element and relations between them
$\overline{6}'m'2$, 622, $6m'm'$, $6/m'm'm'$	21	7	$\overline{m}_{1111}=\overline{m}_{2222}$; $\overline{m}_{1122}=\overline{m}_{2211}$; $\overline{m}_{1133}=\overline{m}_{2233}$; $\overline{m}_{1212}=\overline{m}_{1221}=\overline{m}_{2112}=\overline{m}_{2121}=\frac{\overline{m}_{1111}-\overline{m}_{1122}}{2}$; $\overline{m}_{1313}=\overline{m}_{2323}=\overline{m}_{3113}=\overline{m}_{3223}$; $\overline{m}_{1331}=\overline{m}_{2332}=\overline{m}_{3131}=\overline{m}_{3232}$; $\overline{m}_{3311}=\overline{m}_{3322}$; $\overline{m}_{3333}$.
$\overline{6}'m2'$, $62'2'$, $6mm$, $6/m'mm$	18	5	$\overline{m}_{1112}$; $\overline{m}_{1121}$; $\overline{m}_{1211}=\overline{m}_{2111}=\frac{-\overline{m}_{1112}-\overline{m}_{1121}}{2}$; $\overline{m}_{1222}=\overline{m}_{2122}=\frac{\overline{m}_{1112}+\overline{m}_{1121}}{2}$; $\overline{m}_{1323}=\overline{m}_{3123}$; $\overline{m}_{1332}=\overline{m}_{3132}$; $\overline{m}_{2212}=-\overline{m}_{1121}$; $\overline{m}_{2221}=-\overline{m}_{1112}$; $\overline{m}_{2313}=\overline{m}_{3213}=-\overline{m}_{1323}$; $\overline{m}_{2331}=\overline{m}_{3231}=-\overline{m}_{1332}$; $\overline{m}_{3312}$; $\overline{m}_{3321}=-\overline{m}_{3312}$.
$\overline{6}'$, 6, $6/m'$	39	12	$\overline{m}_{1111}=\overline{m}_{2222}$; $\overline{m}_{1112}$; $\overline{m}_{1121}$; $\overline{m}_{1122}=\overline{m}_{2211}$; $\overline{m}_{1133}=\overline{m}_{2233}$; $\overline{m}_{1211}=\overline{m}_{2111}=\frac{-\overline{m}_{1112}-\overline{m}_{1121}}{2}$; $\overline{m}_{1212}=\overline{m}_{2112}=\frac{\overline{m}_{1111}-\overline{m}_{1122}}{2}$; $\overline{m}_{1221}=\overline{m}_{2121}=\frac{\overline{m}_{1111}-\overline{m}_{1122}}{2}$; $\overline{m}_{1222}=\overline{m}_{2122}=\frac{\overline{m}_{1112}+\overline{m}_{1121}}{2}$; $\overline{m}_{1313}=\overline{m}_{2323}=\overline{m}_{3223}=\overline{m}_{3113}$; $\overline{m}_{1323}=\overline{m}_{3123}$; $\overline{m}_{1331}=\overline{m}_{2332}=\overline{m}_{3131}=\overline{m}_{3232}$; $\overline{m}_{1332}=\overline{m}_{3132}$; $\overline{m}_{2212}=-\overline{m}_{1122}$; $\overline{m}_{2221}=-\overline{m}_{1112}$; $\overline{m}_{2313}=\overline{m}_{3213}=-\overline{m}_{1323}$; $\overline{m}_{2331}=\overline{m}_{3231}=-\overline{m}_{1332}$; $\overline{m}_{3311}=\overline{m}_{3322}$; $\overline{m}_{3312}$; $\overline{m}_{3321}=-\overline{m}_{3312}$; $\overline{m}_{3333}$.

$\overline{6}m2, \overline{6}m'2', 6'22', 6'm'm, 6'/mmm'$	16	3	$\overline{m}_{1113} = -\overline{m}_{1223} = -\overline{m}_{2123} = -\overline{m}_{2213}$ $\overline{m}_{1131} = -\overline{m}_{1232} = -\overline{m}_{2132} = -\overline{m}_{2231}$ $\overline{m}_{1311} = \overline{m}_{1322} = -\overline{m}_{2322} = -\overline{m}_{3122} = -\overline{m}_{3121} = -\overline{m}_{3111} = -\overline{m}_{2322} = -\overline{m}_{3222};$
$6'/mm'm$	16	3	$\overline{m}_{1123} = \overline{m}_{1213} = \overline{m}_{2113} = -\overline{m}_{2223}$ $\overline{m}_{1132} = \overline{m}_{1231} = \overline{m}_{2131} = -\overline{m}_{2232}$ $\overline{m}_{1312} = \overline{m}_{1321} = \overline{m}_{2311} = \overline{m}_{3112} = \overline{m}_{3121} = \overline{m}_{3211}$ $= -\overline{m}_{2322} = -\overline{m}_{3222};$
$\overline{3}, \overline{3}m, \overline{3}m', 6/m, 6'/m', 6/mmm, 6/mm'm', 6'/m'mm'$	0	–	No effect, $\overline{m}_{ijkl} \equiv 0;$
$23, m'3$	21	5	$\overline{m}_{1111} = \overline{m}_{2222} = \overline{m}_{3333};$ $\overline{m}_{1122} = \overline{m}_{2233} = \overline{m}_{3311}; \overline{m}_{1133} = \overline{m}_{2211} = \overline{m}_{3322};$ $\overline{m}_{1212} = \overline{m}_{1331} = \overline{m}_{2112} = \overline{m}_{2323} = \overline{m}_{3131} = \overline{m}_{3223};$ $\overline{m}_{1221} = \overline{m}_{1313} = \overline{m}_{2121} = \overline{m}_{2332} = \overline{m}_{3113} = \overline{m}_{3232}.$
$\overline{4}'3m', 432, m'3m'$	21	3	$\overline{m}_{1111} = \overline{m}_{2222} = \overline{m}_{3333};$ $\overline{m}_{1122} = \overline{m}_{2233} = \overline{m}_{3311} = \overline{m}_{1133} = \overline{m}_{2211} = \overline{m}_{3322};$ $\overline{m}_{1212} = \overline{m}_{2112} = \overline{m}_{1221} = \overline{m}_{2121} = \overline{m}_{1313} = \overline{m}_{3113} =$ $\overline{m}_{1331} = \overline{m}_{3131} = \overline{m}_{2323} = \overline{m}_{3223} = \overline{m}_{2332} = \overline{m}_{3232}.$
$\overline{4}3m, 4'32', m'3m$	18	4	$\overline{m}_{1122} = \overline{m}_{2233} = \overline{m}_{3311};$ $\overline{m}_{1133} = \overline{m}_{2211} = \overline{m}_{3322} = -\overline{m}_{1122}.$ $\overline{m}_{1212} = \overline{m}_{1331} = \overline{m}_{2112} = \overline{m}_{2323} = \overline{m}_{3131} = \overline{m}_{3223};$ $\overline{m}_{1221} = \overline{m}_{1313} = \overline{m}_{2121} = \overline{m}_{2332} = \overline{m}_{3113} = \overline{m}_{3232} = -\overline{m}_{1212};$
$m3, m3m, m3m'$	0	–	No effect, $\overline{m}_{ijkl} \equiv 0;$

References

[1] A.K. Tagantsev, P.V. Yudin (Eds.), Flexoelectricity in Solids: From Theory to Applications, World Scientific, New York, 2016.
[2] V.S. Mashkevich, K.B. Tolpygo, Electrical, optic and elastic properties of crystals of diamond type, Zh. Eksp. Teor. Fiz. 31 (1957) 520–525 (Sov.Phys. JETP, 4 (1957) 455–460).
[3] S.M. Kogan, Piezoelectric effect under an inhomogeneous strain and an acoustic scattering of carriers of current in crystals, Solid State Phys. 5 (1963) 2829–2831.
[4] R.D. Mindlin, Polarization gradient in elastic dielectrics, Int. J. Solids Struct. 4 (1968) 637–642.
[5] A.K. Tagantsev, Piezoelectricity and flexoelectricity in crystalline dielectrics, Phys. Rev. B 34 (1986) 5883–5889.
[6] V.L. Indenbom, E.B. Loginov, M.A. Osipov, Flexoelectric effect and the structure of crystals, Kristallografiya 26 (1981) 1157–1960.
[7] E.A. Eliseev, A.N. Morozovska, M.D. Glinchuk, R. Blinc, Spontaneous flexoelectric/flexomagnetic effect in nanoferroics, Phys. Rev. B. 79 (2009) 165433.
[8] N.D. Sharma, C.M. Landis, P. Sharma, Piezoelectric thin-film superlattices without using piezoelectric materials, J. Appl. Phys. 108 (2010) 024304.
[9] P.V. Yudin, A.K. Tagantsev, Fundamentals of flexoelectricity in solids, Nanotechnology 24 (2013) 432001.
[10] L. Shu, X. Wei, T. Pang, X. Yao, C. Wang, Symmetry of flexoelectric coefficients in crystalline medium, J. Appl. Phys. 110 (2011) 104106.
[11] H. Le Quang, Q.-C. He, The number and types of all possible rotational symmetries for flexoelectric tensors, in: Proceedings of the Royal Society of London A: Mathematical, Physical and Engineering Sciences, vol. 467, 2011, pp. 2369–2386.
[12] E.A. Eliseev, A.N. Morozovska, Hidden symmetry of flexoelectric coupling, Phys. Rev. B 98 (2018) 094108.
[13] E.A. Eliseev, A.N. Morozovska, V.V. Khist, V. Polinger, Symmetry of Flexoelectric Response in Ferroics, arXiv, 2019. Preprint arXiv:1903.02305.
[14] A. Kvasov, A.K. Tagantsev, Dynamic flexoelectric effect in perovskites from first-principles calculations, Phys. Rev. B 92 (2015) 054104.
[15] A.K. Tagantsev, G. Gerra, Interface-induced phenomena in polarization response of ferroelectric thin films, J. Appl. Phys. 100 (2006) 051607.
[16] G. Catalan, L.J. Sinnamon, J.M. Gregg, The effect of flexoelectricity on the dielectric properties of inhomogeneously strained ferroelectric thin films, J. Phys. Condens. Matter 16 (2004) 2253–2264.
[17] G. Catalan, B. Noheda, J. McAneney, L.J. Sinnamon, J.M. Gregg, Strain gradients in epitaxial ferroelectrics, Phys. Rev. B 72 (2005) 020102.
[18] M. Gharbi, Z.H. Sun, P. Sharma, K. White, S. El-Borgi, Flexoelectric properties of ferroelectrics and the nanoindentation size-effect, Int. J. Solids Struct. 48 (2011) 249–256.
[19] M.S. Majdoub, P. Sharma, T. Cagin, Enhanced size-dependent piezoelectricity and elasticity in nanostructures due to the flexoelectric effect, Phys. Rev. B 77 (2008) 125424.
[20] E.A. Eliseev, I.S. Vorotiahin, Y.M. Fomichov, M.D. Glinchuk, S.V. Kalinin, Y.A. Genenko, A.N. Morozovska, Defect driven flexo-chemical coupling in thin ferroelectric films, Phys. Rev. B 97 (2018) 024102.
[21] S.V. Kalinin, V. Meunier, Electronic flexoelectricity in low-dimensional systems, Phys. Rev. B 77 (2008) 033403.
[22] A.N. Morozovska, M.D. Glinchuk, Reentrant phase in nanoferroics induced by the flexoelectric and Vegard effects, J. Appl. Phys. 119 (2016) 094109.

[23] N. Balke, B. Winchester, W. Ren, Y. Hao Chu, A.N. Morozovska, E.A. Eliseev, M. Huijben, R.K. Vasudevan, P. Maksymovych, J. Britson, S. Jesse, I. Kornev, R. Ramesh, L. Bellaiche, L.Q. Chen, S.V. Kalinin, Enhanced electric conductivity at ferroelectric vortex cores in $BiFeO_3$, Nat. Phys. 8 (2012) 81–88.
[24] E.A. Eliseev, Y.M. Fomichov, S.V. Kalinin, Y.M. Vysochanskii, P. Maksymovich, A.N. Morozovska, Labyrinthine domains in ferroelectric nanoparticles: manifestation of a gradient-induced morphological phase transition, Phys. Rev. B 98 (2018) 054101.
[25] A.N. Morozovska, Y.M. Fomichov, P. Maksymovych, Y.M. Vysochanskii, E.A. Eliseev, Analytical description of domain morphology and phase diagrams of ferroelectric nanoparticles, Acta Mater. 160 (2018) 109–120.
[26] P. Lukashev, R.F. Sabirianov, Spin density in frustrated magnets under mechanical stress: Mn-based antiperovskites, J. Appl. Phys. 107 (2010) 09E115.
[27] P. Lukashev, R.F. Sabirianov, Flexomagnetic effect in frustrated triangular magnetic structures, Phys. Rev. B 82 (2010) 094417.
[28] Y.-C. Chen, Q. He, F.-N. Chu, Y.-C. Huang, J.W. Chen, W.I. Liang, R.K. Vasudevan, V. Nagarajan, E. Arenholz, S.V. Kalinin, Y.H. Chu, Electrical control of multiferroic orderings in mixed-phase $BiFeO_3$ films, Adv. Mater. 24 (2012) 3070–3075.
[29] I.I. Naumov, A.M. Bratkovsky, Gap opening in graphene by simple periodic inhomogeneous strain, Phys. Rev. B 84 (2011) 245444.
[30] S.R. Spurgeon, J.D. Sloppy, D.M. Kepaptsoglou, P.V. Balachandran, S. Nejati, J. Karthik, A.R. Damodaran, C.L. Johnson, H. Ambaye, R. Goyette, V. Lauter, Thickness-dependent crossover from charge-to strain-mediated magnetoelectric coupling in ferromagnetic/piezoelectric oxide heterostructures, ACS Nano 8 (2013) 894–903.
[31] S. Sidhardh, M.C. Ray, Flexomagnetic response of nanostructures, J. Appl. Phys. 124 (2018) 244101.
[32] A. Ghobadi, Y.T. Beni, H. Golestanian, Size dependent thermo-electro-mechanical nonlinear bending analysis of flexoelectric nano-plate in the presence of magnetic field, Int. J. Mech. Sci. 152 (2019) 118–137.
[33] E.A. Eliseev, M.D. Glinchuk, V. Khist, V.V. Skorokhod, R. Blinc, A.N. Morozovska, Linear magnetoelectric coupling and ferroelectricity induced by the flexomagnetic effect in ferroics, Phys. Rev. B 84 (2011) 174112.
[34] E.A. Eliseev, A.N. Morozovska, M.D. Glinchuk, B.Y. Zaulychny, V.V. Skorokhod, R. Blinc, Surface-induced piezomagnetic, piezoelectric, linear magnetoelectric effects in nanosystems, Phys. Rev. B 82 (2010) 085408.
[35] E.A. Eliseev, Complete symmetry analyses of the surface-induced piezomagnetic, piezoelectric and linear magnetoelectric effects, Ferroelectric 417 (2011) 100–109.
[36] L.A. Shuvalov (Ed.), Modern Crystallography, Vol. IV: Physical Properties of Crystals, In: Springer Series in Solid-State SciencesSpringer-Verlag, Berlin, 1988.
[37] D.B. Litvin, Magnetic physical-property tensors, Acta Crystallogr. A50 (1994) 406–408.
[38] J.-P. Rivera, A short review of the magnetoelectric effect and related experimental techniques on single phase (multi-) ferroics, Eur. Phys. J. B. 71 (2009) 299–313.
[39] V.G. Bar'yakhtar, V.A. L'vov, D.A. Yablonskii, Inhomogeneous magnetoelectric effect, JETP Lett. 37 (1983) 673–675.
[40] B.M. Tanygin, Symmetry theory of the flexomagnetoelectric effect in the magnetic domain walls, J. Magn. Magn. Mater. 323 (2011) 616–619.
[41] A.P. Pyatakov, A.K. Zvezdin, Flexomagnetoelectric interaction in multiferroics, Eur. Phys. J. B. 71 (2009) 419–427.
[42] W. Cochran, Dynamical, scattering and dielectric properties of ferroelectric crystals, Adv. Phys. 18 (72) (1969) 157–192.

[43] J. Hlinka, I. Gregora, V. Vorlıcek, Complete spectrum of long-wavelength phonon modes in $Sn_2P_2S_6$ by Raman scattering, Phys. Rev. B 65 (2002) 064308.
[44] R.M. Yevych, Y.M. Vysochanskii, M.M. Khoma, S.I. Perechinskii, Lattice instability at phase transitions near the Lifshitz point in proper monoclinic ferroelectrics, J. Phys. Condens. Matter 18 (2006) 4047–4064.
[45] G. Shirane, J.D. Axe, J. Harada, J.P. Remeika, Soft ferroelectric modes in lead titanate, Phys. Rev. B 2 (1970) 155–159.
[46] G. Shirane, J.D. Axe, J. Harada, A. Linz, Inelastic neutron scattering from single-domain $BaTiO_3$, Phys. Rev. B 2 (1970) 3651–3657.
[47] G. Shirane, B.C. Frazer, V.J. Minkiewicz, J.A. Leake, A. Linz, Soft optic modes in barium titanate, Phys. Rev. Lett. 19 (1967) 234–235.
[48] S.W.H. Eijt, R. Currat, J.E. Lorenzo, P. Saint-Gregoire, B. Hennion, Y.M. Vysochanskii, Soft modes and phonon interactions in $Sn_2P_2S_6$ studied by neutron scattering, Eur. Phys. J. 5 (1998) 169–178.
[49] S.W.H. Eijt, R. Currat, J.E. Lorenzo, P. Saint-Gregoire, S. Katano, T. Janssen, B. Hennion, Y.M. Vysochanskii, Soft modes and phonon interactions in $Sn_2P_2Se_6$ studied by means of neutron scattering, J. Phys. Condens. Matter 10 (1998) 4811–4844.
[50] P. Zubko, G. Catalan, A. Buckley, P.R.L. Welche, J.F. Scott, Strain-gradient-induced polarization in $SrTiO_3$ single crystals, Phys. Rev. Lett. 99 (2007) 167601.
[51] P. Zubko, G. Catalan, A. Buckley, P.R.L. Welche, J.F. Scott, Erratum: strain-gradient-induced polarization in $SrTiO_3$ single crystals [Phys. Rev. Lett. 99 (2007) 167601], Phys. Rev. Lett. 100 (2008) 199906.
[52] I. Ponomareva, A.K. Tagantsev, L. Bellaiche, Finite-temperature flexoelectricity in ferroelectric thin films from first principles, Phys. Rev. B 85 (2012) 104101.
[53] R. Maranganti, P. Sharma, Atomistic determination of flexoelectric properties of crystalline dielectrics, Phys. Rev. B 80 (2009) 054109.
[54] R.O. Bell, G. Rupprecht, Elastic constants of strontium titanate, Phys. Rev. 129 (1963) 90–94.
[55] A. Schaefer, H. Schmitt, A. Dorr, Elastic and piezoelectric coefficients of TSSG barium titanate single crystals, Ferroelectrics 69 (1986) 253–266.
[56] A. Biancoli, C.M. Fancher, J.L. Jones, D. Damjanovic, Breaking of macroscopic centric symmetry in paraelectric phases of ferroelectric materials and implications for flexoelectricity, Nat. Mater. 14 (2015) 224–229.
[57] P.V. Yudin, R. Ahluwalia, A.K. Tagantsev, Upper bounds for flexocoupling coefficients in ferroelectrics, Appl. Phys. Lett. 104 (2014) 082913.
[58] J. Hong, D. Vanderbilt, First-principles theory of frozen-ion flexoelectricity, Phys. Rev. B 84 (2011) 180101(R).
[59] J. Hong, D. Vanderbilt, First-principles theory and calculation of flexoelectricity, Phys. Rev. B 88 (2013) 174107.
[60] M. Stengel, Unified ab initio formulation of flexoelectricity and strain-gradient elasticity, Phys. Rev. B 93 (2016) 245107.
[61] C.E. Dreyer, M. Stengel, D. Vanderbilt, Current-density implementation for calculating flexoelectric coefficients, Phys. Rev. B 98 (2018) 075153.
[62] M. Stengel, D. Vanderbilt, New functionalities from gradient couplings: flexoelectricity and more, in: APS March Meeting 2018, Bulletin of the American Physical Society, 2018. Abstract id. C09.001.
[63] W. Ma, L.E. Cross, Strain-gradient-induced electric polarization in lead zirconate titanate ceramics, Appl. Phys. Lett. 82 (2003) 3293–3295.
[64] W. Ma, L.E. Cross, Flexoelectricity of barium titanate, Appl. Phys. Lett. 88 (2006) 232902.
[65] W. Ma, L.E. Cross, Flexoelectric effect in ceramic lead zirconate titanate, Appl. Phys. Lett. 86 (2005) 072905.

[66] D. Lee, A. Yoon, S.Y. Jang, J.-G. Yoon, J.-S. Chung, M. Kim, J.F. Scott, T.W. Noh, Giant flexoelectric effect in ferroelectric epitaxial thin films, Phys. Rev. Lett. 107 (2011) 057602.
[67] J. Lu, X. Liang, W. Yu, S. Hu, S. Shen, Temperature dependence of flexoelectric coefficient for bulk polymer polyvinylidene fluoride, J. Phys. D Appl. Phys. 52 (2019) 075302.
[68] X. Wen, D. Li, K. Tan, Q. Deng, S. Shen, Flexoelectret: an electret with a tunable flexoelectriclike response, Phys. Rev. Lett. 122 (2019) 148001.
[69] L.P. Liu, P. Sharma, Flexoelectricity and thermal fluctuations of lipid bilayer membranes: renormalization of flexoelectric, dielectric, and elastic properties, Phys. Rev. E 87 (2013) 032715.
[70] F. Ahmadpoor, P. Sharma, Flexoelectricity in two-dimensional crystalline and biological membranes, Nanoscale 7 (2015) 16555–16570.
[71] M. Stengel, Flexoelectricity from density-functional perturbation theory, Phys. Rev. B 88 (2013) 174106.
[72] A. Abdollahi, F. Vásquez-Sancho, G. Catalan, Piezoelectric mimicry of flexoelectricity, Phys. Rev. Lett. 121 (2018) 205502.
[73] A.H. Rahmati, S. Yang, S. Bauer, P. Sharma, Nonlinear bending deformation of soft electrets and prospects for engineering flexoelectricity and transverse (d31) piezoelectricity, Soft Matter 15 (2019) 127–148.
[74] I.B. Bersuker, Pseudo Jahn-Teller effect in the origin of enhanced flexoelectricity, Appl. Phys. Lett. 106 (2015) 022903.
[75] V. Polinger, I.B. Bersuker, Pseudo Jahn-Teller effect in permittivity of ferroelectric perovskites, J. Phys. Conf. Ser. 833 (2017) 012012.
[76] A.N. Morozovska, Y.M. Vysochanskii, O.V. Varenyk, M.V. Silibin, S.V. Kalinin, E.A. Eliseev, Flexocoupling impact on the generalized susceptibility and soft phonon modes in the ordered phase of ferroics, Phys. Rev. B 92 (2015) 094308.
[77] A.N. Morozovska, E.A. Eliseev, C.M. Scherbakov, Y.M. Vysochanskii, The influence of elastic strain gradient on the upper limit of flexocoupling strength, spatially-modulated phases and soft phonon dispersion in ferroics, Phys. Rev. B 94 (2016) 174112.
[78] A.N. Morozovska, M.D. Glinchuk, E.A. Eliseev, Y.M. Vysochanskii, Flexocoupling-induced soft acoustic mode and the spatially modulated phases in ferroelectrics, Phys. Rev. B 96 (2017) 094111.
[79] E.A. Eliseev, A.N. Morozovska, M.D. Glinchuk, S.V. Kalinin, Missed surface waves in non-piezoelectric solids, Phys. Rev. B 96 (2017) 045411.
[80] A.N. Morozovska, E.A. Eliseev, Size effect of soft phonon dispersion law in nanosized ferroics, Phys. Rev. B 99 (2019) 115412.

Printed in the United States
By Bookmasters